# Wastewater Characteristics, Treatment and Disposal

# Biological Wastewater Treatment Series

## VOLUME ONE

# Wastewater Characteristics, Treatment and Disposal

*Marcos von Sperling*

*Department of Sanitary and Environmental Engineering*
*Federal University of Minas Gerais, Brazil*

Publishing
LONDON • NEW YORK

**Published by IWA Publishing, Alliance House, 12 Caxton Street, London SW1H 0QS, UK**

Telephone: +44 (0) 20 7654 5500; Fax: +44 (0) 20 7654 5555; Email: publications@iwap.co.uk
Website: **www.iwapublishing.com**

First published 2007
© 2007 IWA Publishing

Copy-edited and typeset by Aptara Inc., New Delhi, India



**Disclaimer**


*British Library Cataloguing in Publication Data*
A CIP catalogue record for this book is available from the British Library

*Library of Congress Cataloguing-in-Publication Data*
A catalogue record for this book is available from the Library of Congress

ISBN: 1 84339 161 9
ISBN 13: 9781843391616

# Contents

# Contents

# Preface

The present series of books has been produced based on the book "*Biological wastewater treatment in warm climate regions*", written by the same authors and also published by IWA Publishing. The main idea behind this series is the subdivision of the original book into smaller books, which could be more easily purchased and used.

The implementation of wastewater treatment plants has been so far a challenge for most countries. Economical resources, political will, institutional strength and cultural background are important elements defining the trajectory of pollution control in many countries. Technological aspects are sometimes mentioned as being one of the reasons hindering further developments. However, as shown in this series of books, the vast array of available processes for the treatment of wastewater should be seen as an incentive, allowing the selection of the most appropriate solution in technical and economical terms for each community or catchment area. For almost all combinations of requirements in terms of effluent quality, land availability, construction and running costs, mechanisation level and operational simplicity there will be one or more suitable treatment processes.

Biological wastewater treatment is very much influenced by climate. Temperature plays a decisive role in some treatment processes, especially the natural-based and non-mechanised ones. Warm temperatures decrease land requirements, enhance conversion processes, increase removal efficiencies and make the utilisation of some treatment processes feasible. Some treatment processes, such as anaerobic reactors, may be utilised for diluted wastewater, such as domestic sewage, only in warm climate areas. Other processes, such as stabilisation ponds, may be applied in lower temperature regions, but occupying much larger areas and being subjected to a decrease in performance during winter. Other processes, such as activated sludge and aerobic biofilm reactors, are less dependent on temperature, as a result of the higher technological input and mechanisation level. The main purpose of this series of books is to present the technologies for urban wastewater treatment as applied to the specific condition of warm temperature, with the related implications in terms of design and operation. There is no strict definition for the range of temperatures that fall into this category, since the books always present how to correct parameters, rates and coefficients for different temperatures. In this sense, subtropical and even temperate climate are also indirectly covered, although most of the focus lies on the tropical climate.

Another important point is that most warm climate regions are situated in developing countries. Therefore, the books cast a special view on the reality of these countries, in which simple, economical and sustainable solutions are strongly demanded. All technologies presented in the books may be applied in developing countries, but of course they imply different requirements in terms of energy, equipment and operational skills. Whenever possible, simple solutions, approaches and technologies are presented and recommended.

Considering the difficulty in covering all different alternatives for wastewater collection, the books concentrate on off-site solutions, implying collection and transportation of the wastewater to treatment plants. No off-site solutions, such as latrines and septic tanks are analysed. Also, stronger focus is given to separate sewerage systems, although the basic concepts are still applicable to combined and mixed systems, especially under dry weather conditions. Furthermore, emphasis is given to urban wastewater, that is, mainly domestic sewage plus some additional small contribution from non-domestic sources, such as industries. Hence, the books are not directed specifically to industrial wastewater treatment, given the specificities of this type of effluent. Another specific view of the books is that they detail biological treatment processes. No physical-chemical wastewater treatment processes are covered, although some physical operations, such as sedimentation and aeration, are dealt with since they are an integral part of some biological treatment processes.

The books' proposal is to present in a balanced way theory and practice of wastewater treatment, so that a conscious selection, design and operation of the wastewater treatment process may be practised. Theory is considered essential for the understanding of the working principles of wastewater treatment. Practice is associated to the direct application of the concepts for conception, design and operation. In order to ensure the practical and didactic view of the series, 371 illustrations, 322 summary tables and 117 examples are included. All major wastewater treatment processes are covered by full and interlinked design examples which are built up throughout the series and the books, from the determination of the wastewater characteristics, the impact of the discharge into rivers and lakes, the design of several wastewater treatment processes and the design of the sludge treatment and disposal units.

The series is comprised by the following books, namely: *(1) Wastewater characteristics, treatment and disposal; (2) Basic principles of wastewater treatment; (3) Waste stabilisation ponds; (4) Anaerobic reactors; (5) Activated sludge and aerobic biofilm reactors; (6) Sludge treatment and disposal.*

Volume 1 (*Wastewater characteristics, treatment and disposal*) presents an integrated view of water quality and wastewater treatment, analysing wastewater characteristics (flow and major constituents), the impact of the discharge into receiving water bodies and a general overview of wastewater treatment and sludge treatment and disposal. Volume 1 is more introductory, and may be used as teaching material for undergraduate courses in Civil Engineering, Environmental Engineering, Environmental Sciences and related courses.

Volume 2 (*Basic principles of wastewater treatment*) is also introductory, but at a higher level of detailing. The core of this book is the unit operations and processes associated with biological wastewater treatment. The major topics covered are: microbiology and ecology of wastewater treatment; reaction kinetics and reactor hydraulics; conversion of organic and inorganic matter; sedimentation; aeration. Volume 2 may be used as part of postgraduate courses in Civil Engineering, Environmental Engineering, Environmental Sciences and related courses, either as part of disciplines on wastewater treatment or unit operations and processes.

Volumes 3 to 5 are the central part of the series, being structured according to the major wastewater treatment processes (*waste stabilisation ponds, anaerobic reactors, activated sludge and aerobic biofilm reactors*). In each volume, all major process technologies and variants are fully covered, including main concepts, working principles, expected removal efficiencies, design criteria, design examples, construction aspects and operational guidelines. Similarly to Volume 2, volumes 3 to 5 can be used in postgraduate courses in Civil Engineering, Environmental Engineering, Environmental Sciences and related courses.

Volume 6 (*Sludge treatment and disposal*) covers in detail sludge characteristics, production, treatment (thickening, dewatering, stabilisation, pathogens removal) and disposal (land application for agricultural purposes, sanitary landfills, landfarming and other methods). Environmental and public health issues are fully described. Possible academic uses for this part are same as those from volumes 3 to 5.

Besides being used as textbooks at academic institutions, it is believed that the series may be an important reference for practising professionals, such as engineers, biologists, chemists and environmental scientists, acting in consulting companies, water authorities and environmental agencies.

The present series is based on a consolidated, integrated and updated version of a series of six books written by the authors in Brazil, covering the topics presented in the current book, with the same concern for didactic approach and balance between theory and practice. The large success of the Brazilian books, used at most graduate and post-graduate courses at Brazilian universities, besides consulting companies and water and environmental agencies, was the driving force for the preparation of this international version.

In this version, the books aim at presenting consolidated technology based on worldwide experience available at the international literature. However, it should be recognised that a significant input comes from the Brazilian experience, considering the background and working practice of all authors. Brazil is a large country with many geographical, climatic, economical, social and cultural contrasts, reflecting well the reality encountered in many countries in the world. Besides, it should be mentioned that Brazil is currently one of the leading countries in the world on the application of anaerobic technology to domestic sewage treatment, and in the post-treatment of anaerobic effluents. Regarding this point, the authors would like to show their recognition for the Brazilian Research Programme on Basic Sanitation (PROSAB), which, through several years of intensive, applied, cooperative research has led to the consolidation of anaerobic treatment and aerobic/anaerobic post-treatment, which are currently widely applied in full-scale plants in Brazil. Consolidated results achieved by PROSAB are included in various parts of the book, representing invaluable and updated information applicable to warm climate regions.

Volumes 1 to 5 were written by the two main authors. Volume 6 counted with the invaluable participation of Cleverson Vitorio Andreoli and Fernando Fernandes, who acted as editors, and of several specialists, who acted as chapter authors: Aderlene Inês de Lara, Deize Dias Lopes, Dione Mari Morita, Eduardo Sabino Pegorini, Hilton Felício dos Santos, Marcelo Antonio Teixeira Pinto, Maurício Luduvice, Ricardo Franci Gonçalves, Sandra Márcia Cesário Pereira da Silva, Vanete Thomaz Soccol.

Many colleagues, students and professionals contributed with useful suggestions, reviews and incentives for the Brazilian books that were the seed for this international version. It would be impossible to list all of them here, but our heartfelt appreciation is acknowledged.

The authors would like to express their recognition for the support provided by the Department of Sanitary and Environmental Engineering at the Federal University of Minas Gerais, Brazil, at which the two authors work. The department provided institutional and financial support for this international version, which is in line with the university's view of expanding and disseminating knowledge to society.

Finally, the authors would like to show their appreciation to IWA Publishing, for their incentive and patience in following the development of this series throughout the years of hard work.

Marcos von Sperling  
Carlos Augusto de Lemos Chernicharo

*December 2006*

# The author

**Marcos von Sperling**
PhD in Environmental Engineering (Imperial College, Univ. London, UK). Associate professor at the Department of Sanitary and Environmental Engineering, Federal University of Minas Gerais, Brazil. Consultant to governmental and private companies in the field of water pollution control and wastewater treatment. marcos@desa.ufmg.br

# 1

# Introduction to water quality and water pollution

## 1.1 INTRODUCTION

Water, because of its properties as a solvent and its capacity to transport particles, incorporates in itself various impurities that characterise the *water quality*.

Water quality is a result of natural phenomena and the acts of human beings. Generally it can be said that *water quality is a function of land use in the catchment area*. This is due to the following factors:

- **Natural conditions**: even with the catchment area preserved in its natural condition, the surface water quality is affected by run off and infiltration resulting from rainfall. The impact of these is dependent on the contact of the water with particles, substances and impurities in the soil. Therefore, the incorporation of suspended solids (e.g. soil particles) or dissolved solids (e.g. ions originating from the dissolution of rocks) occurs even when the catchment area is totally preserved in its natural condition (e.g. occupation of the land with woods and forests). In this case, the soil protection and composition have a great influence.
- **Interference of human beings**: the interference of man manifests itself either in a concentrated form, such as in the discharge of domestic or industrial wastewater, or in a diffused form, such as in the application of fertilisers or pesticides onto the soil. Both contribute to the introduction of

compounds into the water, thus affecting its quality. Therefore, the form in which human beings use and occupy the land has a direct implication in the water quality.

Figure 1.1 presents an example of possible interactions between land use and the presence of factors that modify the water quality in rivers and lakes. Water quality control is associated with a global planning at the whole catchment area level, and not individually, for each impacting source.

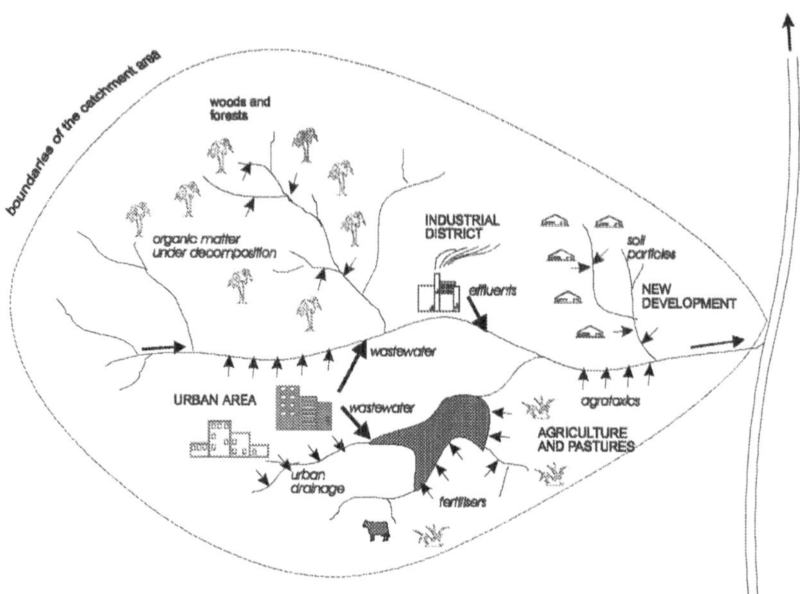



Figure 1.1.　Examples in a catchment area of the interrelation between land use and water quality impacting agents

Separate from the above concept of *existing* water quality, there is the concept of the *desired* water quality. *The desired quality for a water is a function of its intended use*. There are various possible intended uses for a particular water, which are listed in Section 1.2. In summary:

- Existing water quality: function of the land use in the catchment area
- Desired water quality: function of the intended uses for the water

Within the focus of this book, the study of water quality is essential, not only to characterise the consequences of a certain polluting activity, but also to allow the selection of processes and methods that will allow compliance with the desired water uses.

## 1.2 USES OF WATER

The main water uses are:

- *domestic supply*
- *industrial supply*
- *irrigation*
- *animal supply*
- *preservation of aquatic life*
- *recreation and leisure*
- *breeding of aquatic species*
- *generation of electricity*
- *navigation*
- *landscape harmony*
- *dilution and transport of wastes*

In general terms, only the first two uses (domestic supply and industrial supply) are frequently associated with a prior water treatment, in view of their more demanding quality requirements.

There is a direct relation between water use and its required quality. In the above list, the most demanding use can be considered domestic water supply, which requires the satisfaction of various quality criteria. Conversely, the less demanding uses are simple dilution and transportation of wastes, which do not have any specific requirements in terms of quality. However, it must be remembered that multiple uses are usually assigned to water bodies, resulting in the necessity of satisfying diverse quality criteria. Such is the case, for example, of reservoirs constructed for water supply, electricity generation, recreation, irrigation and others.

Besides the cycle of water on Earth (hydrological cycle), there are internal cycles, in which water remains in the liquid state, but has its characteristics modified as a result of its use. Figure 1.2 shows an example of typical routes of water use, composing partial cycles. In these cycles, the water quality is modified at each stage of its journey.

The management of these internal cycles is an essential role in environmental engineering, and includes the planning, design, construction and control of the works necessary for the maintenance of the desired water quality as a function of its intended uses. Therefore, the engineer or scientist must know how to ask for and interpret the results of water quality samples in the various points of the cycle. This book focuses mainly on the aspect of wastewater treatment, and the impact of the discharge of wastewater to receiving bodies is covered in Chapter 3.

## 1.3 WATER QUALITY REQUIREMENTS

Table 1.1 presents in a simplified way the association between the main quality requirements and the corresponding water uses. In cases of water bodies with multiple uses, the water quality must comply with the requirements of the various intended uses. The expression "free" in the table is different from "absolutely free". Zero levels of many contaminants cannot be guaranteed and in most cases are not necessary. The acceptable concentrations are based on risk analysis, a tool that is used for deriving quality guidelines and standards.

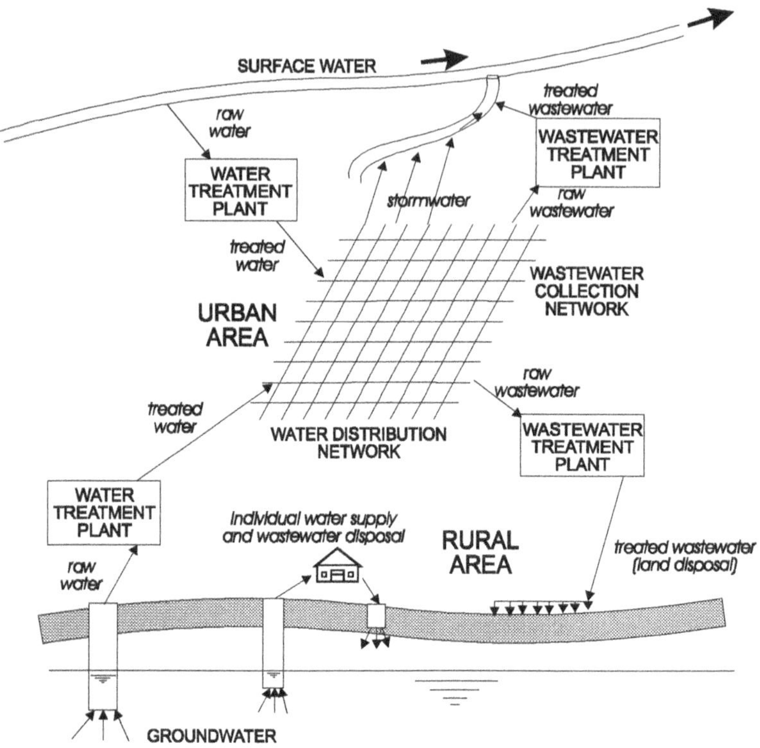


Figure 1.2.    Routes of water use and disposal

- *Raw water*. Initially, water is abstracted from the river, lake or water table, and has a certain quality.
- *Treated water*. After abstraction, water undergoes transformations during its treatment to be able to comply with its intended uses (e.g. public or industrial water supply).
- *Raw wastewater*. The water, after being used, undergoes new transformations in its quality and becomes a liquid waste.
- *Treated wastewater*. Aiming at removing its main pollutants, wastewater undergoes treatment before being discharged into the receiving body. Wastewater treatment is responsible for the new modification in the quality of the liquid.
- *Stormwater*. Rain water flows on the ground, incorporates some pollutants, and is collected at stormwater systems before being discharged into the receiving body.
- *Receiving body*. Stormwater and the effluent from the wastewater treatment plant reach the receiving body where water quality undergoes new modifications, as a result of dilution and self-purification mechanisms.

Table 1.1. Association between water use and quality requirements

| General use | Specific use | Required quality |
|---|---|---|
| Domestic supply | – | – Free from chemical substances harmful to health<br>– Free from organisms harmful to health<br>– Low aggressiveness and hardness<br>– Aesthetically pleasant (low turbidity, colour, taste and odour; absence of macro-organisms) |
| Industrial supply | Water incorporated into the product (e.g. food, drinks, medicines) | – Free from chemical substances harmful to health<br>– Free from organisms harmful to health<br>– Aesthetically pleasant (low turbidity, colour, taste and odour; absence of macro-organisms) |
| | Water that enters into contact with the product | – Variable with the product |
| | Water that does not enter into contact with the product (e.g. refrigeration units, boilers) | – Low hardness<br>– Low aggressiveness |
| Irrigation | Horticulture, products ingested raw or with skin | – Free from chemical substances harmful to health<br>– Free from organisms harmful to health<br>– Non-excessive salinity |
| | Other plantations | – Free from chemical substances harmful to the soil and plantations<br>– Non-excessive salinity |
| Animal water supply | – | – Free from chemical substances harmful to animals health<br>– Free from organisms harmful to animals health |
| Preservation of aquatic life | – | – Variable with the environmental requirements of the aquatic species to be preserved |
| Aquaculture | Animal breeding | – Free from chemical substances harmful to animals, workers and consumers health<br>– Free from organisms harmful to animals, workers and consumers health<br>– Availability of nutrients |
| | Vegetable growing | – Free from chemical substances toxic to vegetables and consumers<br>– Availability of nutrients |

*(continued)*

Table 1.1. (*continued*)

| General use | Specific use | Required quality |
|---|---|---|
| Recreation and leisure | Primary contact (direct contact with the liquid medium – bathing; e.g.: swimming, water-skiing, surfing) | – Free from chemical substances harmful to health<br>– Free from organisms harmful to health<br>– Low levels of suspended solids and oils and grease |
| | Secondary contact (without direct contact with the liquid medium; e.g.: leisure navigation, fishing, contemplative viewing) | – Pleasant appearance |
| Energy generation | Hydroelectric power plants | – Low aggressiveness |
| | Nuclear or thermoelectric power plants (e.g. cooling towers) | – Low hardness |
| Transport | – | – Low presence of course material that could be dangerous to vessels |
| Waste dilution and transportation | – | – |

## 1.4 WATER POLLUTION

*Water pollution* is the addition of substances or energy forms that directly or indirectly **alter** the nature of the water body in such a manner that **negatively affects** its legitimate **uses**.

This definition is essentially practical and, as a consequence, potentially controversial, because of the fact that it associates pollution with negative alterations and with water body uses, concepts that are attributed by human beings. However, this practical view is important, principally when analysing the control measures for pollution reduction

Table 1.2 lists the main pollutants and their source, together with the most representative effects. Chapter 2 covers in detail the main parameters, which characterise the quality of a wastewater (second column in the table). For domestic sewage, which is the main focus of this book, the main pollutants are: *suspended solids, biodegradable organic matter, nutrients* and *pathogenic organisms*. Their impact in the water body is analysed in detail in Chapter 3.

The solution to most of these problems, especially biodegradable organic matter and pathogens, has been reached in many **developed regions**, which are now concentrated on the removal of nutrients and micro-pollutants, together with substantial attention to the pollution caused by storm-water drainage. In **developing**

Table 1.2. Main pollutants, their source and effects

| Pollutant | Main representative parameters | Source | | | | Possible effect of the pollutant |
| | | Wastewater | | Stormwater | | |
| | | Domestic | Industrial | Urban | Agricultural and pasture | |
|---|---|---|---|---|---|---|
| *Suspended solids* | Total suspended solids | XXX | ←→ | XX | X | • Aesthetic problems<br>• Sludge deposits<br>• Pollutants adsorption<br>• Protection of pathogens |
| *Biodegradable organic matter* | Biochemical oxygen demand | XXX | ←→ | XX | X | • Oxygen consumption<br>• Death of fish<br>• Septic conditions |
| *Nutrients* | Nitrogen Phosphorus | XXX | ←→ | XX | X | • Excessive algae growth<br>• Toxicity to fish (ammonia)<br>• Illnesses in new-born infants (nitrate)<br>• Pollution of groundwater |
| *Pathogens* | Coliforms | XXX | ←→ | XX | X | • Water-borne diseases |
| *Non-biodegradable organic matter* | Pesticides Some detergents Others | X | ←→ | X | XX | • Toxicity (various)<br>• Foam (detergents)<br>• Reduction of oxygen transfer (detergents)<br>• Non-biodegradability<br>• Bad odours (e.g.: phenols) |
| *Metals* | Specific elements (As, Cd, Cr, Cu, Hg, Ni, Pb, Zn, etc.) | X | ←→ | X | | • Toxicity<br>• Inhibition of biological sewage treatment<br>• Problems in agriculture use of sludge<br>• Contamination of groundwater |
| *Inorganic dissolved solids* | Total dissolved solids Conductivity | XX | ←→ | | X | • Excessive salinity – harm to plantations (irrigation)<br>• Toxicity to plants (some ions)<br>• Problems with soil permeability (sodium) |

x: *small*　　xx: *medium*　　xxx: *high*　　←→: variable　　*empty*: usually not important

**regions**, the basic pollution problems still need to be dealt with, and the whole array of pollutants needs to be tackled. However, because of scarcity of financial resources in these regions, priorities need to be set (as they have been, in the past, and continue to be, in the developed regions), and the gross pollution by organic matter and contamination by pathogens are likely to deserve higher attention. Naturally, each region has its own specificities, and these need to be taken into account when setting up priorities.

Also from the table, it is seen that it is very difficult to generalise industrial wastewater, because of its variability from process to process and from industry to industry.

In the table, it is also seen that there are two ways in which the pollutant could reach the receiving body (see Figure 1.3):

- *point-source pollution*
- *diffuse pollution*

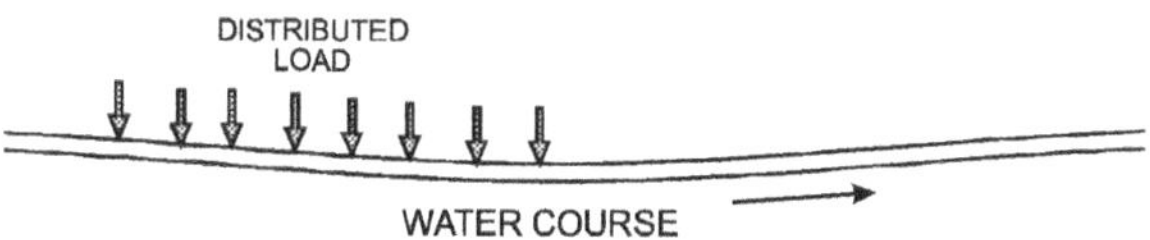


Figure 1.3.   Point-source and diffuse pollution

In *point-source* pollution, the pollutants reach the water body in points concentrated in the space. Usually the discharge of domestic and industrial wastewater generates point-source pollution, since the discharges are through outfalls.

In *diffuse pollution,* the pollutants enter the water body distributed at various locations along its length. This is the typical case of storm water drainage, either in rural areas (no pipelines) or in urban areas (storm water collection system, with multiple discharges into the water body).

The focus of this book is the control of point-source pollution by means of wastewater treatment. In the developing regions, there is practically everything still to be done in terms of the control of point-source pollution originating from cities and industries.

# 2

# Wastewater characteristics

## 2.1 WASTEWATER FLOWRATES

### 2.1.1 Introduction

Wastewater sewerage (collection, treatment and disposal) is accomplished by the following main alternatives (Figure 2.1):

- *Off-site sewerage*
  - Separate sewerage system
  - Combined sewerage system
- *On-site sewerage*

In various countries a **separate sewerage system** is adopted, which separates storm water from sewage, both being transported by independent pipeline systems. In this case, in principle, storm water does not contribute to the wastewater treatment plant (WWTP). In other countries, however, a **combined (unitary) sewerage system** is adopted, which directs sewage and storm water together into the same system (see Figure 2.1). In this case, the pipelines have a larger diameter, to transport not only the sewage flow, but mainly rainwater, and the design of the WWTP has to take into consideration the corresponding fraction of rainwater that is allowed to enter the treatment works. In countries with a warm climate, during the dry season, sewage flows slowly in these large diameter pipes, leading to long detention times which allow decomposition and generation of malodours. In this book, *focus is given to the separate sewerage system*, analysing only the three components listed above. However, the principles for the design of a combined sewerage system, based on dry-weather flow, are the same.

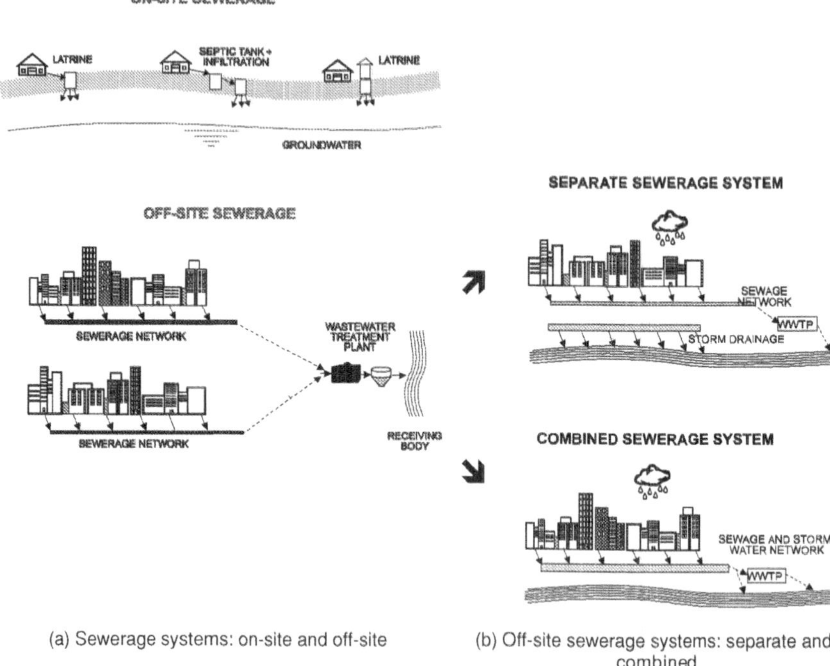



(a) Sewerage systems: on-site and off-site

(b) Off-site sewerage systems: separate and combined

Figure 2.1.   Types of sewerage system

Similarly, the book concentrates on **off-site collection systems** (with a water-borne sewerage collection and transportation network) and does not cover the **on-site systems** (e.g. latrines and septic tanks). These are of great importance and in many cases the best alternative in various regions, being more applicable in locations with a low population density, like rural areas (even though they are also applied in various densely occupied locations, but frequently presenting problems of infiltration in the soil and the resulting contamination of the water table).

Urban wastewater that flows in an off-site sewerage system and contributes to a WWTP is originated from the following three main sources:

- *Domestic sewage* (including residences, institutions and commerce)
- *Infiltration*
- *Industrial effluents* (various origins and types of industry)

For the characterisation of both quantity and quality of the influent to the WWTP, it is necessary to separately analyse each of the three items.

## 2.1.2  Domestic wastewater flow

### 2.1.2.1  Preliminaries

The concept of domestic flow encompasses the sewage originating from homes, as well as commercial activities and institutions that are normally components of the locality. More expressive values originating from significant point sources must be computed separately and added to the global value.

Normally domestic sewage flow is calculated based on the water consumption in the respective locality. The water consumption is usually calculated as a function of the design population and of a value attributed for the average daily per capita water consumption.

It is important to observe that for the design and operation of the sewage treatment works it is not sufficient to consider only the average flow. It is also necessary to quantify the minimum and maximum flowrates, because of hydraulic and process reasons.

This Section describes the population-forecast studies, the estimates of water consumption and the production of wastewater, together with the variations in flow (minimum and maximum flow).

## 2.1.2.2 Population forecast

The population that contributes to the treatment plant is that situated inside the design area served by the sewerage system. However, the design population is only a certain fraction of the total population in this area, because maybe not all the population is connected to the sewerage system. This ratio (population served/total population) is called the **coverage index**. This index can be determined (current conditions) or estimated (future conditions), such as to allow the calculation of the design flow. In the final years of the planning horizon, it is expected that the coverage will be close to 100%, reflecting the improvement and expansion in the collection network. The coverage index is a function of the following aspects:

- *Physical, geographical or topographical conditions of the locality*. It is not always possible to serve all households with the sewerage system. Those not served must adopt other solutions besides the off-site water-borne sewerage system.
- *Adhesion index*. This is the ratio between the population actually connected to the system and the population potentially served by the sewerage system in the streets (not all households are connected to the available system, that is to say, not all adhere to the sewerage system). In some communities, it is compulsory to connect to the collection system, in case it passes in front of the house; in other communities, this is optional.
- *Implementation stages of the sewerage system*. In the initial operating years of the WWTP, maybe not all of the designed collection and transport system has been actually installed, and this affects the initial flow.

For the design of a sewage treatment works it is necessary to know the final population (population at the end of the planning horizon – see Chapter 6 for the concept of planning horizon) as well as the initial population and its evolution with time, in order to allow the definition of implementation stages.

Table 2.1. Population forecast. Methods based on mathematical formulas

| Method | Description | Curve shape | Growth rate | Forecast formula | Coefficients (if regression analysis is not used) |
|---|---|---|---|---|---|
| *Linear growth* | Population growth follows a constant rate. Method used for short-term forecasts. Curve fitting can also be done through regression analysis. | 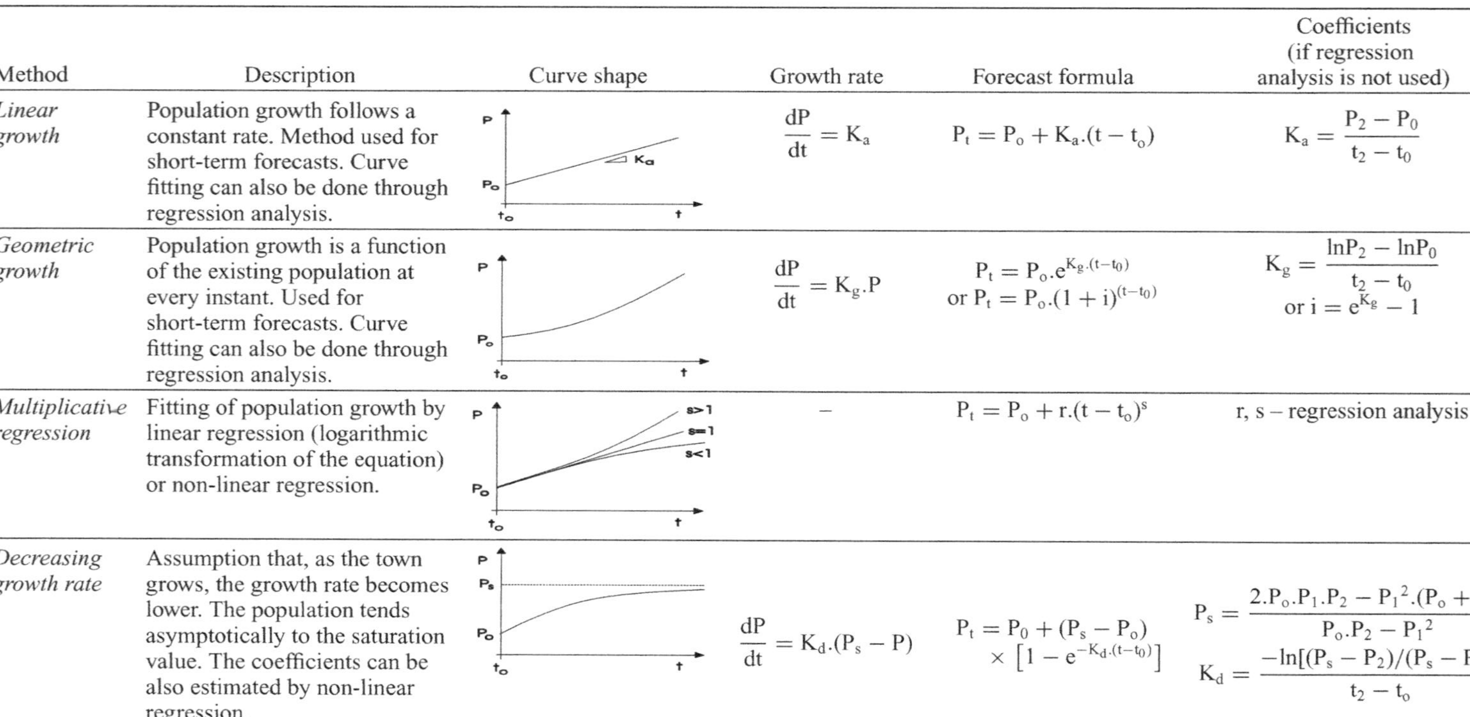 | $\dfrac{dP}{dt} = K_a$ | $P_t = P_o + K_a.(t - t_o)$ | $K_a = \dfrac{P_2 - P_0}{t_2 - t_0}$ |
| *Geometric growth* | Population growth is a function of the existing population at every instant. Used for short-term forecasts. Curve fitting can also be done through regression analysis. | | $\dfrac{dP}{dt} = K_g.P$ | $P_t = P_o.e^{K_g.(t-t_0)}$ <br> or $P_t = P_o.(1 + i)^{(t-t_0)}$ | $K_g = \dfrac{\ln P_2 - \ln P_0}{t_2 - t_0}$ <br> or $i = e^{K_g} - 1$ |
| *Multiplicative regression* | Fitting of population growth by linear regression (logarithmic transformation of the equation) or non-linear regression. | | – | $P_t = P_o + r.(t - t_o)^s$ | $r, s$ – regression analysis |
| *Decreasing growth rate* | Assumption that, as the town grows, the growth rate becomes lower. The population tends asymptotically to the saturation value. The coefficients can be also estimated by non-linear regression. | | $\dfrac{dP}{dt} = K_d.(P_s - P)$ | $P_t = P_0 + (P_s - P_o) \times \left[1 - e^{-K_d.(t-t_0)}\right]$ | $P_s = \dfrac{2.P_o.P_1.P_2 - P_1{}^2.(P_o + P_2)}{P_o.P_2 - P_1{}^2}$ <br> $K_d = \dfrac{-\ln[(P_s - P_2)/(P_s - P_o)]}{t_2 - t_o}$ |

| *Logistic growth* | The population growth follows an S-shaped curve. The population tends asymptotically to a saturation value. The coefficients can be also estimated by non-linear regression. Required conditions: $P_o < P_1 < P_2$ and $P_0.P_2 < P_1^2$. The point of inflexion in the curve occurs at time $t = [t_o - \ln(c)/K_1]$ and with $P_t = P_s/2$. | 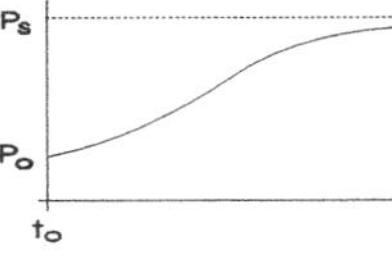 | $$\frac{dP}{dt} = K_1.P.\left(\frac{(P_s - P)}{P_s}\right) \quad P_t = \frac{P_s}{1 + c.e^{K_1.(t-t_o)}} \quad P_s = \frac{2.P_o.P_1.P_2 - P_1^2.(P_o + P_2)}{P_o.P_2 - P_1^2}$$ $$c = (P_s - P_o)/P_o$$ $$K_1 = \frac{1}{t_2 - t_1}.\ln\left[\frac{P_o.(P_s - P_1)}{P_1.(P_s - P_o)}\right]$$ |

*Source:* partly adapted from Qasim (1985)

- $dP/dt$ = population growth rate as a function of time
- $P_o, P_1, P_2$ = population in the years $t_o, t_1, t_2$. The formulas for the decreasing and logistic growth rates require equally-spaced values in time if regression analysis is not employed (inhabitants)
- $P_t$ = population estimated for year t (inhabitants); $P_s$ = saturation population (inhabitants)
- $K_a, K_g, K_d, K_1, i, c, r, s$ = coefficients (obtaining coefficients by regression analysis is preferable as all of the existing data series can be used, and not only $P_o, P_1$ e $P_2$)

The main methods or models used for population forecasts are (Fair et al, 1973; CETESB, 1978; Barnes et al, 1981; Qasim, 1985; Metcalf & Eddy, 1991):

- *linear (arithmetic) growth*
- *geometric growth*
- *multiplicative regression*
- *decreasing growth rate*
- *logistic growth*
- *graphical comparison between similar communities*
- *method of ratio and correlation*
- *prediction based on employment forecast or other utilities forecast*

Tables 2.1 and 2.2 list the main characteristics of the various methods. All of the methods presented in Table 2.1 can also be solved through statistical regression analysis (linear or non-linear). Such methods are found in many commercially available computer programs. Whenever possible it is always best to adopt a regression analysis that allows the incorporation of a largest historical data series instead of two or three 3 points, such as the algebraic methods presented in Table 2.1.

The results of the population forecast must be coherent with the population density in the area under analysis. The population density data are also useful in the computation of flows and loads resulting from a certain area or basin within the town. Typical population density values are presented in Table 2.3. Table 2.4 presents typical saturation population densities, in highly occupied metropolitan areas.

When making population forecasts, the following points must be taken into consideration:

- The population studies are normally very complex. All the variables (unfortunately not always quantifiable) that could interact in the specific locality under study must be analysed. Unexpected events can still occur, which can completely change the predicted trajectory of the population growth. This emphasises the need to establish a realistic value for the planning horizon and for the implementation stages of the WWTP.
- The mathematical sophistication associated with the determination of the coefficients of some forecast equations loses its meaning if it is not based on parallel information, often non-quantifiable, such as social, economical, geographical and historical aspects.
- The common sense of the analyst is very important in the choice of the forecast and in the interpretation of the results. Even though the choice of method is based on the best fit with census data, the extrapolation of the curve requires perception and caution.

Table 2.2. Population forecast based on indirect quantification methods

| Method | Description |
|---|---|
| Graphical comparison | The method involves the graphical fitting of the past population under study. The population data of other similar but larger towns are plotted in a manner that the curves coincide at the current value of the population of the town under study. These curves are used as references for the forecast of the town under study. |
| Ratio and correlation | It is assumed that the town population follows the same trend of the region (physical or political region) in which it is inserted. Based on the census records, the ratio "town population / region population" is calculated and projected for future years. The town population is obtained from the population forecast for the region (made at a planning level by another body) and the calculated ratio. |
| Forecast of employment and utility services | The population is estimated using a job prediction (made by another body). Based on the past population data and people employed, the "job/population" ratio is calculated and projected for future years. The town population is calculated from the forecast of the number of jobs in the town. The procedure is similar to the ratio method. The same methodology can be adopted from the forecast of utility services, such as electricity, water, telephone, etc. The service utility companies normally undertake studies of forecast and expansion of their services with relative reliability. |

*Note:* The forecast of the ratios can be done based on regression analysis
*Source:* Qasim (1985)

Table 2.3. Typical population densities as a function of land use

| Land use | Population density | |
|---|---|---|
| | (inhab/ha) | (inhab/km$^2$) |
| Residential areas | | |
| • single-family dwellings, large lots | 12–36 | 1,200–3,600 |
| • single-family dwellings, small lots | 36–90 | 3,600–9,000 |
| • multiple-family dwellings, small lots | 90–250 | 9,000–25,000 |
| Apartments | 250–2,500 | 25,000–250,000 |
| Commercial areas | 36–75 | 3,600–7,500 |
| Industrial areas | 12–36 | 1,200–3,600 |
| Total (excluding parks and other large-scale equipment) | 25–125 | 2,500–12,500 |

*Source:* adapted from Fair et al (1973) and Qasim (1985) (rounded up values)

Table 2.4. Population densities and average street length per hectare, under saturation conditions, in highly occupied metropolitan areas

| Land use | Saturation population density (inhab/ha) | Average street length (m/ha) |
|---|---|---|
| High standard residential areas, with standard lots of 800 m$^2$ | 100 | 150 |
| Intermediate standard residential areas, with standard lots of 450 m$^2$ | 120 | 180 |
| Popular residential areas, with standard lots of 250 m$^2$ | 150 | 200 |
| Centrally-located mixed residential–commercial areas, with predominance of 3–4 storey buildings | 300 | 150 |
| Centrally-located residential areas, with predominance of 10–12 storey buildings | 450 | 150 |
| Mixed residential–commercial–industrial urban areas, with predominance of commerce and small industries | 600 | 150 |
| Centrally-located residential areas, with predominance of office buildings | 1000 | 200 |

Average data from São Paulo Metropolitan Area, Brazil
*Source:* Alem Sobrinho and Tsutiya (1999)

---

### Example 2.1

Based on the following census records, undertake the population forecast using the methods based on mathematical formulas (Table 2.1). Data:

| Year | Population (inhabitants) |
|---|---|
| 1980 | 10,585 |
| 1990 | 23,150 |
| 2000 | 40,000 |

**Solution:**

**a)   Nomenclature of the years and populations**

According to Table 2.1, there is the following nomenclature:

$t_0 = 1980$  $P_0 = 10,585$ inhab
$t_1 = 1990$  $P_1 = 23,150$ inhab
$t_2 = 2000$  $P_2 = 40,000$ inhab

**b)   Linear (arithmetic) growth**

$$K_a = \frac{P_2 - P_o}{t_2 - t_o} = \frac{40000 - 10585}{2000 - 1980} = 1470.8$$

$$P_t = P_o + K_a.(t - t_o) = 10585 + 1470.8 \times (t - 1980)$$

***Example 2.1 (Continued)***

For example, to calculate the population in the year 2005, t is substituted for 2005 in the above equation. For the year 2010, t = 2010, and so on.

**c)   Geometric growth**

$$K_g = \frac{\ln P_2 - \ln P_o}{t_2 - t_o} = \frac{\ln 40000 - \ln 10585}{2000 - 1980} = 0.0665$$

$$P_t = P_0.e^{K_g.(t-t_o)} = 10585.e^{0.0665\times(t-1980)}$$

**d)   Decreasing growth rate**

$$P_s = \frac{2.P_o.P_1.P_2 - P_1^2.(P_o + P_2)}{P_o.P_2 - P_1^2}$$

$$= \frac{2 \times 10585 \times 23150 \times 40000 - 23150^2 \times (10585 + 40000)}{10585 \times 40000 - 23150^2}$$

$$= 66709$$

The saturation population is, therefore, 66,709 inhabitants.

$$K_d = \frac{-\ln[(P_s - P_2)/(P_s - P_o)]}{t_2 - t_o}$$

$$= \frac{-\ln[66709 - 40000)/(66709 - 10585)]}{2000 - 1980} = 0.0371$$

$$P_t = P_O + (P_s - P_o).\left[1 - e^{-K_d.(t-t_o)}\right]$$

$$= 10585 + (66709 - 10585) \times \left(1 - e^{-0.0371\times(t-1980)}\right)$$

**e)   Logistic growth**

$$P_s = \frac{2.P_o.P_1.P_2 - P_1^2.(P_o + P_2)}{P_o.P_2 - P_1^2}$$

$$= \frac{2 \times 10585 \times 23150 \times 40000 - 23150^2 \times (10585 + 40000)}{10585 \times 40000 - 23150^2}$$

$$= 66709$$

$$c = \frac{(P_s - P_o)}{P_o} = \frac{(66709 - 10585)}{10585} = 5.3022$$

$$K_1 = \frac{1}{t_2 - t_1}.\ln\left[\frac{P_o.(P_s - P_1)}{P_1.(P_s - P_o)}\right]$$

$$= \frac{1}{2000 - 1990}.\ln\left[\frac{10585 \times (66709 - 23150)}{23150 \times (66709 - 10585)}\right] = -0,1036$$

### Example 2.1 (Continued)

$$P_t = \frac{P_s}{1 + c.e^{K_1.(t-t_0)}} = \frac{66709}{1 + 5.3022.e^{-0.1036 \times (t-1980)}}$$

The inflexion in the S-shaped curve occurs at the following year and population:

$$\text{Inflexion time} = t_0 - \frac{\ln(c)}{K_1} = 1980 - \frac{\ln(5.3022)}{-0.1036} = 1996$$

$$\text{Population at inflexion} = \frac{P_s}{2} = \frac{66709}{2} = 33354 \text{ inhab}$$

Before inflexion (year 1996), population growth presented an increasing rate and, after it, a decreasing rate.

**f) Results in table and graphic form**

| Nomenclature | Year | Actual population (census) | Population forecast | | | |
|---|---|---|---|---|---|---|
| | | | Linear | Geometric | Decreasing rate | Logistic |
| $P_0$ | 1980 | 10585 | 10585 | 10585 | 10585 | 10585 |
| $P_1$ | 1990 | 23150 | 25293 | 20577 | 27992 | 23150 |
| $P_2$ | 2000 | 40000 | 40000 | 40000 | 40000 | 40000 |
| – | 2005 | – | 47354 | 55770 | 44525 | 47725 |
| – | 2010 | – | 54708 | 77758 | 48284 | 53930 |
| – | 2015 | – | 62061 | 108414 | 51405 | 58457 |
| – | 2020 | – | 69415 | 151157 | 53998 | 61534 |

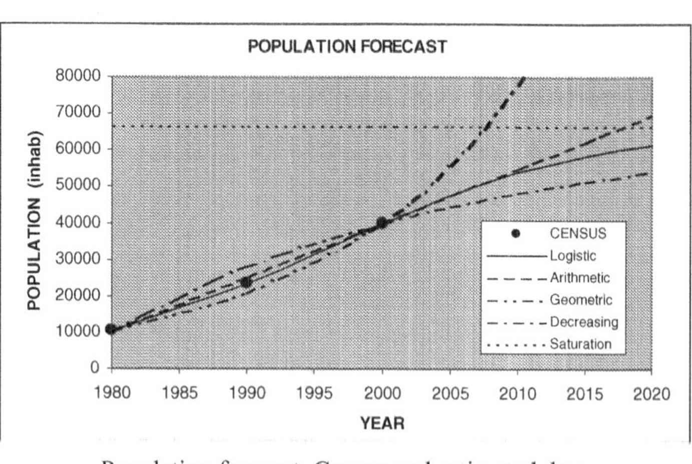


Population forecast. Census and estimated data

### *Example 2.1 (Continued)*

From the graph and table, the following points specific for this data group can be seen:

- The census data (population of the years 1980 to 2000) present an increasing growth rate trend. Visually, it is seen that the decreasing rate model does not fit well.
- The geometric method leads to very high future estimates (that can turn out to be true or not, but that are far away from the other forecasts).
- The logistic and decreasing rate methods tend to the saturation population. (66,709 inhabitants, indicated on the graph)
- In all methods, the calculated population values for the years $P_0$ and $P_2$ are equal to the measured values.
- The population forecast as such is only from year 2000. The years with census data are plotted to permit the visual interpretation of the fit of the curves to the measured data (1980, 1990, 2000).
- The best-fit curve may be chosen from statistical criteria, which give an indication of the prediction error (usually based on the sum of the squared errors), where error or residual is the difference between the estimated and the observed data.
- Spreadsheets may be used, to find the value of the coefficients that lead to the minimum sum of the squared errors (e.g. solver tool in Excel®).

## 2.1.2.3 Average water consumption

As mentioned, the domestic flow is a function of the water consumption. Typical values of per capita water consumption for populations provided with household water connections are presented in Table 2.5.

These values can vary from locality to locality. Table 2.6 presents various factors that influence water consumption. The data listed in Table 2.5 are simply typical average values, being naturally subjected to all the variability resulting from the factors listed in Table 2.6.

Table 2.5. Typical ranges of per capita water consumption

| Community size | Population range (inhabitants) | Per capita water consumption (L/inhab.d) |
|---|---|---|
| Rural settlement | <5,000 | 90–140 |
| Village | 5,000–10,000 | 100–160 |
| Small town | 10,000–50,000 | 110–180 |
| Average town | 50,000–250,000 | 120–220 |
| Large city | >250,000 | 150–300 |

*Note:* in places with severe water shortages, these values may be smaller
*Source:* Adapted from CETESB (1977; 1978), Barnes et al (1981), Dahlhaus & Damrath (1982), Hosang & Bischof (1984)

Table 2.6. Factors that influence water consumption

| Influencing factor | Comment |
|---|---|
| • Water availability | • In locations of water shortage consumption tends to be less |
| • Climate | • Warmer climates induce a greater water consumption |
| • Community size | • Larger cities generally present a larger per capita water consumption (to account for strong commercial and institutional activities) |
| • Economic level of the community | • A higher economic level is associated with a higher water consumption |
| • Level of industrialisation | • Industrialised locations present a higher consumption |
| • Metering of household consumption | • Metering inhibits greater consumption |
| • Water cost | • A higher cost reduces consumption |
| • Water pressure | • High pressure in the distribution system induces greater use and wastage |
| • System losses | • Losses in the water distribution network imply the necessity of a greater water production |

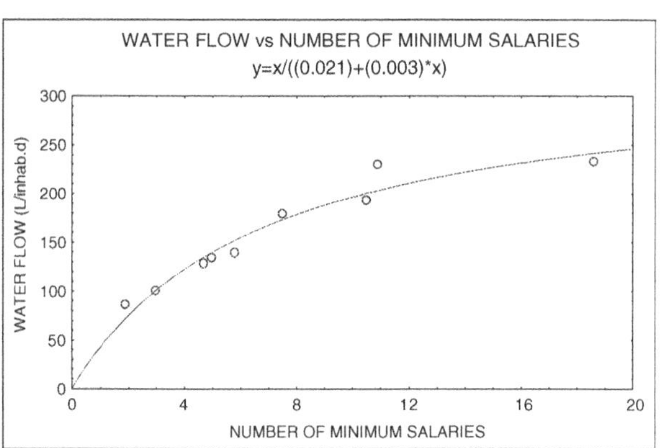


Figure 2.2.   Per capita water consumption as a function of family salary in Belo Horizonte, Brazil

Campos and von Sperling (1996) observed, for predominantly residential sewage originating from nine sub-catchment areas in Belo Horizonte, Brazil, a strong relationship between per capita water consumption and average monthly family income (in number of minimum salaries) (Figure 2.2). Naturally the data are site specific and require caution in their extrapolation to other conditions.

Water consumption data from 45 municipalities in the State of Minas Gerais, Brazil (von Sperling et al, 2002), were investigated by the author. The State of Minas Gerais has many features in common with Brazil, as a whole, and many

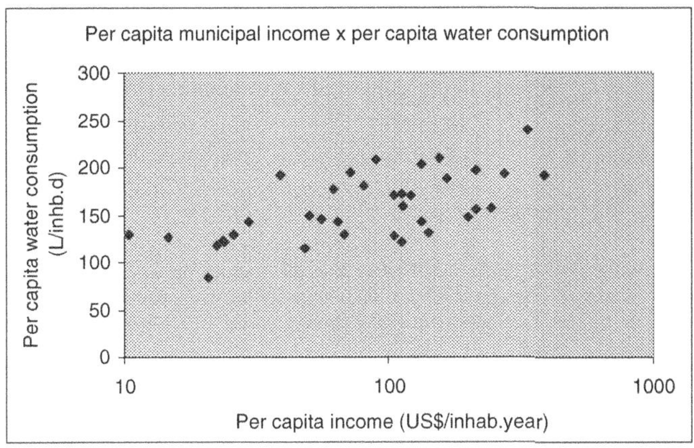


Figure 2.3.   Relationship between per capita water consumption and per capita income.
Data from the state of Minas Gerais, Brazil (von Sperling et al, 2002) (US$1.00 = R$2.50)

other developing countries, because it presents regions with high and low economic level, rainfall and temperature. The range of variation in the data was: per capita water consumption: 84 to 248 l/inhab.d; urban population: 4,000–2,300,000 inhabitants; average per capita income: US$8–1600 per inhabitant per year; mean yearly temperature: 20–26° C; mean yearly rainfall: 300–1750 mm/year. Figure 2.3 presents the relation of the per capita water consumption with per capita income, which was the clearer one. The analysis should be done only in terms of trends and average values, since the correlation coefficient was not high, as a result of a substantial scatter in the data.

Figure 2.4 presents the ranges of variation of the per capita water consumption as a function of the category of the per capita income and rainfall of the 45 municipalities (separation between low and high income: US$110/inhab.year, corresponding to the median of values; separation of high and low rainfall: 1350 mm/year, corresponding to the average value of the State of Minas Gerais). Naturally these values are region specific, but it is believed that a certain extrapolation of trends and ranges can be done, but always judiciously.

Table 2.7 shows ranges of per capita water consumption as a function of income and rainfall, based on the 25 and 75 percentiles presented in Figure 2.4.

Tables 2.8 and 2.9 show the ranges of average water consumption values for various commercial establishments and institutions. This information, which should only be used in the absence of more specific data, is particularly useful in the design of sewage treatment works for small areas, in which the contribution of individual important establishments could have an importance in the general flow calculations.

## 2.1.2.4 Average sewage flow

In general, the production of sewage corresponds approximately to the water consumption. However, the fraction of the sewage that enters the sewerage system can

Table 2.7. Ranges of water consumption values, based on
45 municipalities in the State of Minas Gerais, Brazil

| | Ranges of per capita water consumption (L/inhab.d) | |
|---|---|---|
| Income | Low rainfall | High rainfall |
| Low | 120–165 | 130–190 |
| High | 140–180 | 150–200 |

*Notes:*

- Ranges based on 25 and 75 percentile values from Fig. 2.4
- In larger towns (greater than 200,000 inhabitants), the per capita water consumption was on average approximately 10% higher than in smaller towns
- The ranges present usual values, and it is frequent to observe values outside them

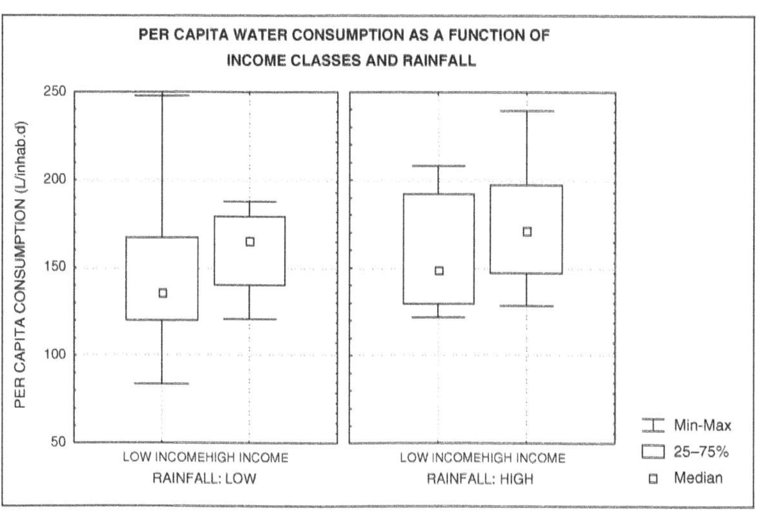


Figure 2.4. Box-and-whisker plot of the per capita water consumption values as a function of categories for per capita income and mean yearly rainfall (45 municipalities in the State of Minas Gerais, Brazil)

be different, due to the fact that part of the water consumed could be incorporated into the storm water system or infiltrate (e.g. watering of gardens and parks). Other influencing factors in a separate sewerage system are: (a) clandestine sewage connections to the storm water system, (b) clandestine connections of storm water into the separate sewerage system and (c) infiltration. The last point is covered separately in Section 2.1.3.

The fraction of the supplied water that enters the sewerage system in the form of sewage is called Return Coefficient (R = sewage flow/water flow). Typical values vary between 60% and 100%, and a value of 80% (R = 0.8) is usually adopted.

Table 2.8. Typical water consumption in some commercial establishments

| Establishment | Unit | Flow range (L/unit.d) |
|---|---|---|
| Airport | Passenger | 8–15 |
| Accommodation (lodging house) | Resident | 80–150 |
| Public toilet | User | 10–25 |
| Bar | Customer | 5–15 |
| Cinema/theatre | Seat | 2–10 |
| Office | Employee | 30–70 |
| Hotel | Guest | 100–200 |
| | Employee | 30–50 |
| Industry (sanitary sewage only) | Employee | 50–80 |
| Snack bar | Customer | 4–20 |
| Laundry – commercial | Machine | 2,000–4,000 |
| Laundry – automatic | Machine | 1,500–2,500 |
| Shop | Toilet | 1,000–2,000 |
| | Employee | 30–50 |
| Department store | Toilet | 1,600–2,400 |
| | Employee | 30–50 |
| | $m^2$ of area | 5–12 |
| Petrol station | Vehicle attended | 25–50 |
| Restaurant | Meal | 15–30 |
| Shopping centre | Employee | 30–50 |
| | $m^2$ of area | 4–10 |

*Source:* EPA (1977), Hosang and Bischof (1984), Tchobanoglous and Schroeder (1985), Qasim (1985), Metcalf & Eddy (1991), NBR-7229/93

Table 2.9. Typical water consumption in some institutional establishments

| Establishment | Unit | Flow range (L/unit.d) |
|---|---|---|
| Rest home | Resident | 200–450 |
| | Employee | 20–60 |
| School | Student | 50–100 |
| – with cafeteria, gymnasium, showers | Student | 40–80 |
| – with cafeteria only | Student | 20–60 |
| – without cafeteria and gymnasium | | |
| Hospital | Bed | 300–1000 |
| | Employee | 20–60 |
| Prison | Inmate | 200–500 |
| | Employee | 20–60 |

*Source:* EPA (1977), Hosang and Bischof (1984), Tchobanoglous and Schroeder (1985), Qasim (1985), Metcalf & Eddy (1991)

The average domestic sewage flow calculation is given by:

$$Q_{d_{av}} = \frac{Pop.L_{pcd}.R}{1000} \quad (m^3/d) \tag{2.2}$$

$$Q_{d_{av}} = \frac{Pop.L_{pcd}.R}{86400} \quad (L/s) \tag{2.3}$$

where:

$Q_{d_{av}}$ = average domestic sewage flow (m³/d or L/s)
$L_{pcd}$ = per capita water consumption (L/inhab.d)
  R = sewage flow/water flow return coefficient

It is important to notice that the water flow to be considered is the flow actually **consumed**, and not the flow produced by the water treatment works. The water flow produced is higher than that consumed due to unaccounted water losses in the distribution system, which can vary typically from 20 to 50%. Thus in a locality where the loss is 30%, for each 100 m³ of water produced, 30 m³ are unaccounted for and only 70 m³ are consumed. Of this 70 m³, around 80% (56 m³/d) return in the form of sewage to the sewerage system.

## 2.1.2.5 Flow variations. Maximum and minimum flows

Water consumption and wastewater generation in a locality vary throughout the day (hourly variations), during the week (daily variations) and throughout the year (seasonal variations).

Figure 2.5 presents typical hourly influent flowrate variations in a WWTP. Two main peaks can be observed: a peak at the beginning of the morning (more pronounced) and a peak at the beginning of the evening (more distributed). The average daily flow corresponds to the line that separates equal areas, below and above the line.

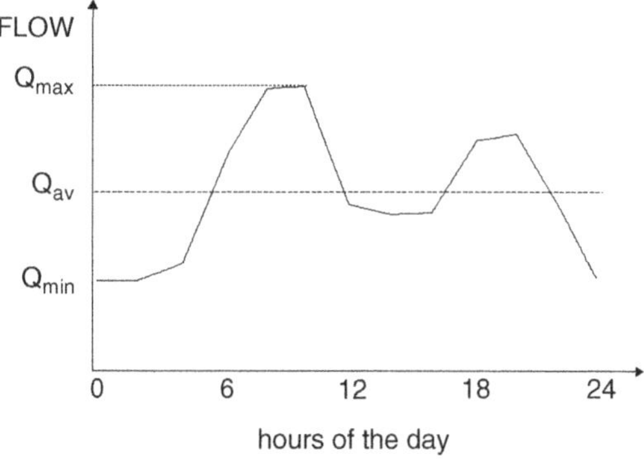


Figure 2.5.   Typical hourly flow variations in the influent to a sewage treatment works

The following coefficients are frequently used to allow the estimation of minimum and maximum **water** flows:

- $K_1 = 1.2$ (peak coefficient for the day with the highest water consumption)
- $K_2 = 1.5$ (peak coefficient for the hour with the highest water consumption)
- $K_3 = 0.5$ (reduction coefficient for the hour with the lowest water consumption)

Table 2.10. Coefficient of hourly variation of sewage flow

| $Q_{max}/Q_{av}$ | $Q_{min}/Q_{av}$ | Author | Reference |
|---|---|---|---|
| $1+(14/(4+P^{0.5}))$ | – | Harmon | Qasim (1985) |
| $5P^{-0.16}$ | $0.2P^{0.16}$ | Gifft | Fair et al (1973) |

*Notes:*
P = population, in thousands
Gifft's formula is indicated for P < 200 (population < 200,000 inhabitants)

Thus, the maximum and minimum water flows can be given by the formulas:

$$Q_{max} = Q_{av} \cdot K_1 \cdot K_2 = 1.8\,Q_{av} \qquad (2.4)$$

$$Q_{min} = Q_{av} \cdot K_3 = 0.5\,Q_{av} \qquad (2.5)$$

If it is possible to carry out flow measurements, to establish the real flow variations, the actual data should be used in the design. The coefficients $K_1$, $K_2$ and $K_3$ are generalised, thus probably not allowing the accurate reproduction of the flow variations in the locality under analysis. Over- or underestimated values affect directly the technical and economical performance of the sewage works design.

When considering hourly variations of **wastewater** flow, it should be taken into consideration that the fluctuations are absorbed and reduced in amplitude along the sewerage system. It is easy to understand that the larger the network (or the population), the lower are the chances of peak flows to overlap simultaneously in the works entrance. Thus the residence time in the sewerage system has a large influence on the absorption of the peak flows. Based on this concept, some authors have developed formulas for correlating the coefficients of variation with population, or with average flow (Table 2.10). As an illustration, the following table presents the calculated coefficients for different populations.

| | $Q_{max}/Q_{av}$ | | $Q_{min}/Q_{av}$ |
|---|---|---|---|
| Population | Harmon | Gifft | Gifft |
| 1,000 | 3.8 | 5.0 | 0.20 |
| 10,000 | 3.0 | 3.4 | 0.14 |
| 100,000 | 2.0 | 2.3 | 0.09 |
| 1,000,000 | 1.4 | - | - |

It can be observed that even the product of the coefficients $K_1$ and $K_2$ utilised for water supply, and frequently adopted as 1.8, could induce an underestimated ratio $Q_{max}/Q_{av}$ for a wide population range.

Table 2.11. Approximate values of infiltration rates in sewerage systems

| Pipe diameter | Type of joint | Groundwater level | Soil permeability | Infiltration coefficient | |
|---|---|---|---|---|---|
| | | | | L/s.km | m³/d.km |
| < 400 mm | Elastic | Below the pipes | Low | 0.05 | 4 |
| | | | High | 0.10 | 9 |
| | | Above the pipes | Low | 0.15 | 13 |
| | | | High | 0.30 | 26 |
| | Non-elastic | Below the pipes | Low | 0.05 | 4 |
| | | | High | 0.50 | 43 |
| | | Above the pipes | Low | 0.50 | 43 |
| | | | High | 1.00 | 86 |
| > 400 mm | – | – | – | 1.00 | 86 |

*Source:* Crespo (1997)


## 2.1.3 Infiltration flow

Infiltration in a sewerage system occurs through defective pipes, connections, joints or manholes. The quantity of infiltrated water depends on various factors, such as the extension of the collection network, pipeline diameters, drainage area, soil type, water table depth, topography and population density (number of connections per unit area) (Metcalf & Eddy, 1991).

When no specific local data are available, infiltration rate is normally expressed in terms of flow per extension of the sewerage system or per area served. The values presented in Table 2.11 can be used as a first estimate, when no specific local data are available (Crespo, 1997).

Metcalf & Eddy (1991) present the infiltration coefficient as a function of the pipe diameter: **0.01 to 1.0 m³/d.km per mm**. For instance, for a pipe diameter of 200 mm, the infiltration rate will range between 2 to 200 m³/d.km.

The length of the network may be measured in the locality by using the map of the location of the sewerage system. In the absence of these data (for instance, for future populations), in preliminary studies of smaller localities, where the population density is usually less, values around 2.5 to 3.5 m of network per inhabitant may be adopted. In medium-size cities this value could be reduced to around 2.0 to 3.0 m/inhab and in densely populated regions, even smaller values may be reached (1.0 to 2.0 m/inhab or even lower). Figure 2.6, based on the 45 municipalities described in Section 2.1.2, presents the ranges of variation of per capita length of sewerage network for two population categories.

Based on the infiltration values per unit length and the per capita sewerage network length, per capita infiltration values may be estimated to range between **8 to 150 L/inhab.d**, excluding the extreme values. In areal terms, based on typical population densities (25 to 125 inhab/ha), infiltration rates between **0.2 and 20 m³/d per ha** of drainage area (**20 to 2000 m³/d.km²**) are obtained. These ranges are very wide, and the designer should analyse carefully the prevailing conditions in

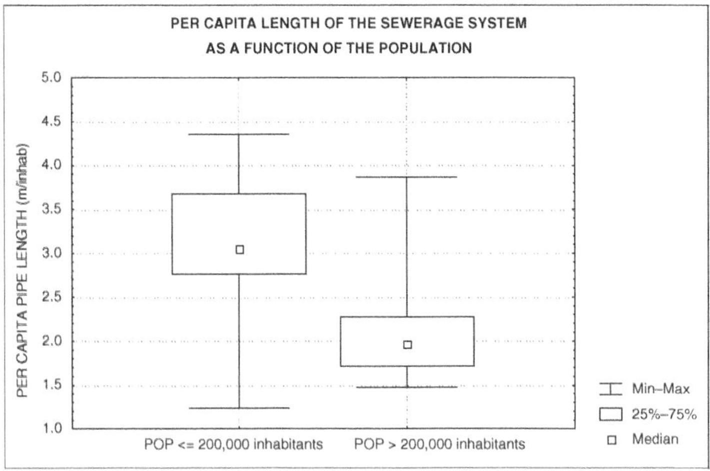


Figure 2.6. Box-and-whisker plot of the per capita length of sewerage network, as a function of two categories of population size (45 municipalities in Minas Gerais, Brazil)

the sewerage network in order to obtain narrower ranges, which could best represent the specific conditions in the community under analysis. The utilisation of good materials and construction procedures helps in reducing the infiltration rates.

In the calculation of the total influent flow to a WWTP, average infiltration values may be used for the computation of average and maximum influent flowrates. For minimum flow conditions, infiltration can be excluded, as a safety measure (in the case of minimum flow, the safety in a design is in the direction of establishing the lowest flow).

## 2.1.4 Industrial wastewater flow

Industrial wastewater flow is a function of the type and size of the industry, manufacturing process, level of recycling, existence of pre-treatment, etc. Even in the case of two industries that manufacture essentially the same product, the wastewater flows can diverge substantially.

If there are large industries contributing to the public sewerage system and subsequently to a WWTP, the adequate evaluation of their respective flows is of great importance. Industrial wastewater has a great influence in the planning and operation of a WWTP. Specific data must be obtained for each significant industry, through industrial surveys, thus allowing the supply of data of interest for the project. With relation to the water consumption and the generation of wastewater, the following information at least must be obtained for the main industries:

- *Water consumption*
  - Total volume consumed (per day or month)
  - Volume consumed in the various stages of the process

- • Internal recirculations
- • Water origin (public supply, wells, etc.)
- • Internal systems of water treatment
- • *Wastewater production*
  - • Total flow
  - • Number of discharge points (with the corresponding industrial process associated with each point)
  - • Discharge pattern (continuous or intermittent; duration and frequency) in each discharge point
  - • Discharge destination (sewerage system, watercourse)
  - • Occasional mixing of wastewater with domestic sewage and storm water

Additionally, whenever possible, effluent flow measurements must be carried out throughout the working day, to record the discharge pattern and variations.

In the event of having no specific information available for the industry, Table 2.12 can be used as a starting point to allow estimation of the probable effluent flow range. These values are presented in terms of water consumption per unit of product manufactured. For simplicity it can be assumed that sewage flow is equal to water consumption. It can be seen from the table that there is a great variety of consumption values for the same type of industry. If there are no specific data available for the industry in question, specific literature references relative to the industrial process in focus must be consulted. The table presented only gives a starting point for more superficial or general studies.

The daily discharge pattern for industrial wastewater does not follow the domestic flow variations, changing substantially from industry to industry. Industrial flow peaks do not necessarily coincide with the domestic peaks, that is to say, the total maximum flow (domestic + industrial) is normally less than the simple sum of the maximum flows.


## 2.2 WASTEWATER COMPOSITION

### 2.2.1 Quality parameters

Domestic sewage contains approximately 99.9% water. The remaining part includes organic and inorganic, suspended and dissolved solids, together with microorganisms. It is because of this 0.1% that water pollution takes place and the wastewater needs to be treated.

The composition of the wastewater is a function of the uses to which the water was submitted. These uses, and the form with which they were exercised, vary with climate, social and economic situation and population habits.

In the design of a WWTP, there is normally no interest in determining the various compounds that make up wastewater. This is due, not only to the difficulty in

Table 2.12. Specific average flows from some industries

| Type | Activity | Unit | Water consumption per unit (m³/unit) (*) |
|---|---|---|---|
| *Food* | Canned fruit and vegetables | 1 tonne product | 4–50 |
| | Sweets | 1 tonne product | 5–25 |
| | Sugar cane | 1 tonne sugar | 0.5 – 10.0 |
| | Slaughter houses | 1 cow or 2,5 pig | 0.5–3.0 |
| | Dairy (milk) | 1000 L milk | 1–10 |
| | Dairy (cheese or butter) | 1000 L milk | 2–10 |
| | Margarine | 1 tonne margarine | 20 |
| | Brewery | 1000 L beer | 5–20 |
| | Bakery | 1 tonne bread | 2–4 |
| | Soft drinks | 1000 L soft drinks | 2–5 |
| *Textiles* | Cotton | 1 tonne product | 120–750 |
| | Wool | 1 tonne product | 500–600 |
| | Rayon | 1 tonne product | 25–60 |
| | Nylon | 1 tonne product | 100–150 |
| | Polyester | 1 tonne product | 60–130 |
| | Wool washing | 1 tonne wool | 20–70 |
| | Dyeing | 1 tonne product | 20–60 |
| *Leather / tanneries* | Tannery | 1 tonne hide | 20–40 |
| | Shoe | 1000 pairs of shoes | 5 |
| *Pulp and paper* | Pulp fabrication | 1 tonne product | 15–200 |
| | Pulp bleaching | 1 tonne product | 80–200 |
| | Paper fabrication | 1 tonne product | 30–250 |
| | Pulp and paper integrated | 1 tonne product | 200–250 |
| *Chemical industries* | Paint | 1 employee | 110 L/d |
| | Glass | 1 tonne glass | 3–30 |
| | Soap | 1 tonne soap | 25–200 |
| | Acid, base, salt | 1 tonne chlorine | 50 |
| | Rubber | 1 tonne product | 100–150 |
| | Synthetic rubber | 1 tonne product | 500 |
| | Petroleum refinery | 1 barrel (117 L) | 0.2–0.4 |
| | Detergent | 1 tonne product | 13 |
| | Ammonia | 1 tonne product | 100–130 |
| | Carbon dioxide | 1 tonne product | 60–90 |
| | Petroleum | 1 tonne product | 7–30 |
| | Lactose | 1 tonne product | 600–800 |
| | Sulphur | 1 tonne product | 8–10 |
| | Pharmaceutical products (vitamins) | 1 tonne product | 10–30 |
| *Manufacturing products* | Precision mechanics, optical, electronic | 1 employee | 20–40 L/d |
| | Fine ceramic | 1 employee | 40 L/d |
| | Machine industry | 1 employee | 40 L/d |
| *Metallurgy* | Foundry | 1 tonne pig iron | 3–8 |
| | Lamination | 1 tonne product | 8–50 |
| | Forging | 1 tonne product | 80 |
| | Electroplating | 1 m³ of solution | 1–25 |
| | Iron and steel plating industry | 1 employee | 60 L/d |
| *Mining* | Iron | 1 m³ mineral taken | 16 |
| | Coal | 1 tonne coal | 2–10 |

* Consumption in m³ per unit produced or L/d per employee
Source: CETESB (1976), Downing (1978), Arceivala (1981), Hosang and Bischof (1984), Imhoff & Imhoff (1985), Metcalf & Eddy (1991), Derísio (1992)

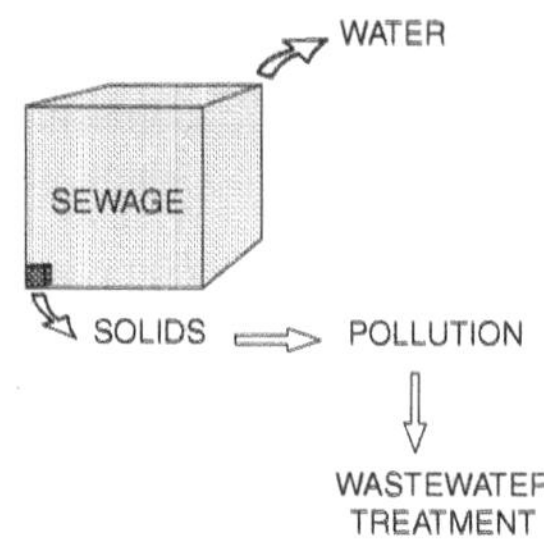



Figure 2.7.   Solids in sewage

undertaking the various laboratory tests, but also to the fact that the results themselves cannot be directly utilised as elements in design and operation. Therefore, many times it is preferable to utilise indirect parameters that represent the character or the polluting potential of the wastewater in question. These parameters define the quality of the sewage, and can be divided into three categories: *physical*, *chemical* and *biological* parameters.

## 2.2.2 Main characteristics of wastewater

Tables 2.13, 2.14 and 2.15 present the main physical, chemical and biological characteristics of domestic sewage.

Table 2.13.  Main physical characteristics of domestic sewage

| Parameter | Description |
|---|---|
| *Temperature* | • Slightly higher than in drinking water<br>• Variations according to the seasons of the years (more stable than the air temperature)<br>• Influences microbial activity<br>• Influences solubility of gases<br>• Influences viscosity of the liquid |
| *Colour* | • Fresh sewage: slight grey<br>• Septic sewage: dark grey or black |
| *Odour* | • Fresh sewage: oily odour, relatively unpleasant<br>• Septic sewage: foul odour (unpleasant), due to hydrogen sulphide gas and other decomposition by-products<br>• Industrial wastewater: characteristic odours |
| *Turbidity* | • Caused by a great variety of suspended solids<br>• Fresher or more concentrated sewage: generally greater turbidity |

*Source:* Adapted from Qasim (1985)

Table 2.14. Main chemical characteristics of domestic sewage

| Parameter | Description |
| --- | --- |
| ***TOTAL SOLIDS*** | *Organic and inorganic; suspended and dissolved; settleable* |
| • *Suspended* | • Part of organic and inorganic solids that are non-filterable |
|   • *Fixed* | • Mineral compounds, not oxidisable by heat, inert, which are part of the suspended solids |
|   • *Volatile* | • Organic compounds, oxidisable by heat, which are part of the suspended solids |
| • *Dissolved* | • Part of organic and inorganic solids that are filterable. Normally considered having a dimension less than $10^{-3}\,\mu\text{m}$. |
|   • *Fixed* | • Mineral compounds of the dissolved solids. |
|   • *Volatile* | • Organic compounds of the dissolved solids |
| • *Settleable* | • Part of organic and inorganic solids that settle in 1 hour in an Imhoff cone. Approximate indication of the settling in a sedimentation tank. |
| ***ORGANIC MATTER*** | *Heterogeneous mixture of various organic compounds. Main components: proteins, carbohydrates and lipids.* |
| *Indirect determination* | |
| • *BOD$_5$* | • Biochemical Oxygen Demand. Measured at 5 days and 20 °C. Associated with the biodegradable fraction of carbonaceous organic compounds. Measure of the oxygen consumed after 5 days by the microorganisms in the biochemical stabilisation of the organic matter. |
| • *COD* | • Chemical Oxygen Demand. Represents the quantity of oxygen required to chemically stabilise the carbonaceous organic matter. Uses strong oxidising agents under acidic conditions. |
| • *Ultimate BOD* | • Ultimate Biochemical Oxygen Demand. Represents the total oxygen consumed at the end of several days, by the microorganisms in the biochemical stabilisation of the organic matter. |
| *Direct determination* | |
| • *TOC* | • Total Organic Carbon. Direct measure of the carbonaceous organic matter. Determined through the conversion of organic carbon into carbon dioxide. |
| ***TOTAL NITROGEN*** | *Total nitrogen includes organic nitrogen, ammonia, nitrite and nitrate. It is an essential nutrient for microorganisms' growth in biological wastewater treatment. Organic nitrogen and ammonia together are called Total Kjeldahl Nitrogen (TKN).* |
| • *Organic nitrogen* | • Nitrogen in the form of proteins, aminoacids and urea. |
| • *Ammonia* | • Produced in the first stage of the decomposition of organic nitrogen. |
| • *Nitrite* | • Intermediate stage in the oxidation of ammonia. Practically absent in raw sewage. |
| • *Nitrate* | • Final product in the oxidation of ammonia. Practically absent in raw sewage. |
| ***TOTAL PHOSPHORUS*** | *Total phosphorus exists in organic and inorganic forms. It is an essential nutrient in biological wastewater treatment.* |
| • *Organic phosphorus* | • Combined with organic matter. |
| • *Inorganic phosphorus* | • Orthophosphates and polyphosphates. |

*(Continued)*

Table 2.14 (*Continued*)

| Parameter | Description |
|---|---|
| **pH** | *Indicator of the acidic or alkaline conditions of the wastewater. A solution is neutral at pH 7. Biological oxidation processes normally tend to reduce the pH.* |
| **ALKALINITY** | *Indicator of the buffer capacity of the medium (resistance to variations in pH). Caused by the presence of bicarbonate, carbonate and hydroxyl ions.* |
| **CHLORIDES** | *Originating from drinking water and human and industrial wastes.* |
| **OILS AND GREASE** | *Fraction of organic matter which is soluble in hexane. In domestic sewage, the sources are oils and fats used in food.* |

*Source:* adapted from Arceivala (1981), Qasim (1985), Metcalf & Eddy (1991)

Table 2.15. Main organisms present in domestic sewage

| Organism | Description |
|---|---|
| *Bacteria* | • Unicellular organisms<br>• Present in various forms and sizes<br>• Main organisms responsible for the stabilisation of organic matter<br>• Some bacteria are pathogenic, causing mainly intestinal diseases |
| *Archaea* | • Similar to bacteria in size and basic cell components<br>• Different from bacteria in their cell wall, cell material and RNA composition<br>• Important in anaerobic processes |
| *Algae* | • Autotrophic photosynthetic organisms, containing chlorophyll<br>• Important in the production of oxygen in water bodies and in some sewage treatment processes<br>• In lakes and reservoirs they can proliferate in excess, deteriorating the water quality |
| *Fungi* | • Predominantly aerobic, multicellular, non-photosynthetic, heterotrophic organisms<br>• Also of importance in the decomposition of organic matter<br>• Can grow under low pH conditions |
| *Protozoa* | • Usually unicellular organisms without cell wall<br>• Majority is aerobic or facultative<br>• Feed themselves on bacteria, algae and other microorganisms<br>• Essential in biological treatment to maintain an equilibrium between the various groups<br>• Some are pathogenic |
| *Viruses* | • Parasitic organisms, formed by the association of genetic material (DNA or RNA) and a protein structure<br>• Pathogenic and frequently difficult to remove in water or wastewater treatment |
| *Helminths* | • Higher-order animals<br>• Helminth eggs present in sewage can cause illnesses |

*Note:* algae are normally not present in untreated wastewater, but are present in the treated effluent from some processes (e.g. stabilisation ponds)
*Source:* Silva & Mara (1979), Tchobanoglous & Schroeder (1985), Metcalf & Eddy (1991), 2003

## 2.2.3  Main parameters defining the quality of wastewater

### 2.2.3.1  Preliminaries

The main parameters predominantly found in domestic sewage that deserve special consideration are:

- *solids*
- *indicators of organic matter*
- *nitrogen*
- *phosphorus*
- *indicators of faecal contamination*

### 2.2.3.2  Solids

All the contaminants of water, with the exception of dissolved gases, contribute to the solids load. In wastewater treatment, the solids can be classified according to (a) their size and state, (b) their chemical characteristics and (c) their settleability:

**Solids in sewage**

- *Classification by size and state*
  - Suspended solids (non-filterable)
  - Dissolved solids (filterable)
- *Classification by chemical characteristics*
  - Volatile solids (organic)
  - Fixed solids (inorganic)
- *Classification by settleability*
  - Settleable suspended solids
  - Non-settleable suspended solids

**a)   Classification by size**

The division of solids by size is above all a practical division. For convention it can be said that particles of smaller dimensions capable of passing through a filter paper of a specific size correspond to the **dissolved solids**, while those with larger dimensions and retained by the filter are considered **suspended solids**. To be more precise, the terms *filterable* (=dissolved) solids and *non-filterable* (=suspended) solids are more adequate. In an intermediate range there are the colloidal solids, which are of importance in water treatment, but are difficult to identify by the simple method of paper filtration. Water analysis results based on typical filter papers show that the major part of colloidal solids is separated as filterable (dissolved) solids.

Sometimes the term *particulate* is used to indicate that the solids are present as suspended solids. In this context, expressions as particulate BOD, COD, phosphorus, etc. are used, to indicate that they are linked to suspended solids. In contrast, soluble BOD, COD and phosphorus are associated with dissolved solids.

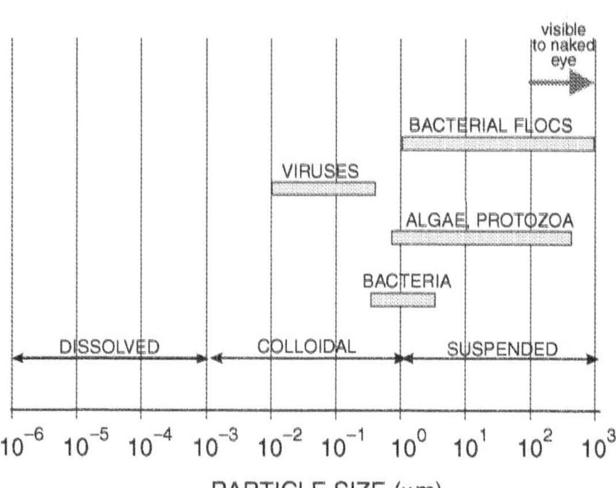


Figure 2.8.  Classification and distribution of solids as a function of size

Figure 2.8 shows the distribution of particles by size. In a general manner, are considered dissolved solids those with a diameter of less than $10^{-3}$ μm, colloidal solids those with a diameter between $10^{-3}$ and $10^{0}$ μm and as suspended solids those with a diameter greater than $10^{0}$ μm.

**b)  Classification by chemical characteristics**

If the solids are submitted to a high temperature (550 °C), the organic fraction is oxidised (volatilised), leaving after combustion only the inert fraction (unoxidised). The *volatile solids* represent an estimate of the *organic* matter in the solids, while the *non-volatile solids* (fixed) represent the *inorganic* or mineral matter. In summary:

$$\text{Total solids} \begin{cases} \nearrow \text{Volatile solids (organic matter)} \\ \searrow \text{Fixed solids (inorganic matter)} \end{cases}$$

**c)  Classification by settleability**

Settleable solids are considered those that are able to settle in a period of 1 hour. The volume of solids accumulated in the bottom of a recipient called an Imhoff Cone is measured and expressed as mL/L. The fraction that does not settle represents the non-settleable solids (usually not expressed in the results of the analysis).

Figure 2.9 shows the typical distribution between the various types of solids present in a raw sewage of average composition.

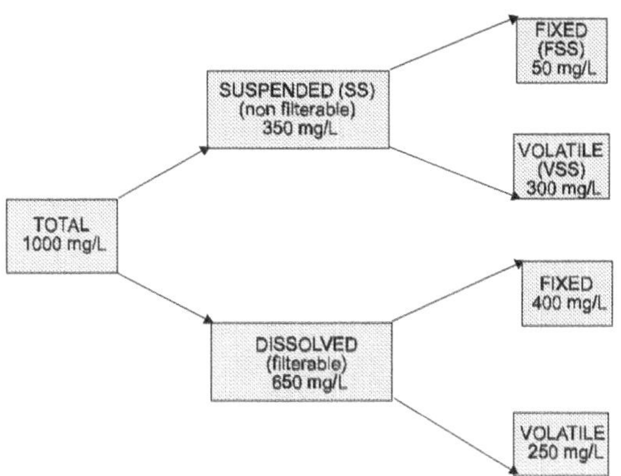



Figure 2.9.   Approximate distribution of the solids in raw sewage
(in terms of concentration)


## 2.2.3.3  Carbonaceous organic matter

The organic matter present in sewage is a characteristic of substantial importance, being the cause of one of the main water pollution problems: consumption of dissolved oxygen by the microorganisms in their metabolic processes of using and stabilising the organic matter. The organic substances present in sewage consist mainly of (Pessoa & Jordão, 1982):

- Protein compounds ($\approx$ 40%)
- Carbohydrates ($\approx$ 25 to $\approx$ 50%)
- Oils and grease ($\approx$ 10%)
- Urea, surfactants, phenols, pesticides and others (lower quantity)

The carbonaceous organic matter (based on organic carbon) present in the influent sewage to a WWTP can be divided into the following main fractions:

**Organic matter in sewage**

- *classification: in terms of form and size*
    - Suspended (particulate)
    - Dissolved (soluble)
- *classification: in terms of biodegradability*
    - Inert
    - Biodegradable

In practical terms it is not usually necessary to classify organic matter in terms of proteins, fats, carbohydrates, etc. Besides, there is a great difficulty in determining in the laboratory the various components of organic matter in wastewater, in view of the multiple forms and compounds in which it can be present. As a result, direct or indirect methods can be adopted for the quantification of organic matter:

- *Indirect methods: measurement of oxygen consumption*
  - Biochemical Oxygen Demand (BOD)
  - Ultimate Biochemical Oxygen Demand ($BOD_u$)
  - Chemical Oxygen Demand (COD)
- *Direct methods: measurement of organic carbon*
  - Total Organic Carbon (TOC)

## a)　Biochemical Oxygen Demand (BOD)

The main ecological effect of organic pollution in a water body is the decrease in the level of dissolved oxygen. Similarly, in sewage treatment using aerobic processes, the adequate supply of oxygen is essential, so that the metabolic processes of the microorganisms can lead to the stabilisation of the organic matter. The basic idea is then to infer the "strength" of the pollution potential of a wastewater by the measurement of the oxygen consumption that it would cause, that is, an indirect quantification of the potential to generate an impact, and not the direct measurement of the impact in itself.

This quantification could be obtained through stoichiometric calculations based on the reactions of oxidation of the organic matter. If the substrate was, for example, glucose ($C_6H_{12}O_6$), the quantity of oxygen required to oxidise the given quantity of glucose could be calculated through the basic equation of respiration. This is the principle of the so-called Theoretical Oxygen Demand (TOD).

In practice, however, a large obstacle is present: the sewage has a great heterogeneity in its composition, and to try to establish all its constituents in order to calculate the oxygen demand based on the chemical oxidation reactions of each of them is totally impractical. Besides, to extrapolate the data to other conditions would not be possible.

The solution found was to measure in the laboratory the consumption of oxygen exerted by a standard volume of sewage or other liquid, in a predetermined time. It was thus introduced the important concept of **Biochemical Oxygen Demand (BOD)**. The BOD represents the *quantity of oxygen required to stabilise, through biochemical processes, the carbonaceous organic matter*. It is an indirect indication, therefore, of the biodegradable organic carbon.

Complete stabilisation takes, in practical terms, various days (around 20 days or more for domestic sewage). This corresponds to the Ultimate Biochemical Oxygen Demand ($BOD_u$). However, to shorten the time for the laboratory test, and to allow a comparison of the various results, some standardisations were established:

- the determination is undertaken on the **5$^{th}$ day**. For typical domestic sewage, the oxygen consumption on the fifth day can be correlated with the final total consumption ($BOD_u$);

- the test is carried out at a temperature of **20°C**, since different temperatures interfere with the bacteria's metabolism, modifying the relation between BOD at 5 days and BOD Ultimate.

The **standard BOD** is expressed as $\mathbf{BOD}_5^{20}$. In this text, whenever the nomenclature BOD is used, implicitly the standard BOD is being assumed.

The BOD test can be understood in this simplified way: on the day of the sample collection, the concentration of dissolved oxygen (DO) in the sample is determined. Five days later, with the sample maintained in a closed bottle and incubated at 20°C, the new DO concentration is determined. This new DO concentration is lower due to the consumption of oxygen during the period. The difference in the DO level on the day zero and day 5 represents the oxygen consumed for the oxidation of the organic matter, being therefore, the $BOD_5$. Thus, for example, a sample from a water body presented the following results (see Figure 2.10):

DO on day 0: 7 mg/L
DO on day 5: 3 mg/L
$BOD_5 = 7 - 3 = 4$ mg/L

**BOD—Biochemical Oxygen Demand**

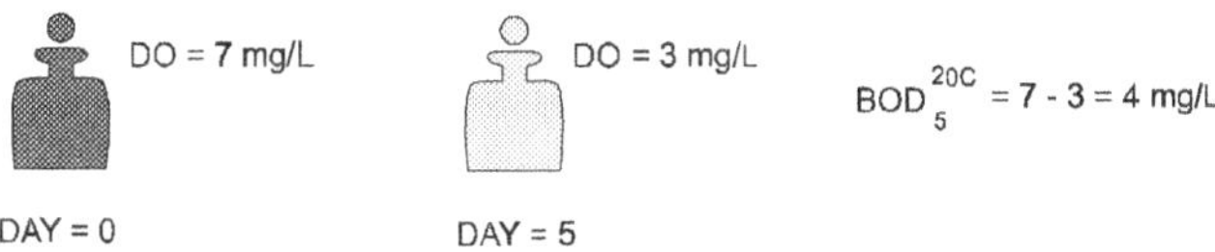


Figure 2.10.   Example of the $BOD_5^{20}$ concept

For sewage, some practical aspects require some adaptations. Sewage, having a large concentration of organic matter, consumes quickly (well before the five days) all the dissolved oxygen in the liquid medium. Thus, it is necessary to make dilutions in order to decrease the concentration of the organic matter, such that the oxygen consumption at 5 days is numerically less than the oxygen available in the sample (the sample is lost if, at day 5, the DO concentration is zero, because it will not be possible to know when the zero concentration was reached). Also it is usually necessary to introduce a seed, containing microorganisms, to allow a faster start of the decomposition process. To measure only the carbonaceous oxygen demand, an inhibitor for nitrification (nitrogenous oxygen demand, associated with the oxidation of ammonia to nitrate) can be added. Domestic sewage has a BOD in the region of 300 mg/L, or that is to say, 1 litre of sewage is associated with the consumption of approximately 300 mg of oxygen, in five days, in the process of the stabilisation of the carbonaceous organic matter.

The main **advantages** of the BOD test are related to the fact that the test allows:

- an approximate indication of the biodegradable fraction of the wastewater;
- an indication of the degradation rate of the wastewater;

- an indication of the oxygen consumption rate as a function of time;
- an approximate determination of the quantity of oxygen required for the biochemical stabilisation of the organic matter present.

However, the following **limitations** may be mentioned (Marais & Ekama, 1976):

- low levels of $BOD_5$ can be found in the case that the microorganisms responsible for the decomposition are not adapted to the waste;
- heavy metals and other toxic substances can kill or inhibit the microorganisms;
- the inhibition of the organisms responsible for the oxidation of ammonia is necessary, to avoid the interference of the oxygen consumption for nitrification (nitrogenous demand) with the carbonaceous demand;
- the ratio of $BOD_u/BOD_5$ varies with the wastewater;
- the ratio of $BOD_u/BOD_5$ varies, for the same wastewater, along the WWTP treatment line;
- the test takes five days, being not useful for operational control of a WWTP.

Despite of the limitations above, the BOD test continues to be extensively used, partly for historical reasons and partly because of the following points:

- the design criteria for many wastewater treatment processes are frequently expressed in terms of BOD;
- the legislation for effluent discharge in many countries, and the evaluation of the compliance with the discharge standards, is normally based on BOD.

Substantial research has been directed towards the substitution of BOD by other parameters. In the area of instrumentation, there are respirometric equipments that make automated measurements of the oxygen consumption, allowing a reduction in the period required for the test. However, universality has not yet been reached regarding the parameter or the methodology.

It is observed that the COD test is being more and more used for design, mathematical modelling and performance evaluation. However, the sanitary engineer must be familiar with the interpretation of the BOD and COD tests and know how to work with the complementary information that they both supply.

The present text utilises BOD in items in which the more consolidated international literature is based on BOD, and it uses COD in the items, usually more recent, in which the literature is based more on COD. In this way, it is easier to compare the design parameters presented in this book with international literature parameters.

**b)   Ultimate Biochemical Oxygen Demand ($BOD_u$)**

The $BOD_5$ corresponds to the oxygen consumption exerted during the first 5 days. However, at the end of the fifth day the stabilisation of the organic material is still not complete, continuing, though at slower rates, for another period of weeks or days.

Table 2.16. Typical ranges for the $BOD_u/BOD_5$ ratio

| Origin | $BOD_u/BOD_5$ |
| --- | --- |
| High concentration sewage | 1.1–1.5 |
| Low concentration sewage | 1.2–1.6 |
| Primary effluent | 1.2–1.6 |
| Secondary effluent | 1.5–3.0 |

*Source:* Calculated using the coefficients presented by Fair et al
(1973) and Arceivala (1981)

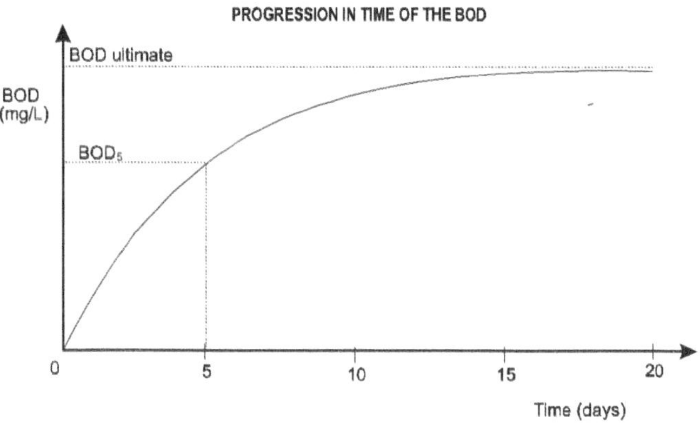


Figure 2.11.   Progression in time of BOD in a sample, showing $BOD_5$ and BOD ultimate

After this, the oxygen consumption can be considered negligible. In this way the
Ultimate Biochemical Oxygen Demand corresponds to the oxygen consumption
until this time, after what there is no significant consumption, meaning that the
organic matter has been practically all stabilised. Figure 2.11 shows the progression
of BOD in time, in a sample analysed along various days.

For domestic sewage, it is considered, in practical terms, that after 20 days
of the test the stabilisation is practically complete. Therefore the $BOD_u$ can be
determined at 20 days. The concept of the test is similar to the standard BOD of 5
days, varying only with the final period of determination of the dissolved oxygen
concentration.

Table 2.16 presents typical ranges of the conversion factor for $BOD_5$ to $BOD_u$
(domestic waste). Such a conversion is important, because various sewage treatment processes are designed using a $BOD_u$ base. Chapter 3 shows how to proceed
with this conversion using a specific formula.

Various authors adopt the ratio $BOD_u/BOD_5$ equal to **1.46**. This means that,
in the case of having a $BOD_5$ of 300 mg/L, the $BOD_u$ is assumed to be equal to
$1.46 \times 300 = 438$ mg/L.

### c)  Chemical Oxygen Demand (COD)

The COD test measures the consumption of oxygen occurring as a result of the chemical oxidation of the organic matter. The value obtained is, therefore, an indirect indication of the level of organic matter present.

The main difference with the BOD test is clearly found in the nomenclature of both tests. The BOD relates itself with the *biochemical* oxidation of the organic matter, undertaken entirely by microorganisms. The COD corresponds to the *chemical* oxidation of the organic matter, obtained through a strong oxidant (potassium dichromate) in an acid medium.

The main **advantages** of the COD test are:

- the test takes only two to three hours;
- because of the quick response, the test can be used for operational control;
- the test results give an indication of the oxygen required for the stabilisation of the organic matter;
- the test allows establishment of stoichiometric relationships with oxygen;
- the test is not affected by nitrification, giving an indication of the oxidation of the carbonaceous organic matter only (and not of the nitrogenous oxygen demand).

The main **limitations** of the COD test are:

- in the COD test, both the biodegradable and the inert fractions of organic matter are oxidised. Therefore, the test may overestimate the oxygen to be consumed in the biological treatment of the wastewater;
- the test does not supply information about the consumption rate of the organic matter along the time;
- certain reduced inorganic constituents could be oxidised and interfere with the result.

For raw domestic sewage, the ratio $COD/BOD_5$ varies between 1.7 and 2.4. For industrial wastewater, however, this ratio can vary widely. Depending on the value of the ratio, conclusions can be drawn about the biodegradability of the wastewater and the treatment process to be employed (Braile & Cavalcanti, 1979):

- *Low $COD/BOD_5$ ratio* (less than 2.5 or 3.0):
  - the biodegradable fraction is high
  - good indication for biological treatment
- *Intermediate $COD/BOD_5$ ratio* (between 2.5 and 4.0):
  - the inert (non-biodegradable) fraction is not high
  - treatability studies to verify feasibility of biological treatment
- *High $COD/BOD_5$ ratio* (greater than 3.5 or 4.0):
  - the inert (non-biodegradable) fraction is high
  - possible indication for physical–chemical treatment

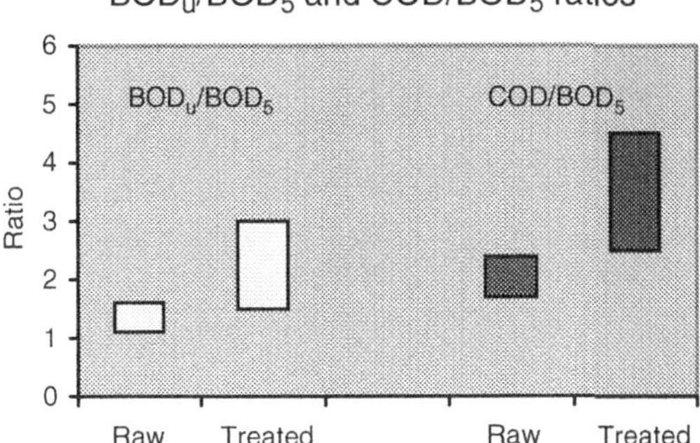

Figure 2.12.   Ranges of values of the ratios $BOD_u/BOD_5$ and $COD/BOD_5$ for raw sewage and biologically treated sewage

The $COD/BOD_5$ ratio also varies as the wastewater passes along the various units of the treatment works. The tendency is for the ratio to increase, owing to the stepwise reduction of the biodegradable fraction, at the same time that the inert fraction remains approximately unchanged. In this way, the final effluent of the biological treatment has values of the $COD/BOD_5$ ratio usually higher than 3.0.

### d)   Total Organic Carbon (TOC)

In this test the organic carbon is **directly** measured, in an instrumental test, and not indirectly through the determination of the oxygen consumed, like the three tests above. The TOC test measures all the carbon released in the form of $CO_2$. To guarantee that the carbon being measured is really organic carbon, the inorganic forms of carbon (like $CO_2$, $HCO_3^-$ etc) must be removed before the analysis or be corrected when calculated (Eckenfelder, 1980). The TOC test has been mostly used so far in research or in detailed evaluations of the characteristics of the liquid, due to the high costs of the equipment.

### e)   Relationship between the representative parameters of oxygen consumption

In samples of raw and treated domestic sewage, the usual ratios between the main representative parameters of oxygen consumption ($BOD_u/BOD_5$ and $COD/BOD_5$) are shown in Figure 2.12. The following comments can be made:

*   The ratios can never be lower than 1.0.
*   The ratios increase, from the condition of untreated to biologically treated wastewater.
*   The higher the treatment efficiency, the higher the value of the ratio.

Table 2.18. Predominant forms of nitrogen in the water

| Form | Formula | Oxidation state |
|---|---|---|
| Molecular nitrogen | $N_2$ | 0 |
| Organic nitrogen | Variable | Variable |
| Free ammonia | $NH_3$ | $-3$ |
| Ammonium ion | $NH_4^+$ | $-3$ |
| Nitrite ion | $NO_2^-$ | $+3$ |
| Nitrate ion | $NO_3^-$ | $+5$ |

## 2.2.3.4 Nitrogen

In its cycle in the biosphere, nitrogen alternates between various forms and oxidation states, resulting from various biochemical processes. In the aquatic medium, nitrogen can be found in the forms presented in Table 2.18.

Nitrogen is a component of great importance in terms of generation and control of the water pollution, principally for the following aspects:

- *Water pollution*
    - nitrogen is an essential nutrient for algae leading, under certain conditions, to the phenomenon of eutrophication of lakes and reservoirs;
    - nitrogen can lead to dissolved oxygen consumption in the receiving water body due to the processes of the conversion of ammonia to nitrite and this nitrite to nitrate;
    - nitrogen in the form of free ammonia is directly toxic to fish;
    - nitrogen in the form of nitrate is associated with illnesses such as methaemoglobinaemia
- *Sewage treatment*
    - nitrogen is an essential nutrient for the microorganisms responsible for sewage treatment;
    - nitrogen, in the processes of the conversion of ammonia to nitrite and nitrite to nitrate (*nitrification*), which can occur in a WWTP, leads to oxygen and alkalinity consumption;
    - nitrogen in the process of the conversion of nitrate to nitrogen gas (*denitrification*), which can take place in a WWTP, leads to (a) the economy of oxygen and alkalinity (when occurring in a controlled form) or (b) the deterioration in the settleability of the sludge (when not controlled).

The determination of the prevailing form of nitrogen in a water body can provide indications about the stage of pollution caused by an upstream discharge of sewage. If the pollution is recent, nitrogen is basically in the form of organic nitrogen or ammonia and, if not recent, in the form of nitrate (nitrite concentrations are normally low). In summary, the distinct forms can be seen in a generalised form presented in Table 2.19 (omitting other sources of nitrogen apart from sewage).

Table 2.19. Relative distribution of the forms of nitrogen under different conditions

| Condition | Prevailing form of nitrogen |
| --- | --- |
| Raw wastewater | • Organic nitrogen<br>• Ammonia |
| Recent pollution in a water course | • Organic nitrogen<br>• Ammonia |
| Intermediate stage in the pollution of a water course | • Organic nitrogen<br>• Ammonia<br>• Nitrite (in lower concentrations)<br>• Nitrate |
| Remote pollution in a water course | • Nitrate |
| Effluent from a treatment process without nitrification | • Ammonia |
| Effluent from a treatment process with nitrification | • Nitrate |
| Effluent from a treatment process with nitrification/ denitrification | • Low concentrations of all forms of nitrogen |

*Note:* organic nitrogen + ammonia = TKN (Total Kjeldahl Nitrogen)

In *raw domestic sewage*, the predominant forms are *organic nitrogen* and *ammonia*. Organic nitrogen corresponds to amina groups. Ammonia is mainly derived from urea, which is rapidly hydrolysed and rarely found in raw sewage. These two, together, are determined in the laboratory by the Kjeldahl method, leading to the *Total Kjeldahl Nitrogen* (**TKN**). Most of the TKN in domestic sewage has physiological origin. The other forms of nitrogen are usually of lesser importance in the influent to a WWTP. In summary:

$$\text{TKN} = \text{ammonia} + \text{organic nitrogen} \quad \textit{(prevailing form in domestic sewage)}$$
$$\text{TN} = \text{TKN} + NO_2^- + NO_3^- \quad \textit{(total nitrogen)}$$

The distribution of ammonia in the raw sewage can be represented schematically as shown in Figure 2.13. It is seen that the fraction of the oxidised nitrogen $NO_x$ (nitrite + nitrate) is negligible in raw sewage. TKN can be further subdivided in a *soluble* fraction (dominated by ammonia) and a *particulate* fraction (associated with the organic suspended solids – nitrogen participates in the constitution of practically all forms of particulate organic matter in sewage).

Ammonia exists in solution in the form of the ion ($NH_4^+$) and in a free form, not ionised ($NH_3$), according to the following dynamic equilibrium:

$$NH_3 + H^+ \leftrightarrow NH_4^+ \qquad (2.6)$$
$$\text{free ammonia} \quad \text{ionised ammonia}$$

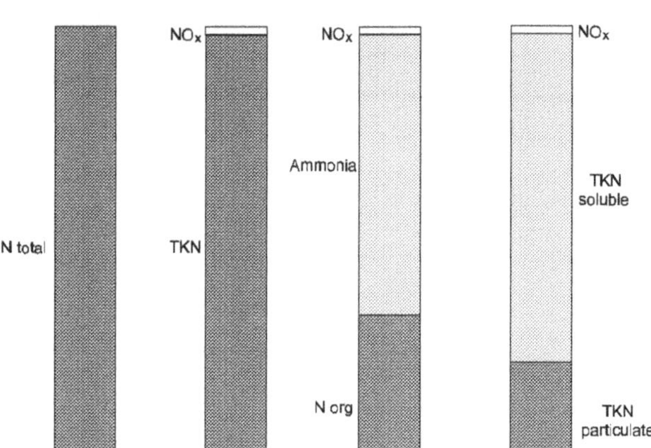


Figure 2.13.   Distribution of nitrogen forms in untreated domestic sewage (adapted from IAWQ, 1995)

The relative distribution has the following values, as a function of the pH values.

**Distribution between the forms of ammonia**

- pH < 8     Practically all the ammonia is in the form of $NH_4^+$
- pH = 9.5   Approximately 50% $NH_3$ and 50% $NH_4^+$
- pH > 11    Practically all the ammonia in the form of $NH_3$

In this way it can be seen that, in the usual range of pH, near neutrality, the ammonia present is practically in the ionised form. This has important environmental consequences, because free ammonia is toxic to fish even in low concentrations. The temperature of the liquid also influences this distribution. At a temperature of 25 °C, the proportion of free ammonia relative to the total ammonia is approximately the double compared with a temperature of 15 °C.

The following equation allows the calculation of the proportion of free ammonia within total ammonia as a function of temperature and pH (Emerson et al, 1975):

$$\frac{\text{Free NH}_3}{\text{Total ammonia}}(\%) = \left\{1 + 10^{0.09018+[2729.92/(T+273.20)]-pH}\right\}^{-1} \times 100 \qquad (2.6)$$

where:

   T = liquid temperature (°C)

Application of Equation 2.6 leads to the values of the ammonia distribution presented in Table 2.20 and illustrated in Figure 2.14.

Table 2.20.  Proportion of free and ionised ammonia within total ammonia, as a function of pH and temperature

| pH | T = 15 °C | | T = 20 °C | | T = 25 °C | |
|---|---|---|---|---|---|---|
| | % $NH_3$ | % $NH_4^+$ | % $NH_3$ | % $NH_4^+$ | % $NH_3$ | % $NH_4^+$ |
| 6.50 | 0.09 | 99.91 | 0.13 | 99.87 | 0.18 | 99.82 |
| 7.00 | 0.27 | 99.73 | 0.40 | 99.60 | 0.57 | 99.43 |
| 7.50 | 0.86 | 99.14 | 1.24 | 98.76 | 1.77 | 98.23 |
| 8.00 | 2.67 | 97.33 | 3.82 | 96.18 | 5.38 | 94.62 |
| 8.50 | 7.97 | 92.03 | 11.16 | 88.84 | 15.25 | 84.75 |
| 9.00 | 21.50 | 78.50 | 28.43 | 71.57 | 36.27 | 63.73 |
| 9.50 | 46.41 | 53.59 | 55.68 | 44.32 | 64.28 | 35.72 |

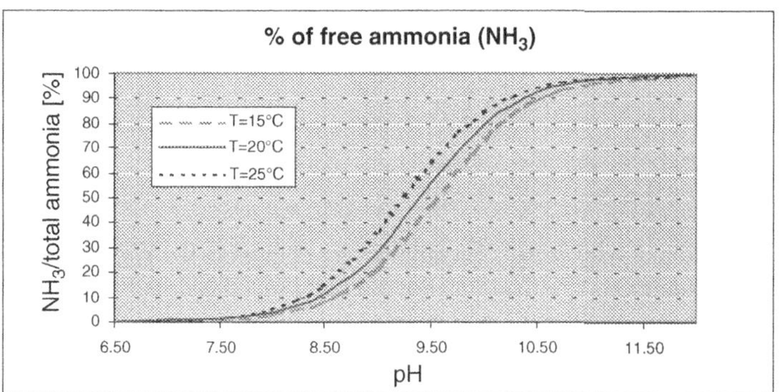


Figure 2.14.  Percentage of free ammonia ($NH_3$) within total ammonia, as a function of pH and temperature

In a watercourse or in a WWTP, the ammonia can undergo subsequent transformations. In the process of **nitrification** the ammonia is oxidised to nitrite and the nitrite to nitrate. In the process of **denitrification** the nitrates are reduced to nitrogen gas.

## 2.2.3.5 *Phosphorus*

Total phosphorus in domestic sewage is present in the form of **phosphates**, according to the following distribution (IAWQ, 1995):

- **inorganic** (polyphosphates and orthophosphates) – main source from detergents and other household chemical products
- **organic** (bound to organic compounds) – physiological origin

Phosphorus in detergents is present, in raw sewage, in the form of soluble polyphosphates or, after hydrolysis, as orthophosphates. *Orthophosphates* are directly available for biological metabolism without requiring conversion to simpler

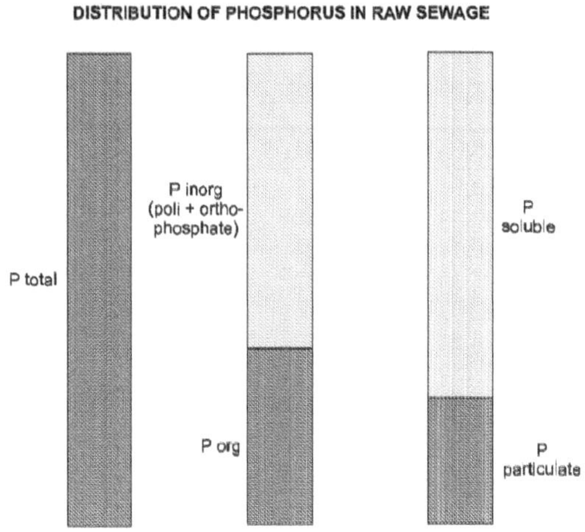

Figure 2.15.    Distribution of phosphorus forms in untreated domestic sewage (IAWQ, 1995)

forms. The forms in which orthophosphates are present in the water are pH dependent, and include $PO_4^{3-}$, $HPO_4^{2-}$, $H_2PO_4^-$, $H_3PO_4$. In typical domestic sewage the prevailing form is $HPO_4^{-2}$. *Polyphosphates* are more complex molecules, with two or more phosphorus atoms. Polyphosphates are converted into orthophosphates by hydrolysis, which is a slow process, even though it takes place in the sewerage collection system itself. Mathematical models for wastewater treatment processes usually consider that both forms of phosphate are represented by orthophosphates since after hydrolysis they will effectively be present as such. Phosphorus in detergents can account for up to 50% of the total phosphorus present in domestic sewage.

Another way of fractionating phosphorus in wastewater is with respect to its form as solids (IAWQ, 1995):

- **soluble phosphorus** (predominantly inorganic) – mainly polyphosphates and orthophosphates (inorganic phosphorus), together with a small fraction corresponding to the phosphorus bound to the soluble organic matter in the wastewater
- **particulate phosphorus** (all organic) – bound to particulate organic matter in the wastewater

Figure 2.15 illustrates the fractionation of phosphorus in untreated domestic sewage.

The importance of phosphorus is associated with the following aspects:

- Phosphorus is an essential nutrient for the growth of the microorganisms responsible for the stabilisation of organic matter. Usually domestic sewage

has sufficient levels of phosphorus, but a lack may occur in some industrial wastewaters;
- Phosphorus is an essential nutrient for the growth of algae, eventually leading, under certain conditions, to the eutrofication of lakes and reservoirs.

## 2.2.3.6 Pathogenic organisms and indicators of faecal contamination

### a) Pathogenic organisms

The list of organisms of importance in water and wastewater quality was presented in Table 2.15. Most of these organisms play various essential roles, mainly related to the transformation of the constituents in the biogeochemical cycles. Biological wastewater treatment relies on these organisms, and this aspect is covered in many parts of this book.

Another important aspect in terms of the biological quality of a water or wastewater is that related to the disease transmission by pathogenic organisms. The major groups of pathogenic organisms are: (a) *bacteria*, (b) *viruses*, (c) *protozoans* and (d) *helminths*.

**Water-related disease** is defined as any significant or widespread adverse effects on human health, such as death, disability, illness or disorders, caused directly or indirectly by the condition, or changes in the quantity or quality of any water (Grabow, 2002). A useful way of classifying the water-related diseases is to group them according to the mechanism by which they are transmitted (*water borne, water hygiene, water based, water related*). Table 2.21 presents the main four categories, with a summary description and the main preventive strategies to be employed. Table 2.22 details the **faecal–oral transmission diseases** (*water borne and water hygiene*), with the main pathogenic agents and symptoms. Faecal–oral diseases are of special interest for the objectives and theme of this book, since they are associated with proper excreta and wastewater treatment and disposal.

The number of pathogens present in the sewage of a certain community varies substantially and depends on: (a) socio-economic status of the population; (b) health requirements; (c) geographic region; (d) presence of agroindustries; (e) type of treatment to which the sewage was submitted.

### b) Indicator organisms

The detection of pathogenic organisms, mainly bacteria, protozoans and viruses, in a sample of water is difficult, because of their low concentrations. This would demand the examination of large volumes of the sample to detect the pathogenic organisms. The reasons are due to the following factors:

- in a population, only a certain fraction suffers from water-borne diseases;
- in the faeces of these inhabitants, the presence of pathogens may not occur in high proportions;

Table 2.21. Mechanisms of transmission of water-related infections

| Mechanism | Description | Transmission mode | Main diseases | Preventive strategy |
| --- | --- | --- | --- | --- |
| Water borne | • Ingestion of contaminated water | • Faecal–oral (pathogen in *water* is ingested by man or animal) | • Diarrhoeas and dysenteries (amoebic dysentery, balantidiasis, *Campylobacter* enteritis, cholera, *E. coli* diarrhoea, giardiasis, cryptosporidiosis, rotavirus diarrhoea, salmonellosis, bacillary dysentery, yersiniosis)<br>• Enteric fevers (typhoid, paratyphoid)<br>• Poliomyelitis<br>• Hepatitis A<br>• Leptospirosis<br>• Ascariasis<br>• Trichuriasis | • Protect and treat drinking water<br>• Avoid use of contaminated water |
| Water hygiene | • Infections of the intestinal tract, due to low availability of water and poor hygiene | • Various faecal-oral routes (e.g. food, bad hygiene) | • Similar to above (water borne) | • Supply water in sufficient quantity<br>• Promote personal, domestic and food hygiene |
| | • Infections of the skin or eyes, due to low availability of water and poor hygiene | • Lack of water and poor hygiene create conditions for their transmission<br>• Non-faecal route (cannot be water borne) | • Infectious skin diseases (e.g. skin sepsis, scabies, fungal infections)<br>• Infectious eye diseases (e.g. trachoma)<br>• Others (e.g. louse-borne typhus) | |

| Water based | • Pathogen (helminth) spends part of its life cycle in a water snail or other aquatic animal | • Pathogen penetrates the skin or is ingested | • Schistosomiasis<br>• Guinea worm<br>• Clonorchiasis<br>• Diphyllobothriasis<br>• Fasciolopsiasis<br>• Paragonimiasis<br>• Others | • Avoid contact with contaminated water<br>• Protect water sources from excreta<br>• Adopt adequate solutions for excreta or wastewater disposal<br>• Combat intermediate host |
| --- | --- | --- | --- | --- |
| Water related | • Insects which breed in water or bite near water | • Insect bites man or animal | • Malaria<br>• Sleeping sickness<br>• Filariasis<br>• River blindness<br>• Mosquito-borne virus (e.g. yellow fever, dengue)<br>• Others | • Combat insect vectors<br>• Destroy breeding sites of insects<br>• Avoid contact with breeding sites<br>• Adopt individual protection (e.g. sprays, nets) |

*Source:* Cairncross and Feachem (1990), Heller & Möller (1995), van Buuren et al (1995), Heller (1997)

Table 2.22. Main water-borne and water hygiene (faecal oral transmission) diseases, according to pathogenic organism

| Organism | Disease | Causal agent | Symptoms / manifestation |
| --- | --- | --- | --- |
| Bacteria | Bacillary dysentery (shigellosis) | *Shigella dysenteriae* | Severe diarrhoea |
| | *Campylobacter* enteritis | *Campylobacter jejuni, Campylobacter coli* | Diarrhoea, abdominal pain, malaise, fever, nausea, vomiting |
| | Cholera | *Vibrio cholerae* | Extremely heavy diarrhoea, dehydration, high death rate |
| | Gastroenteritis | *Escherichia coli* – enteropathogenic | Diarrhoea |
| | Leptospirosis | *Leptospira* – various species | Jaundice, fever |
| | Paratyphoid fever | *Salmonella* – various species | Fever, diarrhoea, malaise, headache, spleen enlargement, involvement of lymphoid tissues and intestines |
| | Salmonella | *Salmonella* – various species | Fever, nausea, diarrhoea |
| | Typhoid fever | *Salmonella typhi* | High fever, diarrhoea, ulceration of small intestine |
| Protozoan | Amoebic dysentery | *Entamoeba histolytica* | Prolonged diarrhoea with bleeding, abscesses of the liver and small intestine |
| | Giardiasis | *Giardia lamblia* | Mild to severe diarrhoea, nausea, indigestion, flatulence |
| | Cryptosporidiosis | *Cryptosporidium* | Diarrhoea |
| | Balantidiasis | *Balantidium coli* | Diarrhoea, dysentery |
| Viruses | Infectious hepatitis | Hepatitis A virus | Jaundice, fever |
| | Respiratory disease | Adenovirus – various types | Respiratory illness |
| | Gastroenteritis | Enterovirus, Norwalk, rotavirus, etc. – various species | Mild to strong diarrhoea, vomiting |
| | Meningitis | Enterovirus | Fever, vomiting, neck stiffness |
| | Poliomyelitis (infantile paralysis) | *Poliomyelitis virus* | Paralysis, atrophy |
| Helminths | Ascariasis | *Ascaris lumbricoides* | Pulmonary manifestations, nutritional deficiency, obstruction of bowel or other organ |
| | Trichuriasis | *Trichuris trichiura* | Diarrhoea, bloody mucoid stools, rectal prolapse |

*Source:* Benenson (1985), Tchobanoglous and Schroeder (1985), Metcalf & Eddy (1991)

- after discharge to the receiving body or sewerage system, there is still a high dilution of the contaminated waste;
- sensitivity and specificity of the tests for some pathogens;
- broad spectrum of pathogens.

In this sense, the final concentration of *pathogens* per unit volume in a water body may be considerably low, making detection through laboratory examination highly difficult.

This obstacle is overcome through the search for *indicator organisms of faecal contamination*. These organisms are predominantly non-pathogenic, but they give a satisfactory indication of whether the water is contaminated by human or animal faeces, and, therefore, of its potential to transmit diseases.

The organisms most commonly used with this objective are bacteria of the **coliform group**. The following are the main reasons for the use of the coliform group as indicators of faecal contamination:

- Coliforms are present in *large quantities in human faeces* (each individual excretes on average $10^{10}$ to $10^{11}$ cells per day) (Branco and Rocha, 1979). About 1/3 to 1/5 of the weight of human faeces consist of bacteria from the coliform group. All individuals eliminate coliforms, and not only those who are ill, as is the case with pathogenic organisms. Thus the probability that the coliforms will be detected after the sewage discharge is much higher than with pathogenic organisms.
- Coliforms present a *slightly higher resistance* in the water compared with the majority of enteric pathogenic bacteria. This characteristic is important, because they would not be good indicators of faecal contamination if they died faster than pathogenic organisms, and a sample without coliforms could still contain pathogens. On the other hand, if their mortality rate were much lower than that of pathogenic microorganisms, the coliforms would not be useful indicators, since their presence could unjustifiably make suspect a sample of purified water. These considerations apply mainly to pathogenic bacteria, since other microorganisms can present a higher resistance compared to coliforms.
- The *removal mechanisms* for coliforms from water bodies, water treatment plants and WWTP are the same mechanisms used for pathogenic bacteria. In this way the removal of pathogenic bacteria is usually associated with the removal of coliforms. Other pathogenic organisms (such as protozoan cysts and helminth eggs), however, can be removed by different mechanisms.
- The bacteriological techniques for coliform detection are *quick and economic* compared with those for pathogens.

The indicators of faecal contamination most commonly used are:

- *total coliforms (TC)*
- *faecal coliforms (FC)* or *thermotolerant coliforms*
- *Escherichia coli (EC)*

The group of **total coliforms (TC)** constitutes a large group of bacteria that have been isolated in water samples and in polluted and non polluted soils and plants, as well as from faeces from humans and other warm-blooded animals. This group was largely used in the past as an indicator, and continues to be used in some areas, although the difficulties associated with the occurrence of non-faecal bacteria are a problem (Thoman and Mueller, 1987). There is no quantifiable relation between TC and pathogenic microorganisms. The total coliforms could be understood in a simplified way as "environmental" coliforms, given their possible occurrence in non-contaminated water and soils, thus representing other free-living organisms, and not only the intestinal ones. For this reason, total coliforms should not be used as indicators of faecal contamination in surface waters. However, in the specific case of potable water supply, it is expected that treated water should not contain total coliforms. These, if found, could suggest inadequate treatment, post contamination or excess of nutrients in the treated water. Under these conditions, total coliforms could be used as indicators of the water treatment efficiency and of the integrity of the water distribution system (WHO, 1993).

**Faecal coliforms (FC)** are a group of bacteria predominantly originated from the intestinal tract of humans and other animals. This group encompasses the genus *Escherichia* and, to a lesser degree, species of *Klebsiella, Enterobacter* and *Citrobacter* (WHO, 1993). The test for FC is completed at a high temperature, aiming at suppressing bacteria of non-faecal origin (Thoman and Mueller, 1987). However, even under these conditions, the presence of non-faecal (free-living) bacteria is possible, although in lower numbers compared with the total coliforms test. As a result, even the test for faecal coliforms does not guarantee that the contamination is really faecal. For this reason, recently the faecal coliforms have been preferably denominated **thermotolerant coliforms**, because of the fact that they are resistant to the high temperatures of the test, but are not necessarily faecal. *Whenever in the present book reference is made to faecal coliforms (traditional in the literature and in the environmental legislation in various countries), it should be understood, implicitly, the more appropriate terminology of thermotolerant coliforms.*

***Escherichia coli* (EC)** is the main bacterium of the faecal (thermotolerant) coliform group, being present in large numbers in the faeces from humans and animals. It is found in wastewater, treated effluents and natural waters and soils that are subject to recent contamination, whether from humans, agriculture, wild animals and birds (WHO, 1993). Its laboratory detection is very simple, principally by recent fluorogenic methods. Different from total and faecal coliforms, *E. coli* is the only that gives **guarantee of exclusively faecal contamination**. For this reason, there is a current tendency in using predominantly *E. coli* as indicator of faecal contamination. However, its detection **does not guarantee that the contamination is from human origin**, since *E. coli* can also be found in other animal faeces. There are some types of *E. coli* that are pathogenic, but this does not invalidate its concept as bacterial indicators of faecal contamination.

The detection of faecal contamination, **exclusively human**, requires the use of complementary biochemical tests, which are not usually undertaken in routine analysis.

Table 2.23. Application of total coliforms, thermotolerant coliforms and *E. coli* as indicators of faecal contamination

| Item | Sample | Total coliform | Faecal (thermotolerant) coliforms | *E. coli* |
|---|---|---|---|---|
| Guarantee that the contamination is of faecal origin | Water bodies reasonably clean<br>Water bodies polluted by sewage | Low<br>Reasonable | Reasonable<br>High | Total<br>Total |
| Guarantee that the faecal contamination is exclusively human | Water bodies reasonably clean<br>Water bodies polluted mainly by domestic sewage | None<br>Reasonable | None<br>High | None<br>High |
| Proportion of *E. coli* in the total count of coliforms | Water bodies reasonably clean<br>Water bodies polluted by domestic sewage<br>Domestic sewage | Variable<br>Reasonable to high<br>Very high | Variable<br>High<br>Very high | –<br>–<br>– |

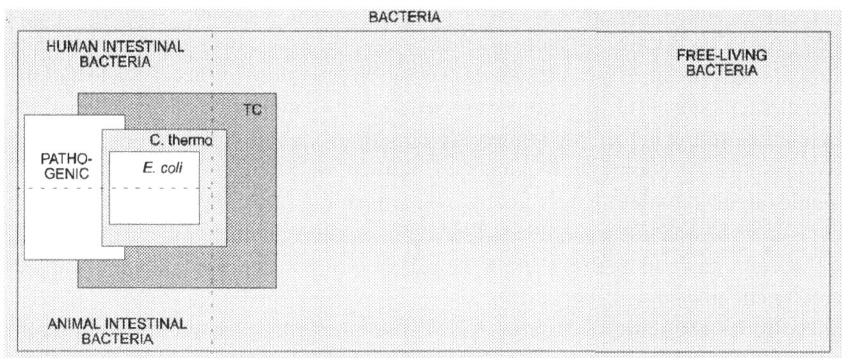


Figure 2.16.   Schematic representation of bacteria and indicators of faecal contamination

Figure 2.16 illustrates the relative distribution of the indicator, pathogenic and other forms of bacteria. Table 2.23 synthesises the application of the three groups of indicators discussed above.

In **sewage**, *E. coli* is the predominant organism within the group of faecal (thermotolerant) coliforms, and the faecal (thermotolerant) coliforms are the predominant group within the total coliforms. For **water bodies**, when doing the interpretation of the tests for indicators of faecal contamination, it is very important to carry out a sanitary survey of the catchment area. This survey helps in establishing the origin of the faecal contamination (presence of domestic sewage discharges or wastes from animals), complementing the information supplied by the laboratory tests.

For the objectives of this book (wastewater treatment), the characterisation of the *faecal origin* is not so important, since it is already accepted that the wastewater will contain faecal matter and organisms. The indicator organisms are used, in this case, as *indicators of the pathogen removal **efficiency** in the wastewater treatment process*. The pathogenic organisms that can be represented are *bacteria* and *viruses*, since they are removed by the same mechanisms of the coliform bacteria. Protozoan cysts and helminth eggs, which are mainly removed by physical mechanisms, such as sedimentation and filtration, are not well represented by coliform bacteria as indicators of treatment efficiency.

There are various other indicator organisms proposed in the literature, each with its own advantages, disadvantages and applicability. Below some of these organisms are briefly discussed.

**Faecal streptococci**. The group of faecal streptococci comprises two main genera: *Enterococcus* and *Streptococcus*. The genus ***Enterococcus*** encompasses many species, the majority of them of faecal human origin; however, some species are from animal origin. All *Enterococcus* present high tolerance to adverse environmental conditions. The genus ***Streptococcus*** comprises the species *S. bovis* and *S. equinus*, which are abundant in animal faeces. Faecal streptococci seldom multiply in polluted waters, and are more resistant than *E. coli* and coliform bacteria (WHO, 1993). Because of these characteristics, they have been used as indicators for *bathing waters*. In the past, the ratio between the values of faecal coliforms and faecal streptococci (FC/FS ratio) was used to give an indication of the origin of the contamination, whether predominantly human or animal. High values of FC/FS would suggest predominantly human contamination, whereas low values of FC/FS would suggest predominantly animal contamination. More recent evidences indicate, however, that these relations are not applicable in a large number of situations, giving unreliable indications about the real origin or the contamination in various catchment areas.

**Sulphite-reducing clostridia**. *Clostridium perfringens* is the most representative species in this group, being normally present in faeces, although in much smaller numbers than *E. coli*. However, it is not exclusively of faecal origin and can be derived from other environmental sources. Clostridial spores can survive in water much longer than organisms of the coliform group and will also resist disinfection. Their presence in disinfected waters may indicate deficiencies in treatment and that disinfectant-resistant pathogens could have survived treatment. Because of its longevity, it is best regarded as indicating intermittent or remote contamination. However, false alarms may also result from its detection, which makes it of special value, but not particularly recommended for routine monitoring of water distribution systems (WHO, 1993).

**Bacteriophages**. For the indication of the presence of viruses, bacteriophages may be representative, owing to their similarities with the enteric human viruses. Bacteriophages are specific viruses that infect bacteria, for example the coliphages, which infect *E. coli*. Coliphages are not present in high numbers in fresh human or animal faeces, but may be abundant in sewage, owing to their fast reproduction rate resulting from the attack to bacterial cells (Mendonça, 2000). Their significance is as indicators of sewage contamination and, because of their greater persistence compared with bacterial indicators, as additional indicators of treatment efficiency or for groundwater protection.

**Helminth eggs**. For helminths, there are no substituting indicators, and helminth eggs are determined directly in laboratory tests. However, the eggs of nematodes, such as *Ascaris*, *Trichuris*, *Necator americanus* and *Ancilostoma duodenale* may be used as indicators of other helminths (cestodes, trematodes and other nematodes), which are removed in water and wastewater treatment by the same mechanism (e.g. sedimentation), being thus indicators of treatment efficiency. Helminth eggs are an important parameter when assessing the use of water or treated wastewater for irrigation, in which workers may have direct contact with contaminated water and consumers may eat the irrigated vegetable uncooked or unpeeled. Helminth eggs may be removed by physical operations, such as sedimentation, which takes place, for instance, in stabilisation ponds. Eggs may be viable or non-viable, and viability may be altered by specific disinfection processes.

This topic is under constant development, and the present text does not aim to go deeper into specific items, covering only the more general and simplified concepts.

## 2.2.4 Relationship between load and concentration

Before presenting the typical concentrations of the main pollutants in sewage, it is important to be clear about the concepts of per capita, load and constituent concentration.

**Per capita load** represents the average contribution of each individual (expressed in terms of pollutant mass) per unit time. A commonly used unit is grams per inhabitant per day (g/inhab.d). For example, when the BOD contribution is 54 g/inhab.d, it is equivalent to saying that every individual discharges 54 grams of BOD on average, per day.

The influent **load** to a WWTP corresponds to the quantity of pollutant (mass) per unit time. In this way, import relations are

$$\boxed{\text{load} = \text{population} \times \text{per capita load}} \tag{2.7}$$

$$\text{load (kg/d)} = \frac{\text{population (inhab)} \times \text{per capita load (g/inhab.d)}}{1000 \, (\text{g/kg})} \tag{2.8}$$

or

$$\boxed{\text{load} = \text{concentration} \times \text{flow}} \tag{2.9}$$

$$\text{load (kg/d)} = \frac{\text{concentration (g/m}^3) \times \text{flow (m}^3/\text{d})}{1000 \, (\text{g/kg})} \tag{2.10}$$

Note: $g/m^3 = mg/L$

The **concentration** of a wastewater can be obtained through the rearrangement of the same dimensional relations:

$$\boxed{\textbf{concentration} = \textbf{load/flow}} \qquad (2.11)$$

$$\text{concentration (g/m}^3) = \frac{\text{load (kg/d)} \times 1000 \text{ (g/kg)}}{\text{flow (m}^3/\text{d})} \qquad (2.12)$$

---

**Example 2.2**

Calculate the total nitrogen load in the influent to a WWTP, given that:

- concentration = 45 mgN/L
- flow = 50 L/s

**Solution:**

Expressing flow in m$^3$/d, :

$$Q = \frac{50 \text{ L/s} \times 86400 \text{ s/d}}{1000 \text{ L/m}^3}$$

The nitrogen load is:

$$\text{load} = \frac{45 \text{ g/m}^3 \times 4320 \text{ m}^3/\text{d}}{1000 \text{ g/kg}} = 194 \text{ kgN/d}$$

*b) In the same works, calculate the total phosphorus concentration in the influent, given that the influent load is 40 kgP/d.*

$$\text{concentration} = \frac{40 \text{ kg/d} \times 1000 \text{ g/kg}}{4320 \text{ m}^3/\text{d}} = 9.3 \text{ gP/m}^3 = 9.3 \text{ mgP/L}$$

---

## 2.2.5 Characteristics of domestic sewage

The typical quantitative physical–chemical characteristics of predominantly domestic sewage in developing countries can be found in a summarised form in Table 2.24.

Campos and von Sperling (1996) verified, for essentially domestic sewage in nine sub-catchment areas in the city of Belo Horizonte, Brazil, relationships between per capita BOD load and BOD concentration with the average family income.

Table 2.24. Physical–chemical characteristics of raw domestic sewage in developing countries

| Parameter | Per capita load (g/inhab.d) | | Concentration (mg/L, except pH) | |
| --- | --- | --- | --- | --- |
| | Range | Typical | Range | Typical |
| *TOTAL SOLIDS* | 120–220 | 180 | 700–1350 | 1100 |
| *Suspended* | 35–70 | 60 | 200–450 | 350 |
| • *Fixed* | 7–14 | 10 | 40–100 | 80 |
| • *Volatile* | 25–60 | 50 | 165–350 | 320 |
| *Dissolved* | 85–150 | 120 | 500–900 | 700 |
| • *Fixed* | 50–90 | 70 | 300–550 | 400 |
| • *Volatile* | 35–60 | 50 | 200–350 | 300 |
| *Settleable* | – | – | 10–20 | 15 |
| *ORGANIC MATTER* | | | | |
| *BOD$_5$* | 40–60 | 50 | 250–400 | 300 |
| *COD* | 80–120 | 100 | 450–800 | 600 |
| *BOD ultimate* | 60–90 | 75 | 350–600 | 450 |
| *TOTAL NITROGEN* | 6.0–10.0 | 8.0 | 35–60 | 45 |
| *Organic nitrogen* | 2.5–4.0 | 3.5 | 15–25 | 20 |
| *Ammonia* | 3.5–6.0 | 4.5 | 20–35 | 25 |
| *Nitrite* | $\approx 0$ | $\approx 0$ | $\approx 0$ | $\approx 0$ |
| *Nitrate* | 0.0–0.3 | $\approx 0$ | 0–2 | $\approx 0$ |
| *PHOSPHORUS* | 0.7–2.5 | 1.0 | 4–15 | 7 |
| *Organic phosphorus* | 0.7–1.0 | 0.3 | 1–6 | 2 |
| *Inorganic phosphorus* | 0.5–1.5 | 0.7 | 3–9 | 5 |
| *pH* | - | - | 6.7–8.0 | 7.0 |
| *ALKALINITY* | 20–40 | 30 | 100–250 | 200 |
| *HEAVY METALS* | $\approx 0$ | $\approx 0$ | $\approx 0$ | $\approx 0$ |
| *TOXIC ORGANICS* | $\approx 0$ | $\approx 0$ | $\approx 0$ | $\approx 0$ |

*Sources:* Arceivala (1981), Jordão & Pessoa (1995), Qasim (1985), Metcalf & Eddy (1991), Cavalcanti et al (2001) and the author's experience.

The higher the income, the higher is the per capita BOD load and the lower is the BOD concentration (Figure 2.17). Family income is expressed as numbers of minimum salaries. The figures are presented in order to show the large influence of economic status, and not to allow direct calculations, since the economical data are region specific.

The typical biological characteristics of domestic sewage, in terms of pathogenic organisms, can be found in Table 2.25.

## 2.2.6 Characteristics of industrial wastewater

### 2.2.6.1 General concepts

The generalisation of typical industrial wastewater characteristics is difficult because of their wide variability from time to time and from industry to industry.

Table 2.25. Microorganisms present in raw domestic sewage in developing countries

| Microorganisms | Per capita load (org/inhab.d) | Concentration (org/100 ml) |
|---|---|---|
| Total coliforms | $10^{10}$–$10^{13}$ | $10^7$–$10^{10}$ |
| Faecal (thermotolerant) coliforms | $10^9$–$10^{12}$ | $10^6$–$10^9$ |
| *E. coli* | $10^9$–$10^{12}$ | $10^6$–$10^9$ |
| Faecal streptococci | $10^7$–$10^{10}$ | $10^4$–$10^7$ |
| Protozoan cysts | $<10^7$ | $<10^4$ |
| Helminth eggs | $10^3$–$10^6$ | $10^0$–$10^3$ |
| Viruses | $10^5$–$10^7$ | $10^2$–$10^4$ |

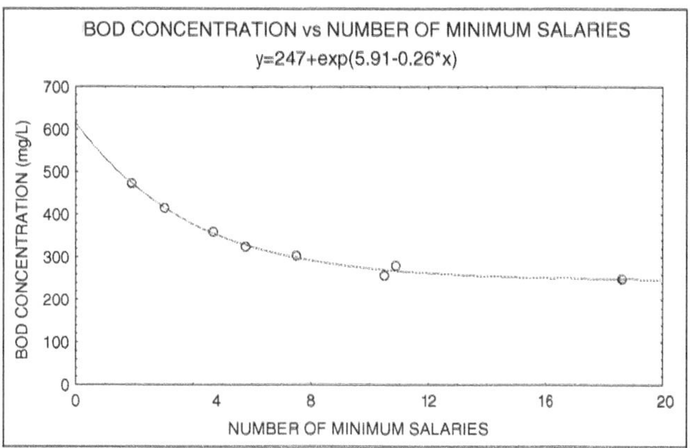


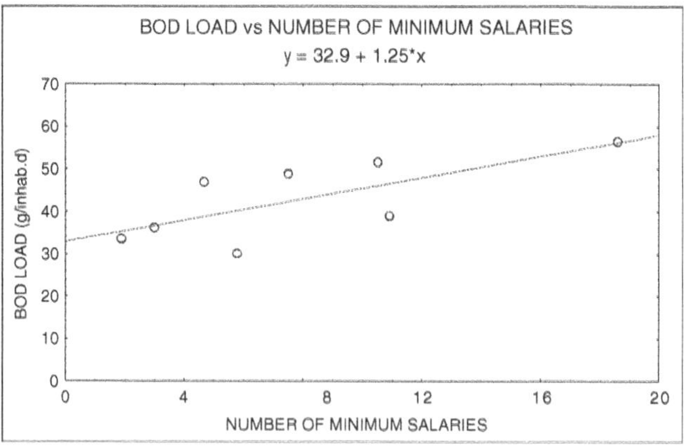


Figure 2.17. BOD concentration and per capita BOD load as a function of family income in domestic sewage from nine catchment areas in Belo Horizonte, Brazil (family income as number of minimum salaries; 1 minimum salary = US$80 at the time of the research)

The following concepts are important in terms of the biological treatment of industrial wastewater:

- **Biodegradability**: capacity of the wastewater to be stabilised through biochemical processes by microorganisms.
- **Treatability**: suitability of the waste to be treated by conventional or existing biological processes.
- **Biodegradable organic matter concentration**: BOD of the wastewater, which can be: (a) higher than in domestic sewage (predominantly biodegradable organic wastewater, treatable through biological processes), or (b) lower than in domestic sewage (predominately inorganic or un-biodegradable wastewater, in which there is less need for BOD removal, but in which the pollutional load can be expressed in terms of other quality parameters)
- **Nutrient availability**: biological wastewater treatment requires a balanced equilibrium between the nutrients C:N:P. This equilibrium is usually found in domestic sewage.
- **Toxicity**: certain industrial wastewaters have toxic or inhibitory constituents that can affect or render biological treatment unfeasible.

Figure 2.18 presents the main options for the treatment and discharge of industrial effluents.

The integration of industrial wastewater with domestic sewage in the public sewerage system, for subsequent combined treatment in a WWTP, may be an interesting alternative. Possible reasons for this alternative would be economy of scale, dilution of undesirable constituents, revenue for the sanitation company for transporting and treating the industrial wastewater, simplification for the industries. However, for this practice to be effective, it is necessary that *previous removal from the industrial effluent* is practised for the constituents that may pose one or more of the following problems:

- Safety risks and problems in the operation of the sewerage collection and interception system.
- Toxicity to the biological treatment.
- Toxicity to the sludge treatment and its final disposal.
- Persistence of contaminants in the effluent of the biological treatment, because of the fact that they have not been removed by the treatment.

The water and sanitation company, to receive the industrial wastewaters, must have specific standards for the discharge of industrial effluents into the public sewerage system.

If a pollutant leads to one of the above problems, the industry must pre-treat the wastewater, in order to place the effluent within the standards of the Sanitation Company for discharge into the public sewerage system.

The industry can opt for complete treatment and discharge the industrial effluent directly into the receiving water body. In this case, the effluents must comply with

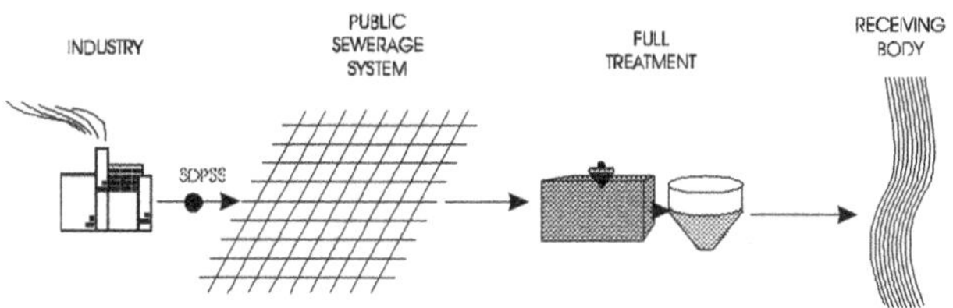


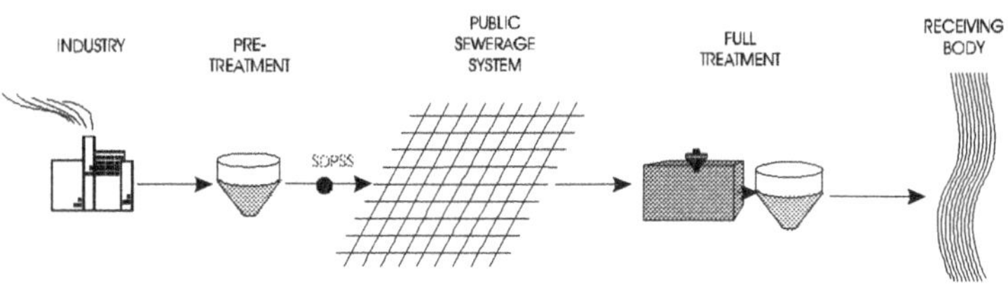


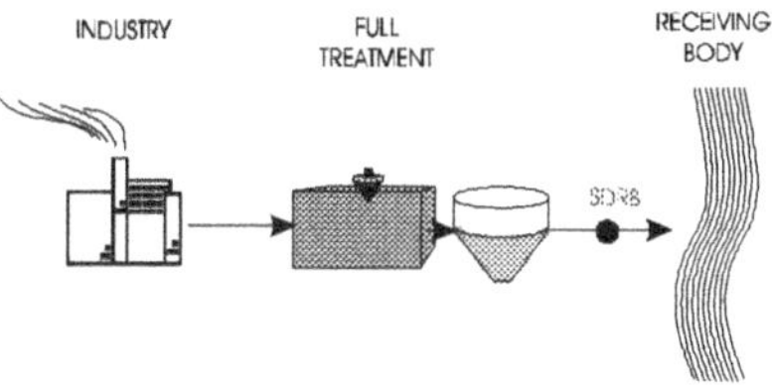


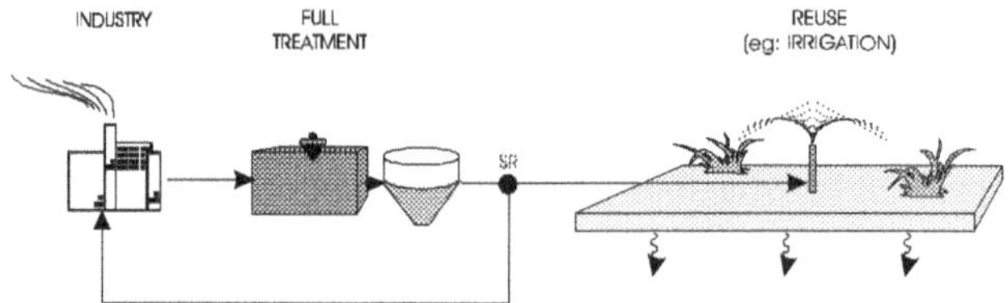


Figure 2.18. Main alternatives for the treatment and discharge of industrial effluents

the legislation established by the Environmental Agency for discharge to receiving bodies.

Another option is for the industry to completely treat the effluent and use it for other purposes (example: irrigation), or recycle it as process water along the production line. Naturally, public health implications need to be well addressed and guidelines or standards for reuse need to be satisfied.

## 2.2.6.2 Pollutants of importance in industrial wastewaters

Industrial effluents, depending on the type of the industrial process, can contain in greater or lesser degrees, the various pollutants described in Section 2.2.3, which are present in domestic sewage (suspended solids, biodegradable organic matter, nitrogen, phosphorus and pathogenic organisms). The present section covers other pollutants, which are not usually found in typical domestic sewage, but which can be of concern in industrial or municipal wastewaters containing a fraction of industrial effluents. The text is based on da Silva et al (2001).

### a) Metals

In the present context, the main implications of metals are:

- Toxicity to human beings and other forms of plant or animal life, as a result of the discharge or disposal of wastewaters to receiving water bodies or land.
- Inhibition to the microorganisms responsible for the biological treatment of wastewater.

In spite of being widely used, the expression "heavy metal" does not have a sole definition, varying from branch to branch of science. From the environmental point of view of this book, heavy metals can be understood as those that, under certain concentrations and exposure time, offer risks to human health and the environment, impairing the activity of living organisms, including those responsible for the biological treatment of wastewater.

The main chemical elements that fit into this category are: Ag, As, Cd, Co, Cr, Cu, Hg, Ni, Pb, Sb, Se and Zn. These elements may be naturally found in soils or waters in variable concentrations, but lower than those ones considered toxic to different living organisms. Among these, As, Co, Cr, Cu, Se and Zn are essential to organisms in certain small quantities, while others have no function in biological metabolism, being toxic to plants and animals.

Most living organisms need only few metals, and in very small doses, characterising the concept of micronutrients, as zinc, magnesium, cobalt and iron. These metals become toxic and dangerous to human health when they exceed certain concentration thresholds. As for lead, mercury and cadmium, these are metals that do not exist naturally in any organism. They do not perform any nutritional or biochemical function in microorganisms, plants or animals. That is, the presence of these metals in living organisms is harmful at any concentration.

Table 2.26. Summary of the sources of contamination and the effects on human health by metals most frequently found in environment

| Metal | Sources of contamination | Effects on health |
|---|---|---|
| Cadmium | Refined flours, cigarettes, odontological materials, steel industry, industrial gaseous effluents, fertilisers, pesticides, fungicides, coffee and tea treated with agrotoxics, ceramics, seafood, bone meal, welding, casting and refining of metals such as zinc, lead and copper. Cadmium derivatives are used in pigments and paintings, batteries, electroplating processes, accumulators, PVC stabilisers, nuclear reactors. | Carcinogenic, causes blood pressure rise and heart swelling. Immunity decreases. Prostate growth. Bone weakening. Joint pains. Anaemia. Pulmonary emphysema. Osteoporosis. Smell loss. Decrease in sexual performance. |
| Lead | Car batteries, paints, fuels, plants treated with agrotoxics, bovine liver, canned foods, cigarettes, pesticides, hair paint, lead-containing gas, newsprint and colour advertisements, fertilisers, cosmetics, air pollution. | Irritability and aggressiveness, indisposition, migraines, convulsions, fatigue, gum bleed, abdominal pains, nausea, muscular weakness, loss of memory, sleeplessness, nightmares, unspecific vascular cerebral accident, alterations of intelligence, osteoporosis, kidney illnesses, anaemia, coagulation problems. It affects the digestive and reproductive system and is a teratogenic agent (causes genetic mutation). |
| Mercury | Thermometers, pesticides and agrotoxics, dental alloy, water, mining, polishers, waxes, jewels, paints, sugar, contaminated tomato and fish, explosives, mercury fluorescent lamps, cosmetic products, production and delivery of petroleum by-products, salt electrolysis cells for chlorine production. | Depressive illness, fatigue, tremors, panic syndrome, paresthesias, lack of motor control, side walking, speech difficulties, loss of memory, loss of sexual performance, stomatitis, loose teeth, pain and paralysis in the edges, headache, anorexia in children, hallucination, vomiting, mastication difficulties, sweating, and pain sense loss. |
| Nickel | Kitchenware, nickel–cadmium batteries, jewels, cosmetics, hydrogenated oils, pottery works, cold permanent wave, welding. | Carcinogenic, may cause: contact dermatitis, gingivitis, skin rash, stomatitis, dizziness, joint pains, osteoporosis and chronic fatigue. |
| Zinc | Metallurgy (casting and refining), lead recycling industries. | Sense of sweetish taste and dryness in the throat, cough, weakness, panalgia, shivering, fever, nausea, vomiting. |
| Chromium | Leather tanning, electroplating. | Dermatitis, cutaneous ulcers, nose inflammation, lung cancer and perforation in the nose septum. |

Table 2.26 (*Continued*)

| Metal | Sources of contamination | Effects on health |
|---|---|---|
| Arsenic | Fuel oil, pesticides and herbicides, metallurgy, sea plants and animals. | Gastrointestinal disturbances, muscular and visceral spasms, nausea, diarrhoea, inflammation of mouth and throat, abdominal pains. |
| Aluminium | Water, processed cheese, white wheat flour, aluminium kitchenware, cosmetics, anti-acids, pesticides and antiperspirant, baker's yeast, salt. | Intestinal constipation, loss of energy, abdominal colics, infantile hyperactivity, loss of memory, learning difficulties, osteoporosis, rickets and convulsions. Related diseases: Alzheimer's and Parkinson's. |
| Barium | Polluted water, agrotoxics, pesticides and fertilisers. | Arterial hypertension, cardiovascular diseases, fatigue and discouragement. |

*Sources:* http://www.rossetti.eti.br; http://www.greenpeace.org.br

In human beings, metals can produce several effects, resulting from their action on molecules, cells, tissues, organs and even the whole system. Besides, the presence of a metal might restrict the absorption of other nutrients essential to the activity of the organism. Metals, because they cannot be metabolised, remain in the organism and carry out their toxic effects, combining with one or more reactive groups, which may be indispensable for normal physiological functions. Depending on the material involved and on the intensity of the intoxication, the effect may range from a topic skin manifestation, pulmonary membrane or digestive tract, to mutagenic, teratogenic or carcinogenic effects, and even death. It is important to emphasise that synergistic effects also need to be taken into account. In most cases, synergistic effects might be far greater than the mere sum of the individual effects.

Although in general metals may be poisonous to plants and animals under the low concentrations in which they may be present in domestic wastewater, no chronic toxicity problems associated with their disposal have been reported. On the other hand, the same could not be said for industrial wastewaters and the resulting sludges (in which metals are concentrated).

Table 2.26 summarises the main sources of contamination from some metals, together with their effects on human health.

In wastewater treatment, limitations associated with metals are mainly related to the inhibition of, or toxicity to, microorganism growth and the incorporation of metals in the sludge. For a certain metal, the maximum allowable load needs to be determined, such that there are no problems with microorganism inhibition, deterioration of effluent quality and impairment to agricultural use of the sludge.

The discharge of a particular industrial wastewater into the public sewerage system will have a variable impact in the WWTP, depending on the dilution factor, the content and type of pollutant, and the treatment process employed. To analyse the impact, it is interesting to perform simulations and to apply a safety factor to the calculated limits. In this way, decisions may be made regarding the acceptance of the effluents into the system, and finally at the WWTP. If the estimated loads are lower than the acceptable limits, the discharge may be accepted. Conversely, if the limits are exceeded, pre-treatment may be required, or no further admissions to the public systems may be accepted. A check must be made on the system to verify whether the biological process is being inhibited, or whether the treated effluent and the sludge to be reused are outside the limits established by the environmental agency. The control must be centred on the industrial discharges, since domestic sewage may not be prevented to be discharged to the public network system.

### b)   Toxic and dangerous organic compounds

Like the section on metals, this text is also based on da Silva et al (2001). Toxic and dangerous organic compounds, even though they usually *do not represent a concern in domestic sewage*, may be of concern in municipal wastewaters that receive industrial effluents.

When wastewaters containing toxic organic compounds are disposed of in the receiving water body without adequate treatment, severe damage may occur, both to the aquatic life and to human beings, who use it as a source of water supply. Most of these compounds are very slowly biodegraded, persisting in the environment for a long period. These compounds are able to penetrate the food chain and, even if they are not detectable in the receiving body, they may be present in large quantities in the higher trophic levels, owing to their bioaccumulation characteristics. Another important fact is that, although some compounds do not pose serious health damages when ingested, their metabolites may be more toxic than the original products. Besides, since wastewaters have a complex composition and normally contain more than one organic pollutant, synergistic effects may take place (the combined effect may be higher than the sum of the individually exerted effects).

Several dangerous pollutants are volatile because of their low solubility, low molecular weight and high vapour pressure. Therefore, they may be transferred to the atmosphere in open units in the WWTP, such as aeration tanks, equalisation tanks and clarifiers, and also pumping stations. If adequate control means are not taken, their volatilisation represents a potential health risk to the population and workers who are frequently exposed to it. The structural integrity of the sewerage collection system is also affected, because many compounds are corrosive, inflammable and explosive (methanol, methyl-ethylketone, hexane, benzene, among others).

Other pollutants are adsorbed and concentrated in the biological flocs in the treatment process, and might cause inhibition to sludge digestion or generate sludge with dangerous characteristics which, if not adequately disposed of, could contaminate groundwater.

In some cases, the toxic pollutants are present in such low concentrations, that are not able to inhibit the biological process, but also are very hard to be removed.

Consequently, the treatment plant effluent may still contain these pollutants and, when discharged into the receiving body, may cause damages to the aquatic life and human beings.

There are relatively few data on the behaviour of these dangerous pollutants in WWTPs. The lack of knowledge of their physical, chemical and biochemical characteristics, as well as their inter-relationships in complex wastewaters, makes it extremely difficult to predict their treatability and destination during the treatment processes. More research is required for the identification of many compounds, understanding of their removal mechanisms and development of predictive models.

The main sources of organic compounds are: chemical and plastic industries, mechanical products, pharmaceutical industries, pesticide formulation, casthouses and steel industries, oil industry, laundries and lumber industries.

The most commonly found organic pollutants in industrial effluents are: phenol, methyl chloride, 1,1,1-trichloroethane, toluene, ethyl benzene, trichloroethylene, tetrachloroethylene, chloroform, bis-2-ethyl-hexyl phthalate, 2,4-dimethyl phenol, naphthalene, butylbenzylphthalate, acrolein, xylene, cresol, acetophenone, methyl-sobutyl-acetone, diphenylamine, aniline and ethyl acetate.

## 2.2.6.3 *Population equivalent*

*Population equivalent* (PE) is an important parameter for characterising industrial wastewaters. PE reflects the equivalence between the polluting potential of an industry (commonly in terms of biodegradable organic matter) and a certain population, which produces the same polluting load. For instance, when an industry is said to have a population equivalent of 20,000 habitants, it is the equivalent to saying that the BOD load of the industrial effluent corresponds to the load generated by a community with a population of 20,000 inhabitants. The formula for the calculation of population equivalent based on BOD is:

$$\boxed{\text{PE (population equivalent)} = \frac{\text{BOD load from industry (kg/d)}}{\text{per capita BOD load (kg/inhab.d)}}} \qquad (2.13)$$

In the case of adopting the value frequently used in the international literature for the per capita BOD load of 54 gBOD/inhab.d, PE may be calculated by:

$$\text{PE (population equivalent)} = \frac{\text{BOD load from industry (kg/d)}}{0.054 \text{ (kg/inhab.d)}} \qquad (2.14)$$

When reporting a value of population equivalent, it is important to make clear the per capita load used as a reference (54 gBOD/inhab.d or other value, more applicable to the region under analysis).

---

**Example 2.3**

Calculate the Population Equivalent (PE) of an industry that has the following data:

- flow $= 120\ \text{m}^3/\text{d}$
- BOD concentration $= 2000\ \text{mg/L}$

**Solution:**

The BOD load is:

$$\text{load} = \text{flow} \times \text{concentration} = \frac{120\ \text{m}^3/\text{d} \times 2000\ \text{g/m}^3}{1000\ \text{g/kg}} = 240\ \text{kgBOD/d}$$

The Population Equivalent is:

$$\text{PE} = \frac{\text{load}}{\text{per capita load}} = \frac{240\ \text{kg/d}}{0.054\ \text{kg/hab.d}} = 4{,}444\ \text{inhab}$$

Thus, the wastewater from this industry has a polluting potential (in terms of BOD) equivalent to a population of 4,444 inhabitants.

---

### 2.2.6.4 Characteristics of industrial wastewater

The characteristics of industrial wastewater vary essentially with the type of industry and with the type of industrial process used. Table 2.27 presents the main parameters that should be investigated for the characterisation of the effluents, as a function of the industry type. This table is only a general and initial guide, since there is always the possibility that the effluent from a certain industry has a parameter of importance not listed, or that a certain parameter in the table is not relevant to the industry in consideration.

The present book addresses mainly the treatment of predominantly domestic sewage. In this way, the main parameter of interest is the organic matter, represented by the BOD. Table 2.28 presents general information about the organic pollution generated by certain industries, including the population equivalent and the BOD loads per unit produced. Example 2.4 illustrates the use of the table for the estimation of the BOD in the industrial wastewater entering a WWTP.

---

**Example 2.4**

A slaughterhouse processes 30 heads of cattle and 50 pigs per day. Estimate the characteristics of the effluent.

**Solution:**

Using the table of industrial wastewater characteristics (Table 2.28), and adopting an average value of 3.0 kgBOD/cattle slaughtered (1 cow $\approx$ 2.5 pigs):

### *Example 2.4 (Continued)*

**a)  BOD load produced**

$$-\text{cows:}\quad \frac{3\text{ kgBOD}}{\text{cow}}\cdot\frac{30\text{ cow}}{\text{d}} = 90\text{ kgDBO/d}$$

$$-\text{pigs:}\quad \frac{3\text{ kgDBO}}{2.5\text{ pigs}}\cdot\frac{50\text{ pigs}}{\text{d}} = 60\text{ kgDBO/d}$$

$$-\text{total:}\quad 90 + 60 = 150\text{ kgBOD/d}$$

**b)  Population Equivalent (PE)**

$$PE = \frac{BODload}{per\ capita\ BODload} = \frac{150\text{ kgDBO/d}}{0.054\text{ kgDBO/inhab.d}} = 2.77\text{ inhab}$$

**c)  Wastewater flow**

Using Table 2.28, and adopting an average value of 2.0 m$^3$/cattle slaughtered (or for 2.5 pigs slaughtered):

$$-\text{cows:}\quad \frac{2.0\text{ m}^3}{\text{cow}}\cdot\frac{30\text{ cow}}{\text{d}} = 60\text{ m}^3/\text{d}$$

$$-\text{pigs:}\quad \frac{2.0\text{ m}^3}{2.5\text{ pigs}}\cdot\frac{50\text{ pigs}}{\text{d}} = 40\text{ m}^3/\text{d}$$

$$-\text{total:}\quad 60 + 40 = 100\text{ m}^3/\text{d}$$

**d)  BOD concentration in the wastewater**

$$concentration = \frac{load}{flow} = \frac{150\text{ kg/d BOD}}{100\text{ m}^3/\text{d}}\cdot 1000\text{ g/kg} = 1{,}500\text{ g/m}^3$$

$$= 1{,}500\text{ mg/L}$$

## 2.2.7  General example of the estimation of flows and pollutant loads

### 2.2.7.1  Problem configuration

Determine the characteristics of the sewage that is going to be generated by the following town, until year 20 of operation. The population forecast for the project produced the values presented in the table below.

The coverage (served population / total population) is 60% at the beginning of operation (year 0), reaching, as a target, the value of 100% from year 5.

Table 2.27. Main parameters of importance for industrial effluents as a function of the industry type

| Type | Activity | BOD or COD | SS | Oils & Grease | Phenols | pH | CN⁻ | Metals |
|---|---|---|---|---|---|---|---|---|
| *Food products* | Sugar and alcohol | x | x | | x | x | | |
| | Meat and fish preservation | x | x | | | x | | |
| | Dairies | x | x | x | | x | | |
| | Slaughter houses | x | x | x | | | | |
| | Fruit and vegetable canning | x | x | | | x | | |
| | Milling of grains | x | x | | | | | |
| *Drinks* | Soft drinks | x | x | x | | x | | |
| | Brewery | x | x | x | | x | | |
| *Textiles* | Cotton | x | | | | x | | |
| | Wool | x | | x | | x | | |
| | Synthetics | x | | | | x | | |
| | Dyeing | | | x | x | x | | x |
| *Tanneries* | Vegetable tanning | x | x | x | | x | | |
| | Chromium tanning | x | x | x | | x | | x |
| *Farming* | Breeding of animals in confined spaces | x | x | | | | | x |
| *Paper* | Process. of pulp-cellulose | x | x | | | x | | x |
| | Manuf. pulp and paper | x | x | | | x | | x |
| *Non-metallic mineral products* | Glass and mirrors | | x | x | | x | | x |
| | Glass fibre | x | x | x | x | | | |
| | Cement | | x | x | | x | | |
| | Ceramics | | x | x | | | | x |

| | | | | | | | | |
|---|---|---|---|---|---|---|---|---|
| *Rubber* | Rubber articles | x | x | x | | x | | |
| | Tyres and tubes | x | x | x | | x | | |
| *Chemical product* | Chemical products (various) | | | | x | x | x | x |
| | Photographic laboratory | | | | | | | x |
| | Paints and colouring agents | | | | | | | x |
| | Insecticides | | | | | | x | x |
| | Disinfectants | | | | x | | | x |
| *Plastic* | Plastics and resins | x | x | | x | x | | x |
| *Perfume/soap* | Cosmetics, detergents, soap | x | | x | | | | x |
| *Mechanical* | Production of metal pieces | | | x | x | | | |
| *Metallurgy* | Production of pig iron | x | x | x | x | x | x | x |
| | Steelworks | | x | x | | x | x | x |
| | Electroplating | | x | x | x | x | x | x |
| *Mining* | Extraction activities | | x | | | x | | |
| *Oil derivatives* | Oils and lubricants | x | | x | x | x | | |
| | Asphalt works | | x | x | | | | |
| *Electrical items* | Electrical items | | | | | | x | x |
| *Wood* | Saw mills | | x | | | | | |
| *Personal services* | Laundries | x | | x | | x | | |

Table 2.28. Characteristics of the wastewater from some industries

| Type | Activity | Unit of production | Specific wastewater flow ($m^3$/unit) | Specific BOD load (kg/unit) | BOD population equivalent [inhab/(unit/d)] | BOD concentration (mg/L) |
|---|---|---|---|---|---|---|
| Food | Canning (fruit/vegetables) | 1 t processed | 4–50 | 30 | 500 | 600–7,500 |
| | Pea processing | 1 t processed | 13–18 | 16–20 | 85–400 | 300–1,350 |
| | Tomato processing | 1 t processed | 4–8 | 1–4 | 50–185 | 450–1,600 |
| | Carrot processing | 1 t processed | 11 | 18 | 160–390 | 800–1,900 |
| | Potato processing | 1 t processed | 7.5–16 | 10–25 | 215–545 | 1,300–3,300 |
| | Citrus fruit processing | 1 t processed | 9 | 3 | 55 | 320 |
| | Chicken meat processing | 1 t produced | 15–60 | 4–30 | 70–1600 | 100–2400 |
| | Beef processing | 1 t processed | 10–16 | 1–24 | 20–600 | 200–6,000 |
| | Fish processing | 1 t processed | 5–35 | 3–55 | 300–2300 | 2,700–3,500 |
| | Sweets / candies | 1 t produced | 5–25 | 2–8 | 40–150 | 200–1,000 |
| | Sugar cane | 1 t produced | 0.5–10 | 2.5 | 50 | 250–5,000 |
| | Dairy (without cheese) | 1000 L milk | 1–10 | 1–5 | 20–100 | 300–5,000 |
| | Dairy (with cheese) | 1000 L milk | 2–10 | 5–40 | 100–800 | 500–8,000 |
| | Margarine | 1 t produced | 20 | 30 | 500 | 1,500 |
| | Slaughter house | 1 cow / 2.5 pigs | 0.5–3 | 0.5–5 | 10–100 | 1,000–5,000 |
| | Yeast production | 1 t produced | 150 | 1100 | 21,000 | 7,500 |
| Confined animal breeding | Pigs | live t.d | 0.2 | 2 | 35–100 | 10,000–50,000 |
| | Dairy cattle (milking room) | live t.d | 0.02–0.08 | 0.05–0.10 | 1–2 | 370–2,300 |
| | Cattle | live t.d | 0.15 | 1.6 | 65–150 | 10,000–50,000 |
| | Horses | live t.d | 0.15 | 4–8 | 65–150 | 20,000–50,000 |
| | Poultry | live t.d | 0.38 | 0.9 | 15–20 | 2,000–3,000 |
| Sugar–alcohol | Alcohol distillation | 1 t cane processed | 60 | 220 | 4,000 | 3,500 |
| Drinks | Brewery | 1 $m^3$ produced | 5–20 | 8–20 | 150–350 | 500–4,000 |
| | Soft drinks | 1 $m^3$ produced | 2–5 | 3–6 | 50–100 | 600–2,000 |
| | Wine | 1 $m^3$ produced | 5 | 0.25 | 5 | – |

|  |  |  | | | | |
| --- | --- | --- | --- | --- | --- | --- |
| Textiles | Cotton | 1 t produced | 120–750 | 150 | 2,800 | 200–1,500 |
| | Wool | 1 t produced | 500–600 | 300 | 5,600 | 500–600 |
| | Rayon | 1 t produced | 25–60 | 30 | 550 | 500–1,200 |
| | Nylon | 1 t produced | 100–150 | 45 | 800 | 350 |
| | Polyester | 1 t produced | 60–130 | 185 | 3,700 | 1,500–3,000 |
| | Wool washing | 1 t produced | 20–70 | 100–250 | 2,000–4,500 | 2,000–5,000 |
| | Dyeing | 1 t produced | 20–60 | 100–200 | 2,000–3,500 | 2,000–5,000 |
| | Textile bleaching | 1 t produced | – | 16 | 250–350 | 250–300 |
| Leather and tanneries | Tanning | 1 t hide processed | 20–40 | 20–150 | 1,000–3,500 | 1,000–4,000 |
| | Shoes | 1000 pairs produced | 5 | 15 | 300 | 3,000 |
| Pulp and paper | Pulp | 1 t produced | 15–200 | 30 | 600 | 300 |
| | Paper | 1 t produced | 30–270 | 10 | 100–300 | |
| | Pulp and paper integrated | 1 t produced | 200–250 | 60–500 | 1,000–10,000 | 300–10,000 |
| Chemical industry | Paint | 1 employee | 0.110 | 1 | 20 | 10 |
| | Soap | 1 t produced | 25–200 | 50 | 1000 | 250–2,000 |
| | Petroleum refinery | 1 barrel (117 L) | 0.2–0.4 | 0.05 | 1 | 120–250 |
| | PVC | 1 t produced | 12.5 | 10 | 200 | 800 |
| Non-metallic industry | Glass and by-products | 1 t produced | 50 | – | – | – |
| | Cement (dry process) | 1 t produced | 5 | – | – | – |
| Steelworks | Foundry | 1 t pig iron produced | 3–8 | 0.6–1.6 | 12–30 | 100–300 |
| | Lamination | 1 t produced | 8–50 | 0.4–2.7 | 8–50 | 30–200 |

*Note:* data not filled in (-) means non-significant data or data not obtained; t = metric ton (1000 kg)

In various cases the water consumption is considered equal to the wastewater flow produced

*Sources:* CETESB (1976), Braile and Cavalcanti (1977), Arceivala (1981), Hosang & Bischof (1984), Salvador (1991), Wentzel (without date), Mattos (1998)

The length of the sewerage collection system is estimated to be 30 km for year 0, increasing to 55 km in year 5 (to sustain the increase in the coverage). From this year, it expands at a rate of 1 km per year.

The town has one dairy industry that processes around 5,000 litres of milk per day, for the production of milk, cheese and butter. There are provisions for expansion at year 10, when the production will be doubled.

| Year | Total urban population (inhabitants) | Coverage (%) | Served population (inhabitants) | Length of the collection system (km) | Industrial production (litres of milk per day) |
|---|---|---|---|---|---|
| 0 | 40,000 | 60 | 24,000 | 30 | 5,000 |
| 5 | 47,000 | 100 | 47,000 | 55 | 5,000 |
| 10 | 53,000 | 100 | 53,000 | 60 | 10,000 |
| 15 | 58,000 | 100 | 58,000 | 65 | 10,000 |
| 20 | 62,000 | 100 | 62,000 | 70 | 10,000 |

Owing to lack of time and other conditions during the design period, it was not possible to obtain samples for characterising the actual sewage composition. Assume adequate values for the missing variables and establish suitable hypotheses for the various parameters in the calculations.

## 2.2.7.2 Flow estimation

### a)  Domestic flow

- *Average flow*
  Assume:
  - per capita water consumption: $L_{pcd} = 160$ L/inhab.d (see Tables 2.5 and 2.7)
  - return coefficient (sewage flow/water flow): $R = 0.8$ (see Section 2.1.2.4)

  The average flow for year 0 is (Equation 2.2):

  $$Q_{d_{av}} = \frac{Pop.L_{pcd}.R}{1000} = \frac{24,000 \times 160 \times 0.8}{1000} = 3,072 \text{ m}^3/\text{d} \ (= 35.6 \text{ L/s})$$

  The flows for the other years are calculated in a similar way, changing only the population.

- *Maximum flow*

  Adopting the Harmon formula (Table 2.10), the $Q_{max}/Q_{av}$ ratio is calculated for the population of every year. For year 0:

  $$\frac{Q_{max}}{Q_{av}} = 1 + \frac{14}{4 + \sqrt{P}} = 1 + \frac{14}{4 + \sqrt{24,000/1,000}} = 2.57$$

The values of $Q_{max}$ are obtained by multiplying $Q_{av}$ by the ratio $Q_{max}/Q_{av}$. Therefore, for year 0:

$$Q_{max} = 2.57 \times 35.6 \, \text{l/s} = 91.5 \, \text{L/s}$$

The ratios and flows for the other years are calculated in a similar manner, altering only the value P (population/1000).

- *Minimum flow*

Adopt a $Q_{min}/Q_{av}$ equal to 0.5. The $Q_{min}$ values are obtained by multiplication with the ratio $Q_{min}/Q_{av}$. Therefore, for year 0:

$$Q_{min} = 0.5 \times 35.6 \, \text{L/s} = 17.8 \, \text{L/s}$$

The ratios and the flows for the other years are calculated in a similar manner.

## b) Infiltration flow

Adopt $Q_{inf} = 0.3$ L/s.km for the sewerage system. Consider the resulting flow value for each year occurring only in the average and maximum flows. For year 0:

$$Q_{inf} = 30 \, \text{km}. \, 0.3 \, \text{L/s.km} = 9.0 \, \text{L/s} \, (= 778 \, \text{m}^3/\text{d})$$

The flows for the other years are calculated in a similar manner, remembering that, from year 5, for each year the sewerage system increases by 1 km.

## c) Industrial wastewater flow

Adopt a value of 7 m$^3$ of wastewater produced per 1000 L of milk processed (water consumption being equal to wastewater production) (see Table 2.28).

Consider that for the years 0 and 5, 5,000 L of milk per day are processed and that for the years 10, 15 and 20, 10,000 L/d of milk are processed (given in the problem).

Assume that the maximum flow is 1.5 times the average flow and the minimum flow is 0.5 times the average flow.

For year 0:

- $Q_{av} = 5 \, \text{m}^3 \, \text{milk} \times 7 \, \text{m}^3 \, \text{wastewater/m}^3 \, \text{milk} = 35 \, \text{m}^3/\text{d} \, (= 0.4 \, \text{L/s})$
- $Q_{max} = 1.5 \times Q_{av} = 1.5 \times 0.4 = 0.6 \, \text{L/s}$
- $Q_{min} = 0.5 \times Q_{av} = 0.5 \times 0.4 = 0.2 \, \text{L/s}$

The flows for the other years are calculated in a similar manner.

## d) Total flow

Total flow corresponds to the sum of the domestic, infiltration and industrial flows. Therefore for year 0, the total influent flow to the WWTP is:

Total flow = domestic flow + infiltration flow + industrial flow

- total average flow $= 35.6 + 9.0 + 0.4 = 45.0$ L/s ($= 3{,}888$ m$^3$/d)
- total maximum flow $= 91.5 + 9.0 + 0.6 = 101.1$ L/s ($= 8{,}735$ m$^3$/d)
- total minimum flow $= 17.8 + 0.0 + 0.2 = 18.0$ L/s ($= 1{,}555$ m$^3$/d)

The flows for the other years are calculated in a similar manner .

### 2.2.7.3  BOD load

**a)  Domestic BOD**

Adopt a per capita BOD production of 50 gBOD$_5$/inhab.d (see Table 2.24)
For the population of year 0:

$$\text{Domestic BOD}_5 \text{ load} = 50 \text{ g/inhab.d} \times 24{,}000 \text{ inhab.} = 1.2 \times 10^6 \text{ g/d}$$

$$= 1{,}200 \text{ kg/d}$$

The loads for the other years are calculated in a similar manner.

**b)  Infiltration water BOD**

Consider that the BOD is zero for infiltration water.

**c)  Industrial BOD**

Adopt a value of 25 kg of BOD per 1000 L of milk processed (see Table 2.28).
Consider that for the years 0 and 5, 5,000 L of milk per day are processed and that for the years 10,15 and 20, 10,000 L/d of milk are processed (given in the problem).
For year 0:

$$\text{Industrial BOD}_5 \text{ load} = 25 \text{ kg/1000 L milk} \times 5{,}000 \text{ L milk/d} = 125 \text{ kg/d}$$

The loads for the other years are calculated in a similar manner.

**d)  Total BOD load**

Total BOD load corresponds to the sum of the domestic, infiltration and industrial BOD. Therefore for year 0, the total BOD load is:

$$\text{Total BOD}_5 \text{ load} = \text{domestic BOD}_5 \text{ load} + \text{infiltration BOD}_5 \text{ load}$$

$$+ \text{ industrial BOD}_5 \text{ load}$$

$$\text{Total BOD}_5 \text{ load} = 1{,}200 + 0 + 125 = 1{,}325 \text{ kg/d}$$

The total loads for the other years are calculated in a similar manner.

Table 2.29. Flows, loads and concentrations in the influent to the WWTP

| | Data from the community | | | Wastewater flows (L/s) | | | | | | | | | | | Average BOD load (kg/d) | | | | Popul. equival. | BOD concentration (mg/L) | | | |
| | | Length | Industrial | Domestic flow | | | | Industrial flow | | | Total flow (L/s) | | | Total average flow | | | | | | | | | |
| | Served popul. | sewer. system | product. | | | | | | | | | | | | | | | | | | | | |
| Year | (inhab.) | (km) | (L milk/d) | Min. | Aver. | Max. | Infiltr | Min. | Aver. | Max. | Min. | Aver. | Max. | (m³/d) | Domestic | Infiltr. | Industrial | Total | (industr.) | Domestic | Infiltr. | Industrial | Total |
| 1 | 2 | 3 | 4 | 5 | 6 | 7 | 8 | 9 | 10 | 11 | 12 | 13 | 14 | 15 | 16 | 17 | 18 | 19 | 20 | 21 | 22 | 23 | 24 |
|---|---|---|---|---|---|---|---|---|---|---|---|---|---|---|---|---|---|---|---|---|---|---|---|
| 0 | 24,000 | 30 | 5,000 | 17.8 | 35.6 | 91.5 | 9.0 | 0.2 | 0.4 | 0.6 | 18.0 | 45.0 | 101.1 | 3,888 | 1,200 | 0 | 125 | 1,325 | 2,500 | 391 | 0 | 3,571 | 341 |
| 5 | 47,000 | 55 | 5,000 | 34.8 | 69.6 | 159.4 | 16.5 | 0.2 | 0.4 | 0.6 | 35.0 | 86.5 | 176.5 | 7,477 | 2,350 | 0 | 125 | 2,475 | 2,500 | 391 | 0 | 3,571 | 331 |
| 10 | 53,000 | 60 | 10,000 | 39.3 | 78.5 | 176.0 | 18.0 | 0.4 | 0.8 | 1.2 | 39.7 | 97.3 | 195.2 | 8,409 | 2,650 | 0 | 250 | 2,900 | 5,000 | 391 | 0 | 3,571 | 345 |
| 15 | 58,000 | 65 | 10,000 | 43.0 | 85.9 | 189.5 | 19.5 | 0.4 | 0.8 | 1.2 | 43.4 | 106.2 | 210.2 | 9,179 | 2,900 | 0 | 250 | 3,150 | 5,000 | 391 | 0 | 3,571 | 343 |
| 20 | 62,000 | 70 | 10,000 | 45.9 | 91.9 | 200.1 | 21.0 | 0.4 | 0.8 | 1.2 | 46.3 | 113.7 | 222.4 | 9,820 | 3,100 | 0 | 250 | 3,350 | 5,000 | 391 | 0 | 3,571 | 341 |

Calculations in each column:

*Col 1*: problem data

*Col 2*: problem data

*Col 3*: problem data

*Col 4*: problem data

*Col 5* = *col 6* × 0.5

*Col 6* = *col 2* × (160 L/inhab.d × 0.8) / 86400 s/d

*Col 7* = *col 6* × (1+14/(4+(*col 2*/1000)$^{0.5}$))

*Col 8* = *col 3* × 0.3 L/s.km

*Col 9* = *col 10* × 0.5

*Col 10* = *col 4* × 7 m³ ww/m³ milk × 1000 L/m³ / 86,400 s/d

*Col 11* = *col 10* × 1.5

*Col 12* = *col 5* + *col 9*

*Col 13* = *col 6* + *col 8* + *col 10*

*Col 14* = *col 7* + *col 8* + *col 11*

*Col 15* = *col 13* × 86400 s/d / 1000 L/m³

*Col 16* = *col 2* × 0.050 kg/inhab.d

*Col 17* = 0

*Col 18* = *col 4* × 25 kg/1000 L milk

*Col 19* = *col 16* + *col 17* + *col 18*

*Col 20* = *col 18* / 0.050 kg/inhab.d

*Col 21* = *col 16* × 1000 L/m³ × 1000 g/kg / (*col 6* × 86400 s/d)

*Col 22* = *col 17* × 1000 L/m³ × 1000 g/kg / (*col 8* × 86400 s/d)

*Col 23* = *col 18* × 1000 L/m³ × 1000 g/kg / (*col 10* × 86400 s/d)

*Col 24* = (*col 19* / *col 15*) × 1000 g/kg

### 2.2.7.4 BOD concentration

The BOD concentration is given by the quotient between the BOD load and the wastewater flow (see Equation 2.11). The BOD concentration for the influent to the WWTP for the year 0 is:

$$\text{Concentration} = \text{load/flow} = (1{,}325 \text{ kg/d}) / (3{,}888 \text{ m}^3/\text{d}) = 0.341 \text{ kg/m}^3$$

$$= 341 \text{ g/m}^3 = 341 \text{ mg/L}$$

The BOD concentrations for the other years are calculated in a similar manner.

### 2.2.7.5 Presentation of results

Table 2.29 presents a summary of the various values determined following the proposed criteria. The table can be expanded to include other wastewater characteristics, such as suspended solids, nitrogen and phosphorus. The methodology to be used is the same as for BOD.

# 3

# Impact of wastewater discharges to water bodies

## 3.1 INTRODUCTION

The present chapter covers basic aspects of water quality and water pollution, analysing in more detail three important topics related to the discharge of sewage to receiving water bodies (rivers, lakes and reservoirs):

- Pollution by organic matter (*dissolved oxygen consumption*)
- Contamination by pathogenic microorganisms (*bacterial die-off*)
- Pollution of lakes and reservoirs (*eutrophication*, caused by nitrogen and phosphorus)

In each of these main items, *causes*, *effects*, *control* and *modelling* of the pollution are discussed. Later in the chapter, water quality legislation is discussed, introducing the concepts of *discharge standard* and *quality standard for the water body*. The importance of the chapter is related to the planning of the removal efficiency and effluent quality to be achieved in the WWTP.

## 3.2 POLLUTION BY ORGANIC MATTER AND STREAM SELF PURIFICATION

### 3.2.1 Introduction

The present section covers one of the main problems of water pollution, largely solved in most developed regions but still of great importance in developing regions, that is, the consumption of dissolved oxygen (DO) after sewage discharge.

The introduction of *organic matter* into a water body results, indirectly, in the *consumption of dissolved oxygen*. This occurs as a result of the processes of the stabilisation of the organic matter undertaken by bacteria, which use the oxygen available in the liquid medium for their respiration. As expected, the decrease in the dissolved oxygen concentration in the water body has various implications from the environmental point of view.

The objective of this section is the study of the processes of consumption of dissolved oxygen and of stream self-purification, in which the water body recovers itself, through purely natural mechanisms. Both of these processes are analysed from an ecological viewpoint and, subsequently, more specifically through the mathematical representation of the DO profile in the water body.

In broader terms, the process of **self-purification** is associated with the *re-establishment of the equilibrium of the aquatic ecosystem, after the alterations induced by the effluent discharge*. Within a more specific point of view, the conversion of organic compounds into inert compounds, not deleterious from an ecological viewpoint, is an integral part of the process.

It should be understood that the concept of self-purification presents the same relativity as the concept of pollution (see Chapter 1). Water can be considered purified from one point of view, even if not fully purified in hygienic terms, presenting, for instance, pathogenic organisms. From a pragmatic approach, water could be considered purified when its characteristics are not conflicting anymore with their intended uses in each reach of the watercourse. This is because there is no absolute purification: the ecosystem reaches a new equilibrium, but under conditions that are different from before (upstream), owing to the increase in the concentrations of certain compounds and by-products resulting from the decomposition process. As a consequence, the aquatic community is different, even if at a new equilibrium state.

The knowledge and quantification of the self-purification phenomenon is very important, because of the following objectives:

- *To use the assimilation capacity of the rivers.* From a strictly ecological point of view, it could be argued that ecosystems should remain unaltered. However, from a pragmatic perspective, it can be considered that the capacity of a water body to assimilate discharges, without presenting environmental problems, is a natural resource that can be exploited. This realistic vision is of great importance in developing countries where the lack of financial resources justifies the use of the water course to complement, up to a certain point, the processes that occur in sewage treatment (provided this is done with parsimony and with well-defined and safe technical criteria).
- *To avoid effluent discharges above the assimilative capacity of the water body.* In this way, the assimilative capacity of the water body can be used up to a level that is acceptable and non-detrimental. Beyond this level, no further discharges could be allowed.

## 3.2.2 Ecological aspects of stream self purification

The ecosystem of a water body upstream of the discharge of untreated wastewater is usually in a state of equilibrium. Downstream of the discharge, the equilibrium between the communities is affected, resulting in an initial disorganisation followed by a subsequent tendency towards rearrangement.

In this sense, self-purification can be understood as a phenomenon of **ecological succession**. Along the river, there is a systematic sequence of replacements of a community by another, until a stable community is established, in equilibrium with the local conditions.

The presence or absence of pollution can be characterised by the concept of **species diversity**:

*Ecosystem in natural conditions:*
- High number of species
- Low number of individuals in each species

*Ecosystem under disturbance:*
- Low number of species
- High number of individuals in each species

A reduction in the diversity of the species is due to the fact that the *pollution is selective* for the species: only those that adapt to the new environmental conditions survive and, further, proliferate (resulting in a high number of individuals in these few species). The other species do not resist to the new environmental conditions and perish (leading to a reduction in the total number of species).

Because self-purification is a process that develops with time, and considering the dimension of the river as predominantly longitudinal, the stages of ecological succession can be associated with physically identifiable zones in the river. There are four main zones:

- *Degradation zone*
- *Active decomposition zone*
- *Recovery zone*
- *Clean water zone*

These zones occur downstream of the discharge of a predominantly biodegradable organic wastewater. It should be remembered that upstream of the discharge there is a clean water zone characterised by ecological equilibrium and good water quality. Figure 3.1 presents the trajectory along the four zones of the three main water quality parameters: organic matter, heterotrophic bacteria (feeding on organic matter) and dissolved oxygen.

**DEGRADATION ZONE**

| Characteristic | Description |
| --- | --- |
| *General characteristics* | This zone starts soon after the discharge of wastewater to the water body. The main chemical characteristic is the high concentration of organic matter, still in a complex stage but potentially decomposable. |
| *Aesthetic aspects* | At the discharge point, water is turbid due to the solids present in the sewage. The sedimentation of the solids results in the formation of sludge banks. |
| *Organic matter and dissolved oxygen* | In this zone there is complete disorder, compared to the stable community that existed before. Decomposition of the organic matter, carried out by microorganisms, can have a slow start, depending on the adaptation of the microorganisms to the waste. Normally, in the case of predominantly organic wastewater, the microorganisms present in the wastewater itself are those responsible for the start of the decomposition. Because the decomposition can still be incipient, the oxygen consumption for the respiratory activities of the microorganisms can also be low, allowing sufficient dissolved oxygen for fish. After the adaptation of the microorganisms, the consumption rate of the organic matter becomes high, also implying a high rate of dissolved oxygen consumption. |
| *Microorganisms* | After the adaptation period, bacteria start to proliferate, with a massive predominance of aerobic forms, that is, those that depend on the oxygen available in the medium for their metabolic processes. Bacteria, having an abundance of food in the form of the organic matter introduced by the wastewater and also with sufficient oxygen for their respiration, have excellent conditions for development and reproduction. The quantity of organic matter is at a maximum at the discharge point and, due to the consumption by microorganisms, starts to decrease. |
| *Decomposition by-products* | There is an increase in the levels of carbon dioxide, one of the products of the microbial respiratory process. With the increase in $CO_2$ concentration, which is then converted into carbonic acid in water, water may become more acidic and pH may decrease. |
| *Bottom sludge* | Anaerobic conditions start to prevail in the sludge at the bottom, due to the difficulty in gas exchange with the atmosphere. As a consequence, there is a production of hydrogen sulphide, which is a potential generator of unpleasant smell. |
| *Nitrogen* | Complex nitrogen compounds are still present in high levels, although a large part undergoes conversion to ammonia. |
| *Aquatic community* | There is a substantial reduction in the number of living species, although the number of individuals in each one is extremely high, characterising a disturbed ecosystem. Less adapted forms disappear, while resistant and better-adapted forms prevail. The quantity of coliform bacteria is very high, when the discharge is associated with domestic sewage. Also occurring are protozoans that feed on the bacteria, besides fungi that feed on the organic matter. The presence of algae is rare because of the difficulty in light penetration, owing to the turbidity of the water. An evasion of hydras, sponges, crustaceans, molluscs and fish takes place. |

**ACTIVE DECOMPOSITION ZONE**

| Characteristics | Description |
| --- | --- |
| *General characteristics* | After the initial disturbance, the ecosystem begins to organise itself. Microorganisms, present in large numbers, actively decompose organic matter. The impact reaches the highest levels and water quality is at its worst state. |
| *Aesthetic aspects* | The strongest water colouration can still be observed, together with the dark deposits of sludge at the bottom. |
| *Organic matter and dissolved oxygen* | In this zone the dissolved oxygen reaches its lowest concentration. Depending on the magnitude of the discharge, dissolved oxygen may be completely consumed by the microorganisms. In this situation, anaerobic conditions occur in all the liquid bulk. Aerobic life disappears, giving way to predominantly anaerobic microorganisms. |
| *Microorganisms* | Bacteria begin to reduce in number, mainly due to the reduction in the available food, which has been largely stabilised. Other factors still interact in the decrease of bacteria, such as light, flocculation, adsorption and sedimentation. |
| *Decomposition by-products* | In the event of anaerobic reactions taking place, by-products are, besides carbon dioxide and water, methane, hydrogen sulphide, mercaptans and others, many of them responsible for the generation of bad odours. |
| *Nitrogen* | Nitrogen is still present in the organic form, although the larger part is already in the form of ammonia. At the end of the zone, in the presence of dissolved oxygen, oxidation of ammonia to nitrite may start. |
| *Aquatic community* | The number of enteric bacteria, pathogenic or not, decreases rapidly. This is due to the fact that these bacteria, well adapted to the environmental conditions in the human intestinal tract, do not resist to the new environmental conditions, which are adverse to their survival. The number of protozoans increases, leading to the rising to a new level in the food pyramid, in the ecological succession process. The presence of some macroorganisms occurs along with insect larvae, adapted to survive under the prevailing conditions. However, the macro fauna is still restricted in species. Hydras, sponges, crustaceans, molluscs and fish have not yet returned. |

**RECOVERY ZONE**

| Characteristics | Description |
| --- | --- |
| *General characteristics* | After the intense phase of organic matter consumption and degradation of the aquatic environment, the recovery stage commences. |
| *Aesthetic aspects* | Water is clearer and its general appearance is improved. Sludge deposits at the bottom present a less fine and more granulated texture. There is no release of gases or bad smells. |
| *Organic matter and dissolved oxygen* | Organic matter, intensely consumed in the previous zones, is largely stabilised and transformed into inert compounds. This implies a lower rate of oxygen consumption through bacterial respiration. In parallel with this, atmospheric oxygen is introduced into the liquid mass, increasing the level of dissolved oxygen (oxygen production by atmospheric reaeration is now larger than its consumption for the stabilisation of the organic matter). The anaerobic conditions that eventually occurred in the previous zone are not present anymore, resulting in another change in the aquatic fauna and flora. |

| Characteristics | Description |
| --- | --- |
| *Nitrogen* | Ammonia is converted into nitrite and the nitrite to nitrate. Also the phosphorus compounds are transformed into phosphates. A fertilisation of the medium takes place because of the presence of nitrates and phosphates, which are nutrients for algae. |
| *Algae* | Due to the presence of nutrients and the higher transparency of the water (allowing a larger light penetration), there are conditions for development of algae. With them, there is the production of oxygen by photosynthesis, increasing the levels of dissolved oxygen in the medium. Also as a result of the presence of algae, the food web becomes more diversified. |
| *Aquatic community* | The number of bacteria is now small and, as a result, so is the number of protozoan bacteriophages. Algae are under full development: the first ones to appear are the blue algae (cyanobacteria) on the surface and banks, followed by flagellates, green algae and finally diatoms. Microcrustaceans occur in their maximum number. Molluscs, various worms, dinoflagellates, sponges and insect larvae are present at high numbers. The food chain is more diversified, generating food for the first more tolerant fishes. |

**CLEAN WATER ZONE**

| Characteristics | Description |
| --- | --- |
| *General characteristics* | Water is clean again. Conditions are similar to those upstream of the discharge, at least in respect to dissolved oxygen, organic matter and bacteria levels, and probably pathogenic organisms. |
| *Aesthetic aspects* | The appearance of the water is similar to that before the pollution occurred. |
| *Organic matter and dissolved oxygen* | In the liquid there is a predominance of the completely oxidised and stable forms of inorganic matter, although sludge at the bottom may not be necessarily stabilised. The concentration of dissolved oxygen is close to the saturation level, owing to the low consumption by the microbial population and possibly high production by the algae. |
| *Aquatic community* | Because of the mineralisation that occurred in the previous zone, water is now richer in nutrients than before the pollution. Therefore, the production of algae is higher. There is the re-establishment of the normal food web. Various organisms, including large freshwater crustacea, molluscs and fish are present. Species diversity is high. The ecosystem is now stable and the community reaches its climax again. |

## 3.2.3 Dissolved oxygen balance

### 3.2.3.1 Interacting factors in the DO balance

In ecological terms, the most negative impact of the pollution in a water body caused by organic matter is the decrease in the level of dissolved oxygen, caused by the respiration of microorganisms involved in the purification of the sewage. The impact is extended to all the aquatic community, and each reduction in the level of the dissolved oxygen is selective for certain species.

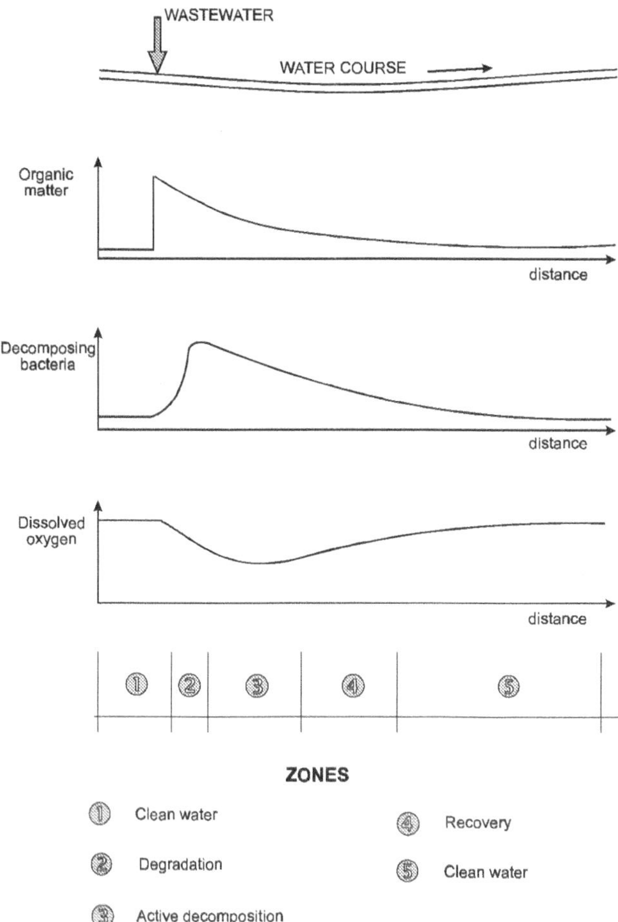


Figure 3.1.   Schematic profile of the concentration of the organic matter, bacteria and dissolved oxygen along the length of the water course, with the indication of the self-purification zones

Dissolved oxygen has been traditionally used for the determination of the degree of pollution and self purification in water bodies. Its measurement is simple and its level can be expressed in quantifiable concentrations, allowing mathematical modelling.

Water is an environment poor of oxygen, by virtue of its low solubility. While in the air its concentration is in the order of 270 mg/L, in water, at normal conditions of temperature and pressure, its concentration is reduced approximately to only 9 mg/L. In this way, any large consumption brings substantial impacts in the DO level in the liquid mass.

Table 3.1. Main interacting mechanisms in the DO balance

| Oxygen consumption | Oxygen production |
|---|---|
| – oxidation of the organic matter (respiration)<br>– benthic demand (sludge at the bottom)<br>– nitrification (ammonia oxidation) | – atmospheric reaeration<br>– photosynthesis |

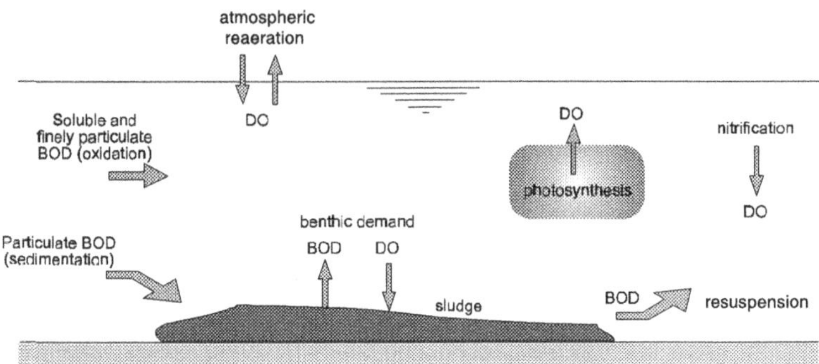


Figure 3.2.   Interacting mechanisms in the dissolved oxygen balance

In the self-purification process there is a *balance* between the sources of consumption and the sources of production of oxygen. When the consumption rate is higher than the production rate, the oxygen concentration tends to decrease, the opposite occurring when the consumption rate is lower than the production rate. The main interacting mechanisms in the DO balance in a water body can be found in Figure 3.2 and Table 3.1. In general, the concentrations of the constituents (such as DO) in a water body change as a result of physical processes of *advection* (transportation by the water as it flows in the river channel) and *dispersion* (transportation due to turbulence and molecular diffusion) and biochemical and physical processes of *conversion* (reaction) (Fig. 3.3). The processes take place in the three dimensions of the water body, although in rivers the longitudinal axis (X) is the prevailing one. The mechanisms listed in Table 3.1 are associated with conversion processes.

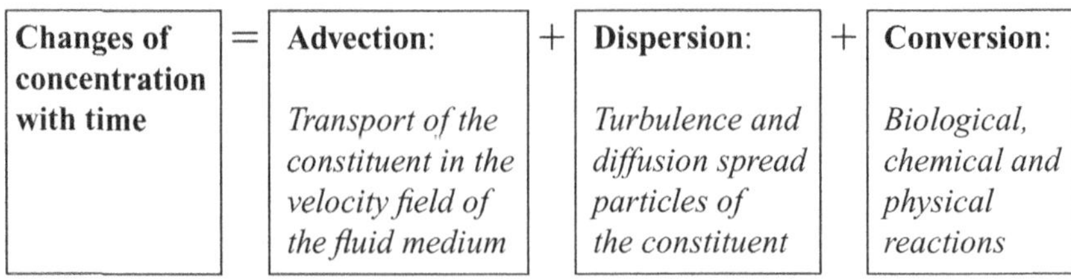

| **Changes of concentration with time** | = | **Advection:**<br><br>*Transport of the constituent in the velocity field of the fluid medium* | + | **Dispersion:**<br><br>*Turbulence and diffusion spread particles of the constituent* | + | **Conversion:**<br><br>*Biological, chemical and physical reactions* |
|---|---|---|---|---|---|---|

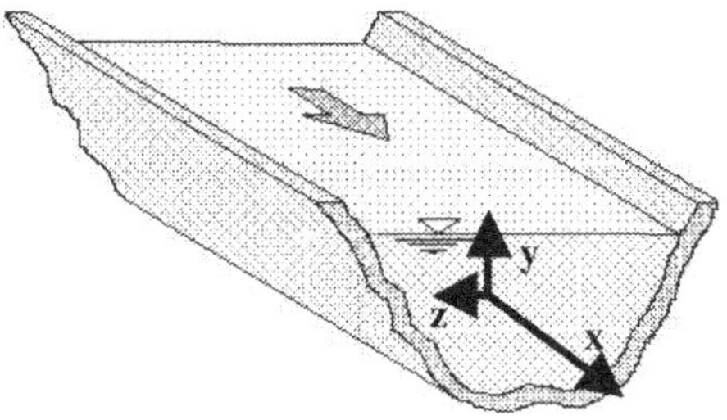

Figure 3.3.   Axes along which spatial and temporal variations of the concentrations of the water constituents take place (in rivers, the X axis is predominant)

# Oxygen consumption

## a)   Oxidation of the organic matter

The organic matter in sewage is present in two forms: *suspended* (*particulate*) and *dissolved* (*soluble*). Settleable suspended solids tend to settle in the water body, forming a sludge layer at the bottom. The dissolved matter, together with the suspended solids of small dimensions (hardly settleable) remains in the liquid mass.

The oxidation of the latter fraction of organic matter corresponds to the main factor in the oxygen consumption. The consumption of DO is due to the respiration of the microorganisms responsible for the oxidation, principally aerobic heterotrophic bacteria. The simplified equation for the stabilisation (oxidation) of organic material is:

$$\text{Organic matter} + O_2 + \text{bacteria} \rightarrow CO_2 + H_2O + \text{bacteria} + \text{energy} \qquad (3.1)$$

Bacteria, in the presence of oxygen, convert the organic matter to simple and inert compounds, such as water and carbon dioxide. As a result, bacteria tend to grow and reproduce, generating more bacteria, while there is availability of food (organic matter) and oxygen in the medium.

## b)   Benthic (sediment) demand

The settled organic matter in suspension, which formed the bottom sludge layer, also needs to be stabilised. A large part of the conversion is completed under anaerobic conditions, because of the difficulty of oxygen to penetrate the sludge layer. This form of conversion, being anaerobic, implies the non- consumption of oxygen.

However, the upper part of the sludge layer, in the order of some millimetres of thickness, still has access to oxygen from the supernatant water. The sludge stabilisation is completed under aerobic conditions in this fine layer, resulting in the consumption of oxygen. Besides, some partial by-products of the anaerobic decomposition may dissolve, cross the aerobic sludge layer and diffuse itself in the bulk of the liquid, exerting an oxygen demand. The oxygen demand originating from these combined factors associated with the sludge is called *benthic* (or *sediment*) demand.

Another factor that can cause oxygen demand is the reintroduction of previously settled organic matter into the bulk of the liquid, caused by the *resuspension* of the sludge layer. This resuspension occurs in occasions of high flows and velocities in the water course. The sludge, not yet completely stabilised, represents a new source of oxygen demand.

The importance of the benthic demand and the resuspension of the sludge in the dissolved oxygen balance depends on a series of simultaneously interacting factors, many of them difficult to quantify.

### c)  Nitrification

Another oxidation process is the one associated with the conversion of ammonia into nitrite and this nitrite into nitrate, in the process of *nitrification*.

The microorganisms involved in this process are chemoautotrophs, which have carbon dioxide as the main carbon source and which draw their energy from the oxidation of an inorganic substrate, such as ammonia.

The transformation of ammonia into nitrite is completed according to the following simplified reaction:

$$\text{ammonia} + O_2 \rightarrow \text{nitrite} + H^+ + H_2O + \text{energy} \qquad (3.2)$$

The transformation of nitrite into nitrate occurs in sequence, in accordance with the following simplified reaction:

$$\text{nitrite} + O_2 \rightarrow \text{nitrate} + \text{energy} \qquad (3.3)$$

It is seen that in both reactions there is oxygen consumption. This consumption is referred to as nitrogenous demand or second-stage demand, because it takes place after the oxidation of most of the carbonaceous matter. This is due to the fact that the nitrifying bacteria have a slower growth rate compared with the heterotrophic bacteria, implying that nitrification also occurs at a slower rate.

## Oxygen production

### a)  Atmospheric reaeration

Atmospheric reaeration is frequently the main factor responsible for the introduction of oxygen into the liquid medium.

Gas transfer is a physical phenomenon, through which gas molecules are exchanged between the liquid and the gas at their *interface*. This exchange results in an increase in the concentration in the liquid gas phase, if this phase is not saturated with gas.

This is what happens in a water body, in which the DO concentration is reduced due to the processes of the stabilisation of the organic matter. As a consequence, DO levels are lower than the saturation concentration, which is given by the solubility of the gas at a given temperature and pressure. In this situation there is an *oxygen deficit*. If there is a deficit, there is the search for a new equilibrium, thus allowing a larger absorption by the liquid mass.

The transfer of oxygen from the gas phase to the liquid phase occurs basically through the following two mechanisms:

- Molecular diffusion
- Turbulent diffusion

In a quiescent water body, *molecular diffusion* prevails. This diffusion may be described as a tendency of any substance to uniformly spread itself about all of the available space. However, this mechanism is very slow and requires a long time for the gas to reach the deepest layers of the water body.

The mechanism of *turbulent diffusion* is much more efficient, because it involves the main factors of an effective aeration: creation of interfaces and renewal of interfaces. The first one is important, because it is through these interfaces that gas exchange occurs. The second one is also significant, because the fast renewal of the interfaces permits that localised saturation points are not formed, besides conducting the dissolved gas to the various depths of the liquid mass, as a result of the mixing.

The diffusion condition to prevail is a function of the hydrodynamic character-istics of the water body. A shallow river with rapids presents excellent conditions for an efficient turbulence. In these conditions, molecular diffusion is negligible. On the other hand, in lakes, molecular diffusion tends to predominate, unless wind promotes good mixing and interface renewal.

**b) Photosynthesis**

Photosynthesis is the main process used by autotrophic organisms to synthesise organic matter, being a characteristic of organisms containing chlorophyll. The process takes place only in the presence of light energy, according to the following simplified equation (there are many intermediate steps):

$$CO_2 + H_2O + \text{light energy} \rightarrow \text{organic matter} + O_2 \qquad (3.4)$$

*Photosynthesis reaction is exactly opposite to the respiration reaction.* While photosynthesis is a process of fixing light energy and forming glucose molecules of high energy potential, respiration is essentially the opposite, that is, release of this energy for subsequent use in metabolic processes (Branco, 1976).

Light dependence controls the distribution of photosynthetic organisms to loca-tions to where light is present. In waters with a certain turbidity, such as from soil particles or suspended solids from waste discharges, the possibility of the presence of algae is smaller and, as a result, so is the photosynthetic activity. This is seen in the first self-purification zones, where the predominance is almost exclusively of heterotrophic organisms. In these zones, respiration surpasses production.

Table 3.2. River water quality parameters modelled by some computer software currently available

| Quality parameter | Program | | | | | | | | | |
|---|---|---|---|---|---|---|---|---|---|---|
| | 1 | 2 | 3 | 4 | 5 | 6 | 7 | 8 | 9 | 10 |
| Temperature | √ | | √ | √ | √ | √ | √ | o.s. | o.s. | |
| Bacteria | | | √ | √ | √ | √ | √ | o.s. | o.s. | |
| DO-BOD | √ | √ | √ | √ | √ | √ | √ | o.s. | o.s. | √ |
| Nitrogen | √ | √ | √ | √ | √ | √ | √ | o.s. | o.s. | √ |
| Phosphorus | √ | √ | √ | √ | √ | √ | √ | o.s. | o.s. | √ |
| Silica | | | √ | | √ | √ | √ | o.s. | o.s. | |
| Phytoplankton | √ | √ | √ | √ | √ | √ | √ | o.s. | o.s. | √ |
| Zooplankton | | | √ | | √ | √ | | o.s. | o.s. | |
| Benthic algae | | | | | √ | √ | √ | o.s. | o.s. | |

1 = QUAL2E (USEPA, 1987); 2 = WASP5 (USEPA, 1988); 3 = CE-QUAL-ICM (US Army Engineer Waterways Experiment Station, 1995); 4 = HEC5Q (US Army Engineer Hydrologic Engineering Centre, 1986); 5 = MIKE11 (Danish Hydraulic Institute, 1992); 6 = ATV Model (ATV, Germany, 1996); 7 = Salmon-Q (HR Wallingford, England, 1994); 8 = DUFLOW (Wageningen Univ., Holland, 1995); 9 = AQUASIM (EAWAG, Switzerland, 1994); 10 = DESERT (IIASA, Austria, 1996); o.s. = open structure (can be modified by the user).
*Source:* Rauch et al, 1998; Shanaham et al, 1998; Somlyódy, 1998

In general, the autotrophs carry out much more synthesis than oxidation, generating a positive balance of organic compounds that constitute an energy reserve for the heterotrophs, besides an excess of oxygen that sustains the respiration of other organisms.

### 3.2.3.2 *Water quality models*

**a)  More complete water quality models**

River quality models have been used since the development of the classic model of DO and BOD by Streeter and Phelps, in 1925. This model represented a milestone in water and environmental engineering. Subsequently, various other models were developed, including the model of Camp (1954), increasing the level of complexity and the number of state and input variables, but at the same time maintaining the same conceptual structure of the classic Streeter–Phelps model. A widely known model, within the relatively recent generation of models, is the QUAL2E model, developed by the United States Environmental Protection Agency (USEPA), which represents in greater depth the cycles of O, N and P in water. There still is another class of models, understood as ecosystem models that represent suspended solids, various algae groups, zooplankton, invertebrates, plants and fish.

The Task Group on River Water Quality Modelling (2001) from IWA (International Water Association) developed a new model (IWA, Scientific and Technical Report 12), with a large number of components and processes, presented in matrix-format. This Task Group also presented an interesting comparison between various models currently available, synthesised in Table 3.2 (Rauch et al, 1998; Shanaham et al, 1998; Somlyódy, 1998).

Naturally that, the larger the number of variables (quality parameters) represented by a model, the larger is the number of kinetic parameters and stoichiometric coefficients to be obtained or adopted and, therefore, the larger is the difficulty in calibrating the model.

For developing countries, with all the large regional diversity of problems and solutions concerning water quality, it is difficult to establish generalisations about the use of models. However it is always important to have in mind that all the water quality models mentioned have been developed in countries that have already practically solved their basic pollution problems, such as the discharge of raw wastewater containing organic matter (domestic and industrial). In these conditions, it is natural that attention is now given to transient events and diffused pollution, for example. Nevertheless, in most of the developing counties the basic problems have not yet been solved, and simpler models still have a large contribution to give for the adequate management of water resources.

## b)   Simplified models

In the present text, for the sake of simplicity, only the two main components in the DO balance are covered, namely:

- *oxygen consumption*: oxidation of organic matter (respiration)
- *oxygen production*: atmospheric reaeration.

Naturally there are cases that justify the inclusion of other components, when it is felt that these are important in the DO balance. However, field and laboratory work for a reliable evaluation of these parameters needs to be undertaken intensively and rigorously, substantially increasing the complexity level of the study. The adoption of sophisticated mathematical models demands the availability of time and financial resources compatible with the proposed formulation, what is frequently not the case in developing countries. Therefore, in the present text, the more simplified version of the model (Streeter–Phelps version) is adopted, allowing an easier identification of occasional problems in its structure and parameter values.

Another important point is that *any user of a sophisticated model should understand well the basic principles of the Streeter–Phelps model*, in order to avoid a blind use of the computer software, without knowing the basic processes that are being represented.

It should be explained that the model described is *restricted to aerobic conditions* in the water body. Under anaerobic conditions, the conversion rate of organic matter is slower, being carried out by a biomass with completely different characteristics. Anaerobic conditions may occur frequently in simulations of the discharge of untreated wastewater to water courses with small dilution capacity.

## c)   Hydraulic representation

In the model structure, the hydraulic regime of the water body must be taken into consideration. There are basically three types of hydraulic models for a water

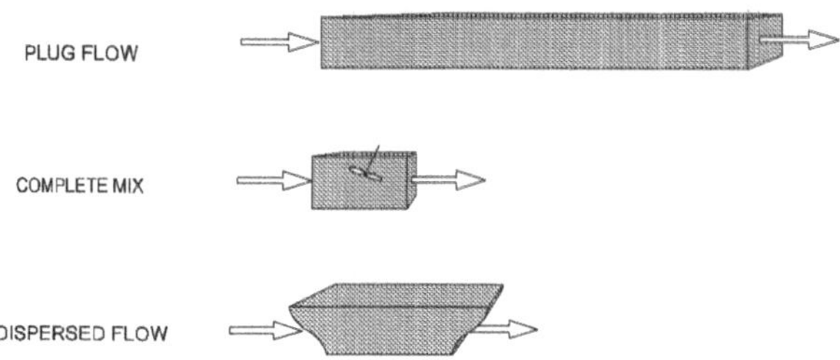


Figure 3.4.　Different hydraulic regimes for a water body: (a) plug flow, (b) complete mix, (c) dispersed flow

body or a reactor (see Figure 3.4):

* *Plug flow*
* *Complete mix*
* *Dispersed flow*

A water body in the ideal *complete-mix* regime is characterised by having the same concentration at all points in the liquid mass. Thus, the effluent concentration is equal at whatever point in the water body. This representation is usually applied to well-mixed lakes and reservoirs. This regime is also called *CSTR* (completely-stirred tank reactor).

A predominantly linear water body, such as a river, can be characterised through the *plug-flow* regime. In the ideal plug flow there are no exchanges between the upstream and downstream sections. Each section functions as a plug, in which the water quality is the same in all points and the community is adapted to the ecological conditions prevailing in each moment. Along with the downstream movement of the plug, the various self-purification reactions take place. Hydraulically, this model is similar to the case in which a tank with water, equal to the plug, remains the same period of time subjected to the same reactions and processes, therefore having the same water quality as that of the plug in the water body (see Figure 3.5).

The two characteristics represented above are for idealised situations. In reality, water bodies present a characteristic of dispersion of the pollutants, which is intermediate between the two extreme situations: total dispersion (completely mixing) and no dispersion (plug flow). Therefore the water bodies or their reaches can be characterised by a dispersion coefficient. High dispersion coefficients are associated with water bodies approaching a completely mixing regime, whereas reduced coefficients are associated with water bodies approaching plug-flow conditions. The *dispersed-flow* regime is particularly relevant with rivers under estuarine influence or with very low flow velocities.

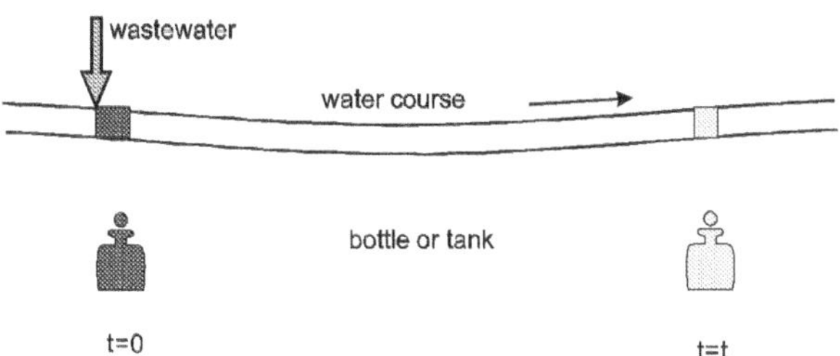


Figure 3.5.   Comparison between a bottle, tank or vessel and a plug in a plug-flow reactor

In the present chapter, the simplified approach of considering a river to be represented by the plug-flow regime is adopted, which is acceptable for most situations.

### 3.2.3.3  The dissolved oxygen profile

It is interesting to analyse the variations (decrease and increase) in the DO concentration along the water course in a graph, which plots the so-called *DO profile* or *DO sag curve*. In this graph, the vertical axis is the DO concentration and the horizontal axis is the distance or travelling time, along which the biochemical transformations take place. From the graph, the following elements can be obtained:

- identification of the consequences of the discharge
- connection of the pollution with the self-purification zones
- relative importance of the consumption and production of oxygen
- critical point of lowest DO concentration
- comparison between the critical DO concentration and the minimum allowable concentration, according with the legislation
- location where the water course returns to the desired conditions

The modelling of these items depends essentially on the understanding of the two main interacting mechanisms in the DO balance: *deoxygenation* and *atmospheric reaeration*. These topics are covered in the following subsections.

## 3.2.4  Kinetics of deoxygenation

### 3.2.4.1  Mathematical formulation

As already seen, the main ecological effect of organic pollution in a water body is the decrease in the levels of dissolved oxygen. This decrease is associated with

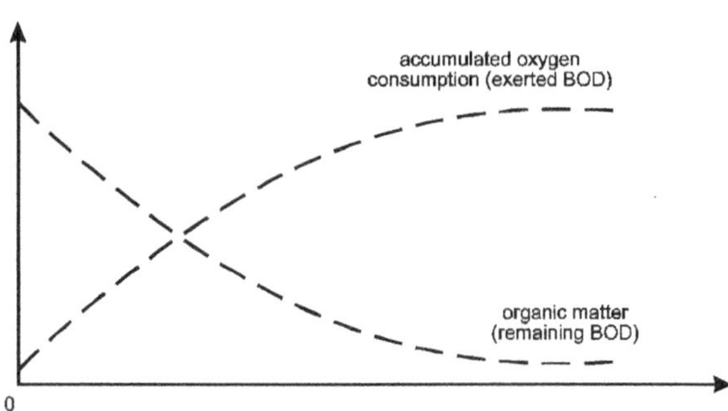


Figure 3.6.   Exerted BOD (oxygen consumed) and remaining BOD (remaining organic matter) along time

the Biochemical Oxygen Demand (BOD), described in Chapter 2. To standardise the results, the concept of the standard BOD is frequently used, being expressed by $BOD_5^{20}$. However, the oxygen consumption in the sample varies with time, that is, the BOD value is different each day. The objective of the present section is to mathematically analyse the progress of oxygen consumption with time.

The concept of BOD, representing both the organic matter concentration and the oxygen consumption, can be understood by the following two distinct angles, both having as units mass of oxygen per unit volume (e.g. $mgO_2/L$):

- *remaining BOD*: concentration of the organic matter remaining in the liquid mass at a given time
- *exerted BOD*: cumulative oxygen consumption for the stabilisation of the organic matter

The progress of BOD with time, according with both concepts, can be seen in Figure 3.6.

The two curves are symmetrical, like mirror images. At time zero, the organic matter is present in its total concentration, while the oxygen consumed is zero. With the passing of time, the remaining organic matter reduces, implying an increase in the accumulated oxygen consumption. After a period of several days, the organic matter has been practically all stabilised (remaining BOD close to zero), while the oxygen consumption has been practically all exerted (BOD almost completely exerted). The understanding of this phenomenon is important, because both curves are an integral part of the DO model.

The kinetics of the reaction of the remaining organic matter (remaining BOD) follows a *first-order reaction*. A first-order reaction is that in which the rate of change of the concentration of a substance is proportional to the first power of the concentration. The first-order reactions are of fundamental importance in environmental engineering, since many reactions are modelled according with this kinetics.

The equation of the progress of the remaining BOD with time can be expressed by the following differential equation:

$$\frac{dL}{dt} = -K_1.L \tag{3.5}$$

where:

$L$ = remaining BOD concentration (mg/L)
$t$ = time (day)
$K_1$ = deoxygenation coefficient (day$^{-1}$)

The interpretation of Equation 3.5 is that the oxidation rate of the organic matter (dL/dt) is proportional to the organic matter concentration still remaining (L), at any given time t. Therefore, the larger the BOD concentration, the faster is the deoxygenation. After a certain time, in which BOD has been reduced by stabilisation, the reaction rate will be lower, as a result of the lower concentration of organic matter.

The *deoxygenation coefficient* $K_1$ is a parameter of great importance in the modelling of dissolved oxygen, being discussed in the next section.

The integration of Equation 3.5, between the limits of $L = L_0$ and $t = 0$ and $t = t$, leads to:

$$\boxed{L = L_0.e^{-K_1.t}} \tag{3.6}$$

where:

$L$ = remaining BOD at any given time t (mg/L)
$L_0$ = remaining BOD in t = 0 (mg/L)
$t$ = time (d)

Attention should be given to the fact that, in many references, this equation is written in a decimal form (base 10), and not in base e. Both forms are equivalent, provided the coefficient is expressed in the compatible base ($K_1$ *base e* = 2.3 × $K_1$ *base 10*). In the present text, the values of the coefficients are expressed in **base e.**

In terms of *oxygen consumption*, the quantification of the exerted BOD is important. This is obtained through Equation 3.6, leading to:

$$\boxed{y = L_0.(1 - e^{-K_1.t})} \tag{3.7}$$

where:

$y$ = exerted BOD at a time t (mg/L). Note that $y = L_0 - L$.
$L_0$ = remaining BOD, at t = 0 (as defined above), or exerted BOD (when t = $\infty$).
Also called *ultimate BOD demand*, by the fact that it represents the total BOD at the end of the stabilisation process (mg/L).

### Example 3.1

The interpretation of a laboratory analysis of a river water sample taken downstream from a sewage discharge leads to the following values: (a) coefficient of deoxygenation: $K_1 = 0.25$ d$^{-1}$; (b) ultimate demand $L_0 = 100$ mg/L. Calculate the exerted BOD at days 1, 5 and 20.

**Solution:**

Using Equation 3.7, where $y = L_0 \cdot (1 - e^{-K_1 \cdot t})$:

- For t = 1 day:

$$y_1 = 100\,(1 - e^{-0.25 \times 1}) = 22 \text{ mg/L}$$

- For t = 5 days:

$$y_5 = 100\,(1 - e^{-0.25 \times 5}) = 71 \text{ mg/L} \;(= BOD_5)$$

- For t = 20 days:

$$y_{20} = 100\,(1 - e^{-0.25 \times 20}) = 99 \text{ mg/L}$$

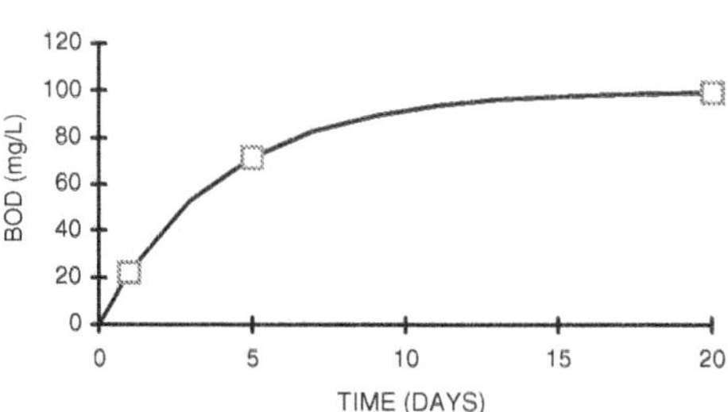


It is seen that at day 20 the BOD has been practically all exerted ($y_{20}$ practically equal to $L_0$).

The ratio between $BOD_5$ and the ultimate demand $L_0$ is: $71/100 = 0.71$. Therefore, at day 5, approximately 71% of the oxygen consumption have been exerted or, in other words, 71% of the total organic matter (expressed in terms of BOD) has been stabilised. Inversely, the $L_0/BOD_5$ ratio is equal to $100/71 = 1.41$.

## 3.2.4.2 The deoxygenation coefficient $K_1$

The coefficient $K_1$ depends on the characteristics of the organic matter, besides temperature and presence of inhibitory substances. Treated effluents, for example, have a lower degradation rate due to the fact that the larger part of the easily biodegradable organic matter has already been removed in the treatment plant, leaving only the slowly biodegradable fraction in the effluent. Average values of $K_1$ can be found present in Table 3.3.

There are mathematical and statistical processes that can be utilised for the determination of the deoxygenation coefficient, in case there are samples from the water under investigation. The input data for these methods are the values of the exerted BOD at various days, typically days 1, 2, 3, 4 and 5 or 1, 3, 5, 7 and 9. The laboratory tests must include, not only the BOD at 5 days, but also the BOD for all the other days, so that the rate of deoxygenation can be estimated. Nitrification must be inhibited in the BOD test, especially for the sequence that goes up to nine days.

The determination is not trivial, because there are two parameters to be simultaneously determined: $K_1$ and $L_0$. *Non-linear regression analysis* can be used, fitting Equation 3.7 to the various pairs of t and BOD to obtain the values of the parameters $K_1$ and $L_0$. In the present book, it is sufficient to use the values of $K_1$ obtained from the table of typical values (Table 3.3). It should be noted that, especially in the case of shallow rivers receiving untreated sewage, the deoxygenation may be higher than that determined in the laboratory, due to biofilm respiration at the river bottom.

Figure 3.7 illustrates the influence of the value of $K_1$, through the trajectories of the cumulative oxygen consumption of two samples with different values of $K_1$ but with the same ultimate demand value ($L_0 = 100$ mg/L). The sample with the higher $K_1$ ($0.25$ d$^{-1}$) presents a faster oxygen consumption rate compared with the sample with the lower $K_1$ ($0.10$ d$^{-1}$). Values of BOD close to the ultimate demand are reached in less time with the sample with the greater $K_1$.

The importance of the coefficient $K_1$ and the relativity of the BOD$_5$ concept can be analysed through the following example (see Figure 3.8). Two distinct samples present the same value of BOD$_5$ ($100$ mg/L). Apparently, one could conclude that the impact in terms of dissolved oxygen consumption is the same in the two

Table 3.3. Typical values of $K_1$ (base e, 20 °C)

| Origin | $K_1$ (day$^{-1}$) |
| --- | --- |
| Water course receiving concentrated raw sewage | 0.35–0.45 |
| Water course receiving raw sewage of a low concentration | 0.30–0.40 |
| Water course receiving primary effluent | 0.30–0.40 |
| Water course receiving secondary effluent | 0.12–0.24 |
| Water course with clean water | 0.09–0.21 |

*Source:* Adapted from Fair et al (1973) and Arceivala (1981)

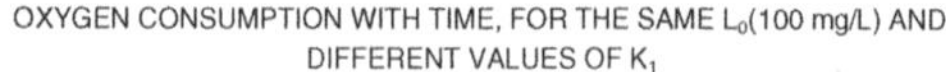

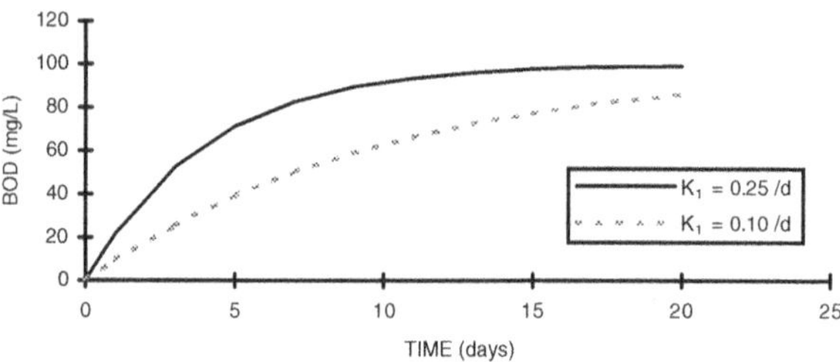



Figure 3.7.   Trajectory of the oxygen consumption for different values of $K_1$ and same values of ultimate BOD

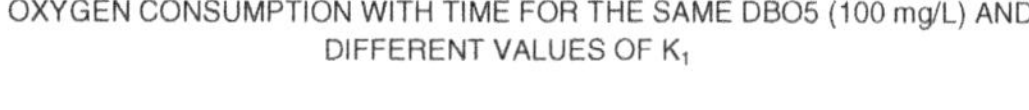

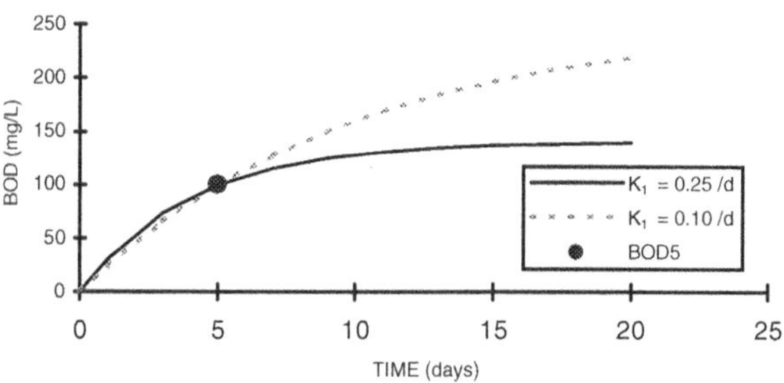



Figure 3.8.   Influence of the coefficient $K_1$ on the progression of BOD. Two samples with the same BOD value at 5 days (100 mg/L), but with different values of $K_1$ and, hence, different values of the ultimate BOD.

situations. However, if the progression of BOD is measured through various days, different BOD values can be observed for all the days, with the exception of the fifth day. This is due to the fact that the coefficients of deoxygenation are distinct in the two samples. The first presents a slower stabilisation rate ($K_1 = 0.10$ day$^{-1}$), implying a high ultimate BOD, still not reached on day 20. The second sample presents a higher $K_1$ ($K_1 = 0.25$ day$^{-1}$), and the demand is practically satisfied by day 20.

These considerations emphasise the aspect that the interpretation of the BOD data must always be associated with the concept of the coefficient of deoxygenation and, consequently, the rate of oxidation of the organic matter. This comment is of greater importance with industrial wastewaters, which are capable of presenting a large variability with regards to biodegradability or to the stabilisation rate of the organic matter.

### 3.2.4.3 Influence of temperature

Temperature has a great influence on the microbial metabolism, affecting, as a result, the stabilisation rates of organic matter. The empirical relation between temperature and the deoxygenation coefficient can be expressed in the following form:

$$\boxed{K_{1_T} = K_{1_{20}}.\theta^{(T-20)}}$$

(3.8)

where:

$K_{1_T} = K_1$ at a temperature $T(d^{-1})$
$K_{1_{20}} = K_1$ at a temperature $T = 20\,°C\ (d^{-1})$
$T$ = liquid temperature $(°C)$
$\theta$ = temperature coefficient $(-)$

A value usually employed for $\theta$ in this reaction is **1.047**. The interpretation of this value with relation to Equation 3.8 is that the value of $K_1$ increases 4.7% for every $1\,°C$ increment in the temperature of the water.

Also to be commented is that changes in the temperature affect $K_1$, but do not alter the value of the ultimate demand $L_0$.

## 3.2.5 Kinetics of reaeration

### 3.2.5.1 Mathematical formulation

The theory of gas transfer is covered in detail in Chapter 11. In the present chapter only the essential concepts necessary for the understanding of the atmospheric reaeration phenomenon are presented. When water is exposed to a gas, a continuos exchange of molecules occurs between the liquid and gas phases. As soon as the solubility concentration of the gas in the liquid phase is reached, both flows start to be equal in magnitude, such that there is no overall change of the gas concentration in both phases. This dynamic equilibrium defines the **saturation concentration** $(C_s)$ of the gas in the liquid phase.

However, in case that there is the consumption of dissolved gas in the liquid phase, the main transfer flux is in the gas-liquid direction, in order to re-establish the equilibrium. The atmospheric reaeration process takes place according to this concept. The oxygen consumption in the stabilisation of the organic matter makes the DO concentration to be below the saturation level. As a result, there is a greater flux of atmospheric oxygen to the liquid mass (Figure 3.9).

The kinetics of reaeration can also be characterised by a first-order reaction (similarly to the deoxygenation), according to the following equation:

$$\frac{dD}{dt} = -K_2.D$$

(3.9)

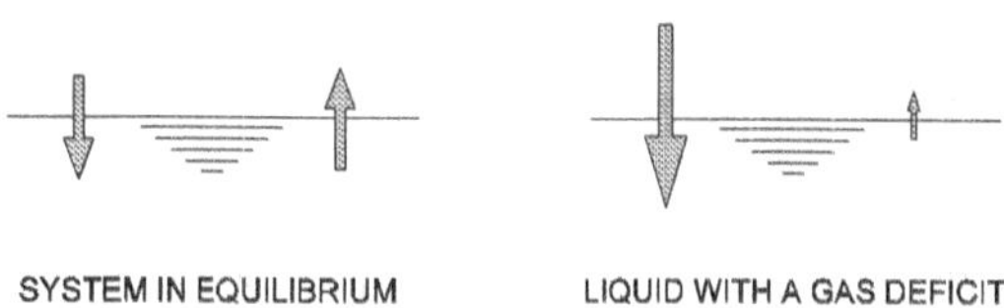



Figure 3.9.   Gas exchanges in a system in equilibrium and in a liquid with a dissolved gas deficit

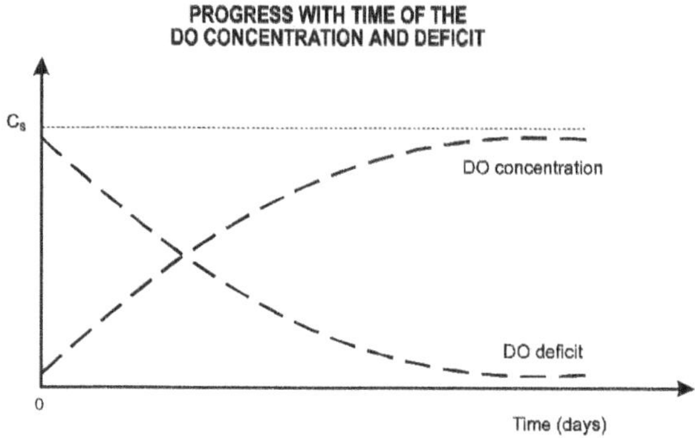



Figure 3.10.   Temporal progress of the concentration and deficit of dissolved oxygen

where:

$D =$ dissolved oxygen deficit, that is, the difference between the saturation concentration $(C_s)$ and the existing concentration at a time t (C) (mg/L)

$t =$ time (d)

$K_2 =$ reaeration coefficient (base e) $(d^{-1})$

Through Equation 3.9 it is seen that the absorption rate of oxygen is directly proportional to the existing deficit. The larger the deficit, the greater the gas transfer rate.

Integration of Equation 3.9, with $D_0$ when $t = 0$, leads to:

$$D = D_0.e^{-K_2.t} \qquad (3.10)$$

where:

$D_0 =$ initial oxygen deficit (mg/L)

The temporal progress of the deficit $(D = C_s - C)$ and the DO concentration (C) can be seen in Figure 3.10. The deficit and concentration curves are symmetrical and like mirror images. With the increase of the DO concentration with time due to the reaeration, the deficit decreases at the same rate.

Table 3.4. Typical values for $K_2$ (base e, 20 °C)

| Water body | $K_2$ (day$^{-1}$) | |
| --- | --- | --- |
|  | Deep | Shallow |
| Small ponds | 0.12 | 0.23 |
| Slow rivers, large lakes | 0.23 | 0.37 |
| Large rivers with low velocity | 0.37 | 0.46 |
| Large rivers with normal velocity | 0.46 | 0.69 |
| Fast rivers | 0.69 | 1.15 |
| Rapids and waterfalls | >1.15 | >1.61 |

*Source:* Fair et al (1973), Arceivala (1981)

## 3.2.5.2 *The reaeration coefficient $K_2$*

In a sample of deoxygenated water, the value of the coefficient $K_2$ can be determined through statistical methods. These methods are based on regression analysis, using either the original Equation 3.10, or some logarithmic transformation of it. The input data are the DO values at various times t. The output data are the saturation concentration $C_s$ and the coefficient $K_2$. In a water body, however, the experimental determination of $K_2$ is very complex, being outside the scope of the present text.

The value of the coefficient $K_2$ has a larger influence on the results of the DO balance than the coefficient $K_1$, because of the fact that the ranges of variation of $K_1$ are narrower. There are three methods for estimating the value of the coefficient $K_2$ in the river under study:

- average tabulated values
- values as a function of the hydraulic characteristics of the water body
- values correlated with the flow of the water body

**a)   Average tabulated values of $K_2$**

Some researchers, studying water bodies with different characteristics, proposed average values for $K_2$ based on a qualitative description of the water body (Table 3.4).

*Shallower* and *faster* water bodies tend to have a *larger reaeration coefficient,* due, respectively, to the greater ease in mixing along the depth and the creation of more turbulence on the surface (see Figure 3.11). The values in Table 3.4 can be used in the absence of specific data from the water body. It must be taken into consideration that *the values from this table are usually lower than those obtained by the other methods* discussed below. However, there are indications that, in some situations, the tabulated values result in better fitting to measured DO data than those obtained from hydraulic formula.

**b)   $K_2$ values as a function of the hydraulic characteristics of the water body**

Other researchers correlated the reaeration coefficient $K_2$ with the hydraulic variables of the water body. Various field techniques were employed in their studies, such as tracers, equilibrium disturbance, mass balance and others.

**INFLUENCE OF PHYSICAL CHARACTERISTICS ON COEFFICIENT K$_2$**

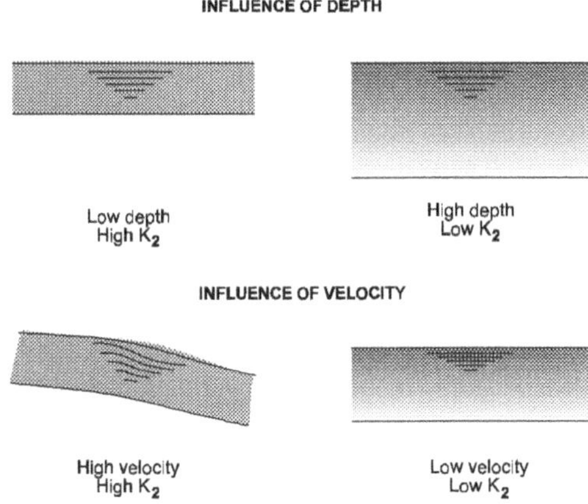



Figure 3.11.   Influence of the physical characteristics of the water body on the coefficient K$_2$

The literature presents various formulas, conceptual and empirical, relating K$_2$ with the depth and the velocity of the water body. Table 3.5 and Figure 3.12 present three of the main formulas, with application ranges that are complementary.

If there are natural cascades with free water falls, other formulations for the estimation of the atmospheric reaeration may be used. Von Sperling (1987) obtained the following empirical formula, based on the study of some waterfalls in Brazil:

$$C_e = C_0 + K.(C_s - C_0) \tag{3.11}$$

$$K = 1 - 1.343.H^{-0.128}.(C_s - C_0)^{-0.093} \tag{3.12}$$

where:
    $C_e$ = effluent (downstream) DO concentration (mg/L)
    $C_0$ = influent (upstream) DO concentration (mg/L)
    $K$ = efficiency coefficient (−)
    $C_s$ = DO saturation concentration (mg/L)
    $H$ = height of each free fall (m)

## c)   K$_2$ values correlated with the river flow

Another approach for estimating K$_2$ is through the correlation with the river flow. This can be justified by the fact that the depth and the velocity are intimately associated with flow.

Table 3.5. Values of the coefficient $K_2$, according with models based on hydraulic data (base e, 20 °C)

| Researcher | Formula | Application range |
|---|---|---|
| O'Connor & Dobbins (1958) | $3.73.v^{0.5}H^{-1.5}$ | $0.6\ m \leq H < 4.0\ m$<br>$0.05\ m/s \leq v < 0.8\ m/s$ |
| Churchill et al (1962) | $5.0.v^{0.97}H^{-1.67}$ | $0.6\ m \leq H < 4.0\ m$<br>$0.8\ m/s \leq v < 1.5\ m/s$ |
| Owens et al (cited by Branco, 1976) | $5.3.v^{0.67}H^{-1.85}$ | $0.1\ m \leq H < 0.6\ m$<br>$0.05\ m/s \leq v < 1.5\ m/s$ |

*Notes:*
- v: velocity of the water body (m/s)
- H: height of the water column (m)
- Ranges of applicability adapted and slightly modified from Covar (EPA, 1985), for simplicity

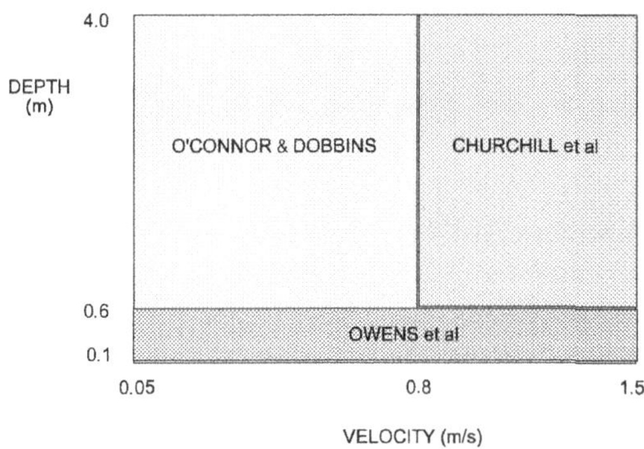


Figure 3.12.   Approximate applicability ranges of three hydraulic formulas for estimating $K_2$ (modified from Covar, cited in EPA, 1985)

The procedure is based on the determination of $K_2$ using the hydraulic formula (section b above), for each pair of values of v and H from historical records in the river. Subsequently, a regression analysis is performed between the resulting values of $K_2$ and the corresponding flow values Q. The relation between $K_2$ and Q may be expressed as $K_2 = m.Q^n$, where m and n are coefficients.

The advantage of this form of expression is that the reaeration coefficient may be calculated for any flow conditions (by interpolation or even some extrapolation), especially minimum flows, independently from depth and velocity values.

### 3.2.5.3 Influence of temperature

The influence of temperature is felt in two different ways:

- *an increase in temperature reduces the solubility of oxygen in the liquid medium (decrease of the saturation concentration $C_s$)*
- *an increase in temperature accelerates the oxygen absorption processes (increase of $K_2$)*

These factors act in opposite directions. The increase in $K_2$ implies an increase in the reaeration rate. However, a reduction in the saturation concentration corresponds to a decrease in the oxygen deficit D, resulting in a reduction in the reaeration rate. The overall influence on the reaeration rate depends on the magnitude of each variation but is frequently not substantial.

The influence of the temperature on the saturation concentration is discussed in Section 3.2.7k.

The influence of temperature on the reaeration coefficient can be expressed in the traditional form (Equation 3.13):

$$\boxed{K_{2_T} = K_{2_{20}}.\theta^{(T-20)}} \qquad (3.13)$$

where:

$K_{2_T} = K_2$ at a temperature $T(d^{-1})$
$K_{2_{20}} = K_2$ at a temperature $T = 20\,^{\circ}C\ (d^{-1})$
$\quad T =$ liquid temperature $(^{\circ}C)$
$\quad \theta =$ temperature coefficient $(-)$

A value frequently used for the temperature coefficient $\theta$ is **1.024**.


## 3.2.6 The DO sag curve

### 3.2.6.1 Mathematical formulation of the model

In 1925 the researchers Streeter and Phelps established the mathematical bases for the calculation of the dissolved oxygen profile in a water course. The structure of the model proposed by them (known as the **Streeter–Phelps model**) is classical within environmental engineering, setting the basis for all the other more sophisticated models that succeeded it. For the relatively simple situation in which only the deoxygenation and the atmospheric reaeration are taken into account in the DO balance, the rate of change of the oxygen deficit with time can be expressed by the following differential equation, originated from the interaction of the deoxygenation and reaeration equations previously seen:

Rate of change of the DO deficit = DO consumption − DO production $\qquad$ (3.14)

$$\frac{dD}{dt} = K_1.L - K_2.D \qquad (3.15)$$

Integration of this equation leads to:

$$D_t = \frac{K_1.L_0}{K_2 - K_1}.(e^{-K_1.t} - e^{-K_2.t}) + D_0.e^{-K_2.t} \qquad (3.16)$$

This is the general equation that expresses the variation of the oxygen deficit as a function of time. The DO concentration curve ($DO_t$ or $C_t$) can be obtained directly from this equation, knowing this:

$$DO_t = C_s - D_t \qquad (3.17)$$

Thus:

$$\boxed{C_t = C_s - \left\{ \frac{K_1.L_0}{K_2 - K_1}.(e^{-K_1.t} - e^{-K_2.t}) + (C_s - C_0).e^{-K_2.t} \right\}} \qquad (3.18)$$

In the DO sag curve, one point is of fundamental importance: the point in which the DO concentration reaches its lowest value. This is called *critical time*, and the DO concentration, the *critical concentration*. The knowledge of the critical concentration is very important, because it is based on it that the need and efficiency of the wastewater treatment will be established. The treatment must be implemented with a BOD removal efficiency which is sufficient to guarantee that the critical DO concentration is higher than the minimum value required by legislation (standard for the water body).

The DO sag curve as a function of time (or of the distance) is S-shaped, as shown in Figure 3.13. In the curve, the main points are identified: the DO concentration in the river and the critical DO concentration.

### 3.2.6.2 Model equations

***a)  DO concentration and deficit in the river immediately after mixing with the wastewater***

$$\boxed{C_0 = \frac{Q_r.DO_r + Q_w.DO_w}{Q_r + Q_w}} \qquad (3.19)$$

$$\boxed{D_0 = C_s - C_0} \qquad (3.20)$$

where:

$C_0$ = initial oxygen concentration, immediately after mixing (mg/L)

$D_0$ = initial oxygen deficit, immediately after mixing (mg/L)

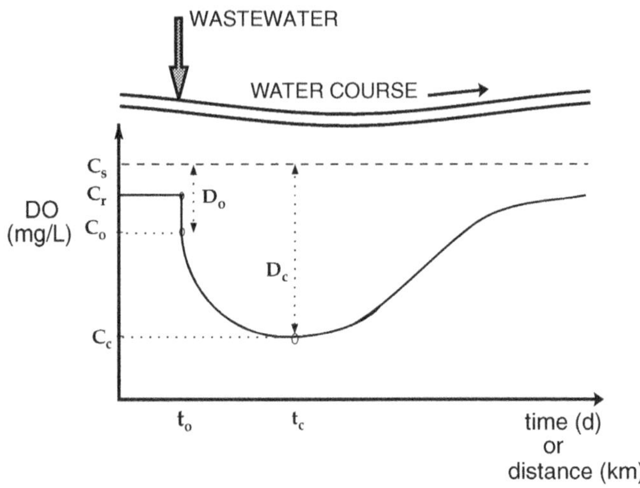


Figure 3.13.   Characteristic points in the DO sag curve

$C_s$ = oxygen saturation concentration (mg/L)
$Q_r$ = river flow upstream of the wastewater discharge (m³/s)
$Q_w$ = wastewater flow (m³/s)
$DO_r$ = dissolved oxygen concentration in the river, upstream of discharge (mg/L)
$DO_w$ = dissolved oxygen concentration in the wastewater (mg/L)

It can be observed that the value of $C_0$ is obtained through the weighted average between the flows and the DO levels in the river and the wastewater.

***b)   BOD₅ and ultimate BOD concentrations in the river immediately after mixing with the wastewater***

$$BOD_{5_0} = \frac{(Q_r.BOD_r + Q_w.BOD_w)}{Q_r + Q_w} \qquad (3.21)$$

$$L_0 = BOD_{5_0}.K_T = \frac{(Q_r.BOD_r + Q_w.BOD_w)}{Q_r + Q_w}.K_T \qquad (3.22)$$

where:
$BOD_{5_0}$ = BOD₅ concentration, immediately after mixing (mg/L)
$L_0$ = ultimate oxygen demand ($BOD_u$), immediately after mixing (mg/L)
$BOD_r$ = BOD₅ concentration in the river (mg/L)
$BOD_w$ = BOD₅ concentration in the wastewater (mg/L)
$K_T$ = coefficient for transforming BOD₅ to the ultimate $BOD_u$ (−)

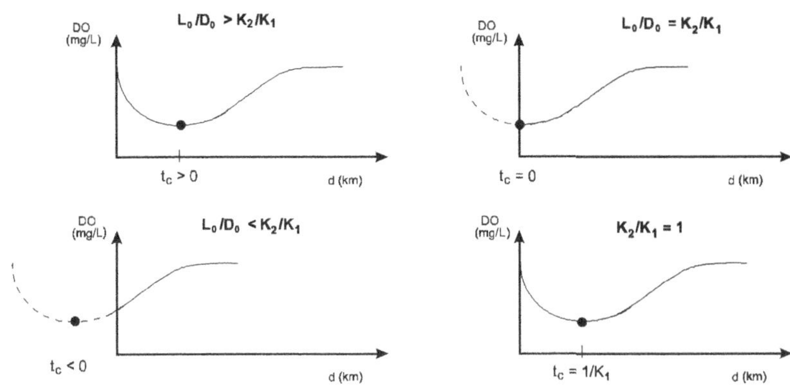



Figure 3.14.   Relation between the critical time and the terms $(L_0/D_0)$ and $(K_2/K_1)$

$$K_T = \frac{BOD_u}{BOD_5} = \frac{1}{1 - e^{-5.K_1}} \tag{3.23}$$

The value of $L_0$ is also obtained through the weighted average between the flows and the biochemical oxygen demands of the river and of the wastewater.

### c)   DO profile as a function of time

$$C_t = C_s - \left\{ \frac{K_1.L_0}{K_2 - K_1}.(e^{-K_1.t} - e^{-K_2.t}) + (C_0 - C_0).e^{-K_2.t} \right\} \tag{3.24}$$

In the event that a negative DO concentration ($C_t < 0$) is calculated, even though mathematically possible, there is no physical meaning. In this case, anaerobic conditions (DO $= 0$ mg/L) have been reached and the Streeter−Phelps model is no longer valid.

### d)   Critical time (time when the minimum DO concentration occurs)

$$t_c = \frac{1}{K_2 - K_1}.\ln\left\{ \frac{K_2}{K_1}.\left[ 1 - \frac{D_0.(K_2 - K_1)}{L_0.K_1} \right] \right\} \tag{3.25}$$

The following situations can occur when using the critical time formula, depending on the relation between $(L_0/D_0)$ and $(K_2/K_1)$ (see Figure 3.14):

- **$L_0/D_0 > K_2/K_1$**
  *Critical time is positive.* From the mixing point there will be a decrease in the DO concentration, leading to a critical deficit that is higher than the initial deficit.

- **$L_0/D_0 = K_2/K_1$**
  *Critical time is equal to zero*, that is, it occurs exactly in the mixing point. The initial deficit is equal to the critical deficit. The water course has a good regenerating capacity for the discharge received, and will not suffer a drop in DO level.
- **$L_0/D_0 < K_2/K_1$**
  *Critical time is negative*. This indicates that, from the mixing point, the dissolved oxygen concentration tends to increase. The initial deficit is the largest observed deficit. In terms of DO, the water course presents a self-purification capacity that is higher than the degeneration capacity of the wastewater. In practical terms, the critical time can be considered equal to zero, with the occurrence of the lowest DO values at the mixing point.
- **$K_2/K_1 = 1$**
  The application of the critical time formula leads to a *mathematical indetermination*. The limit when $K_2/K_1$ tends to 1 leads to a critical time equal to $1/K_1$.

*e)   Critical deficit and concentration of dissolved oxygen*

$$D_c = \frac{K_1}{K_2} . L_0 . e^{K_1 . t_c} \tag{3.26}$$

$$C_c = C_s - D_c \tag{3.27}$$

*f)   BOD removal efficiency required in the wastewater treatment*

The Streeter–Phelps model still permits the calculation of the maximum allowable BOD load of the sewage, which will lead to the critical DO concentration being equal to the minimum permitted by the legislation. Such procedure involves some iterations because, for each alteration of the maximum permissible load, there is a modification of the critical time. However, in a real situation, with more than one discharge point, this approach becomes not very practical. What is usually done its to consider BOD removal efficiencies which are compatible with the existing or available wastewater treatment processes, and to recalculate the DO profile for each new condition. *The most economic situation is usually that in which the minimum DO concentration is only marginally higher than the minimum permitted by legislation.*

### 3.2.7 Input data for the DO model

The following input data are necessary for the utilisation of the Streeter–Phelps model (see Figure 3.15):

- river flow, upstream of the discharge ($Q_r$)
- wastewater flow ($Q_w$)

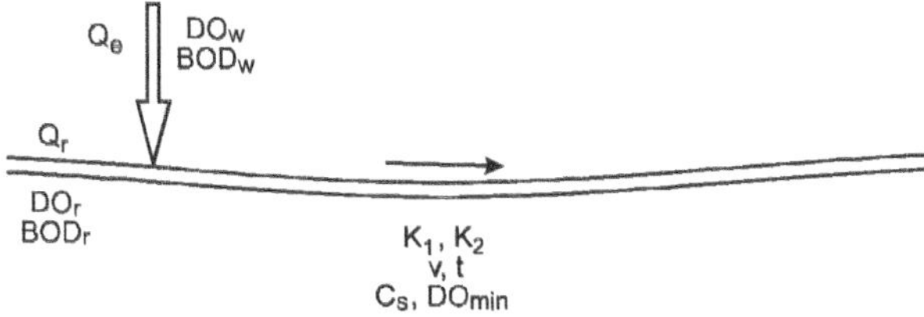


Figure 3.15.  Input data required for the Streeter–Phelps model

- dissolved oxygen in the river, upstream of the discharge ($DO_r$)
- dissolved oxygen in the wastewater ($DO_w$)
- $BOD_5$ in the river, upstream of the discharge ($BOD_r$)
- $BOD_5$ of the wastewater ($BOD_w$)
- deoxygenation coefficient ($K_1$)
- reaeration coefficient ($K_2$)
- velocity of the river (v)
- travelling time (t)
- saturation concentration of DO ($C_s$)
- minimum dissolved oxygen permitted by legislation ($DO_{min}$)

## a) *River flow ($Q_r$)*

The flow of the receiving body is a variable of extreme importance in the model, having a large influence on the simulation results. Therefore it is necessary to obtain the most precise flow value, whenever possible.

The use of the DO model can be with any of the following flows, depending on the objectives:

- flow observed in a certain period
- mean flow (annual average, average in the rainy season, average in the dry season)
- minimum flow

The *observed flow* in a certain period is used for model calibration (adjusting the model coefficients), so that the simulated data are as close as possible to the observed (measured) data in the water body during the period under analysis.

The *mean flow* is adopted when the simulation of the average prevailing conditions is desired, such as during the year, rainy months or dry months.

The *minimum flow* is utilised for the planning of catchment areas, the evaluation of the compliance with environmental standards of the water body and for the allocation of pollutant loads. Therefore, the determination of the required efficiencies for the treatment of various discharges must be determined in the **critical conditions**. These critical conditions in the receiving body occur exactly in the minimum flow period, when the dilution capacity is lower.

The *critical flow* must be calculated from the historical flow measurement data from the water course. The analysis of methods to estimate minimum flows is outside the scope of the present text, being well covered in hydrology books. Usually a minimum flow with a return period of 10 years and a duration of the minimum of 7 days ($Q_{7,10}$), is adopted. This can be understood as a value that may repeat itself in a probability of every 10 years, consisting of the lowest average obtained in 7 consecutive days. Therefore, in each year of the historical data series the 365 average daily flows are analysed. In each year a period of 7 days is selected, which resulted in the lowest average flow (average of 7 values). With the values of the lowest 7-day average for every year, an statistical analysis is undertaken, allowing interpolation or extrapolation of the value for a return period of 10 years.

Adoption of the 10-year return period in the $Q_{7,10}$ concept leads to small flows and frequently to the requirement of high BOD removal efficiencies, the cost of which should always be borne in mind, especially in developing countries. For these countries, probably a shorter return period would be more realistic, especially considering that the current condition is probably already of a polluted river.

Another approach is the utilisation of percentiles, such as a 90%ile value ($\mathbf{Q_{90}}$). In this concept, 90% of the flow values are greater than the critical flow, and only 10% are lower than it. This approach usually leads to critical flows that are greater than those based on $Q_{7,10}$.

Under any flow conditions, the utilisation of the concept of the *specific discharge* ($L/s.km^2$) is helpful. Knowing the drainage area at the discharge point and adopting a value for the specific discharge, the product of both leads to the flow of the water course. The specific discharge values can vary greatly from region to region, as a function of climate, topography, soil, etc., For $Q_{7,10}$ conditions, the following ranges are typical: (a) arid regions: probably less than $1,0$ $L/s.km^2$; (b) regions with good water resources availability: maybe higher than $3,0$ $L/s.km^2$; and (c) in intermediate regions: values close to $2.0$ $L/s.km^2$.

### b)    *Wastewater flow ($Q_w$)*

Wastewater flows considered in DO modelling are usually *average flows*, without coefficients for the hour and day of highest consumption. The sewage flow is obtained through conventional procedures, using data from population, per capita water consumption, infiltration, specific contribution (in the case of industrial wastes), etc. The calculation is detailed in Chapter 2.

### c)    *Dissolved oxygen in the river, upstream of the discharge point ($DO_r$)*

The dissolved oxygen level in a water body, upstream of a waste discharge, is a result of the upstream activities in the catchment area.

Ideally, historical data should be used in this analysis. When doing so, coherence is required: if the simulation is for a dry period, only samples pertaining to the dry period should be considered.

In case that it is not possible to collect water samples at this point, the DO concentration can be estimated as a function of the approximate pollution level of the water body. If there are *few indications of pollution*, a $DO_r$ value of 80% to 90% of the oxygen saturation value (see item k below) can be adopted.

In the event that the water body is already well polluted upstream of the discharge, a sampling campaign is justified, or even an upstream extension of the boundaries of the studies should be considered, in order to include the main polluting points. In such a situation, the value of $DO_r$ will be well below the saturation level.

### d) *Dissolved oxygen in the wastewater ($DO_w$)*

In sewage, the dissolved oxygen levels are normally nihil or close to zero. This is due to the large quantity of organic matter present, implying a high consumption of oxygen by the microorganisms. Therefore the DO of **raw sewage** is usually adopted as **zero** in the calculations.

In case that the wastewater is **treated**, the following considerations could be made:

- *Primary treatment.* Primary effluents can be assumed as having DO equal to zero.
- *Anaerobic treatment.* Anaerobic effluents also have DO equal to zero.
- *Activated sludge* and *biofilm reactors.* Effluents from these systems undergo a certain aeration at the effluent weir on the secondary sedimentation tanks, enabling DO to increase to 2 mg/L or slightly more. If the discharge outfall is long, this oxygen could be consumed as a result of the remaining BOD from the treatment.
- *Facultative or maturation ponds.* Effluents from facultative or maturation ponds can have during day time DO levels close to saturation, or even higher, due to the production of pure oxygen by the algae. At night the DO levels decrease. For the purpose of the calculations, $DO_w$ values around 4 to 6 mg/L can be adopted.
- *Effluents subjected to final reaeration.* Treatment plant effluents may be subject to a final reaeration stage, in order to increase the level of dissolved oxygen. A simple system is composed by cascade aeration, made up of a sequence of steps in which there is a free fall of the liquid. DO values may raise a few milligrams per litre, depending on the number and height of the steps. Sufficient head must be available for the free falls. Gravity aeration should not be used directly for anaerobic effluents, due to the release of $H_2S$ in the gas-transfer operation. Section 11.10 presents the methodology for calculating the increase in DO.

Table 3.6. BOD$_5$ as a function of the water body characteristics

| River condition | BOD$_5$ of the river (mg/L) |
|---|---|
| Very clean | 1 |
| Clean | 2 |
| Reasonably clean | 3 |
| Doubtful | 5 |
| Bad | >10 |

*Source:* Klein (1962)

### e)   BOD$_5$ in the river, upstream of discharge (BOD$_r$)

BOD$_5$ in the river, upstream of the discharge, is a function of the wastewater discharges (point or diffuse sources) along the river down to the mixing point. The same considerations made for DO$_r$ about sampling campaigns and the inclusion of upstream polluting points are also valid here.

Klein (1962) proposed the classification presented in Table 3.6, in the absence of specific data.

### f)   BOD$_5$ in the wastewater (BOD$_w$)

The BOD$_5$ concentration in **raw** domestic sewage has an average value in the order of 300 mg/L. The BOD can also be estimated through the quotient between the BOD load (calculated from the population and the per capita BOD contribution) and the wastewater flow (domestic sewage + infiltration). For more details, see Section 2.2.5.

In case there are industrial discharges of importance, particularly from agro-industries and others with high content of organic matter in the effluent, they must be included in the calculation. These values can be obtained by sampling or through literature data. See Section 2.2.6.

For a treated wastewater, of course the BOD removal efficiency must be taken into account, since treatment is the main environmental control measure to be adopted. In this case, the BOD$_5$ in the wastewater is:

$$\boxed{BOD_{tw} = \left(1 - \frac{E}{100}\right) \cdot BOD_{rw}}$$
(3.28)

where:

$BOD_{tw}$ = BOD$_5$ of the treated wastewater (mg/L)
$BOD_{rw}$ = BOD$_5$ of the raw wastewater (mg/L)
$E$ = BOD$_5$ removal efficiency of the treatment (%)

Table 4.9 presents typical ranges of BOD removal efficiency from various wastewater treatment systems. An overview of these systems can be found in

Chapter 4. Various other chapters of this book are dedicated to the detailed description of these systems.

### g)  *Deoxygenation coefficient ($K_1$)*

The deoxygenation coefficient can be obtained following the criteria presented in Section 3.2.4.2. It must be noted that water bodies that receive biologically treated wastewater have a lower value of $K_1$ (see Table 3.3). For liquid temperatures different from $20\,^{\circ}C$, the value of $K_1$ must be corrected (seer Section 3.2.4.3).

### h)  *Reaeration coefficient ($K_2$)*

The reaeration coefficient can be obtained following the procedures outlined in Section 3.2.5.2. For liquid temperatures different from $20\,^{\circ}C$, the value of $K_2$ must be corrected (see Section 3.2.5.3).

### i)  *Velocity of the water body (v)*

The velocity of the liquid mass in the water course may be estimated using the following methods:

- direct measurement in the water course
- data obtained from flow-measuring points
- use of hydraulic formulas for open channels
- correlation with flow

In DO simulations that can be done under any flow conditions, obtaining the velocity through the last two methods is more convenient. In other words, it is important that the velocity is coherent with the flow under consideration, since in dry periods the velocities are usually lower, with the opposite occurring in wet periods.

The hydraulic formulas are presented in pertinent literature. The most adequate friction factor should be chosen as a function of the river bed characteristics (see Chow, 1959).

The flow-correlation method should follow a methodology similar to the one described in Item 3.2.5.2c for the reaeration coefficient. The model to be obtained can have the form $v = cQ^d$, where c and d are coefficients obtained from regression analysis.

### j)  *Travel time (t)*

In the Streeter–Phelps model, the theoretical travel time that a particle takes to complete a certain river reach is a function of the velocity and the distance. This is because the model assumes a plug-flow regime and does not consider the effects of dispersion. Therefore, knowing the distances and having determined the velocities in each reach, the residence time is obtained directly from the relation:

$$\boxed{t = \frac{d}{v \cdot 86{,}400}} \qquad (3.29)$$

where:

$t$ = travel (residence) time (d)
$d$ = distance (m)
$v$ = velocity in the water body (m/s)
86,400 = number of seconds per day (s/d)

### k)   *DO saturation concentration ($C_s$)*

The saturation concentration of the oxygen can be calculated based on theoretical considerations, or through the use of empirical formulas. The value of $C_s$ is a function of water temperature and altitude:

*   *The increase in temperature reduces the saturation concentration* (the greater agitation of molecules in the water tends to make the dissolved gases pass to the gas phase)
*   *The increase in altitude reduces the saturation concentration* (the atmospheric pressure is lower, thus exerting a lower pressure for the gas to be dissolved in the water).

There are some empirical formulas in the literature (the majority based on regression analysis) that directly supply the value of $C_s$ (mg/L) as a function of, for example, the temperature $T$ (°C). A formula frequently employed is (Pöpel, 1979):

$$C_s = 14.652 - 4.1022 \times 10^{-1}.T + 7.9910 \times 10^{-3}.T^2 - 7.7774 \times 10^{-5}.T^3$$

(3.30)

The influence of the altitude can be computed by the following relation (Qasim, 1985):

$$f_H = \frac{C_s{}'}{C_s} = \left(1 - \frac{\text{Altitude}}{9450}\right)$$

(3.31)

where:

$f_H$ = correction factor for altitude, for the DO saturation concentration (−)
$C_s'$ = DO saturation concentration at the altitude H (mg/L)
Altitude = altitude (m above sea level)

Salinity also affects the solubility of oxygen. The influence of dissolved salts can be computed by the following empirical formula (Pöpel, 1979):

$$\boxed{\gamma = 1 - 9 \times 10^{-6} \cdot C_{sal}}$$

(3.32)

where:

$\gamma$ = solubility reduction factor ($\gamma = 1$ for pure water)
$C_{sal}$ = dissolved salts concentration (mg Cl⁻/L)

Table 3.7. Saturation concentration for oxygen in clean water (mg/L)

| Temperature (°C) | Altitude (m) | | | |
|---|---|---|---|---|
| | 0 | 500 | 1000 | 1500 |
| 10 | 11.3 | 10.7 | 10.1 | 9.5 |
| 11 | 11.1 | 10.5 | 9.9 | 9.3 |
| 12 | 10.8 | 10.2 | 9.7 | 9.1 |
| 13 | 10.6 | 10.0 | 9.5 | 8.9 |
| 14 | 10.4 | 9.8 | 9.3 | 8.7 |
| 15 | 10.2 | 9.7 | 9.1 | 8.6 |
| 16 | 10.0 | 9.5 | 8.9 | 8.4 |
| 17 | 9.7 | 9.2 | 8.7 | 8.2 |
| 18 | 9.5 | 9.0 | 8.5 | 8.0 |
| 19 | 9.4 | 8.9 | 8.4 | 7.9 |
| 20 | 9.2 | 8.7 | 8.2 | 7.7 |
| 21 | 9.0 | 8.5 | 8.0 | 7.6 |
| 22 | 8.8 | 8.3 | 7.9 | 7.4 |
| 23 | 8.7 | 8.2 | 7.8 | 7.3 |
| 24 | 8.5 | 8.1 | 7.6 | 7.2 |
| 25 | 8.4 | 8.0 | 7.5 | 7.1 |
| 26 | 8.2 | 7.8 | 7.3 | 6.9 |
| 27 | 8.1 | 7.7 | 7.2 | 6.8 |
| 28 | 7.9 | 7.5 | 7.1 | 6.6 |
| 29 | 7.8 | 7.4 | 7.0 | 6.6 |
| 30 | 7.6 | 7.2 | 6.8 | 6.4 |

Table 3.7 presents the saturation concentrations for oxygen in clean water at different temperatures and heights.

***l)   Minimum allowable dissolved oxygen concentration in the water body (DO$_{min}$)***

The minimum levels of dissolved oxygen that need to be maintained in the water body are stipulated by the legislation applicable in the country or region. In the absence of specific legislation, it is usual to try to maintain DO concentrations in the water body equal to or above 5.0 mg/L.

## 3.2.8 Measures to control water pollution by organic matter

When analysing the possible pollution control strategies for a water body, it is fundamental to have a regionalised view of the catchment area as a whole, aiming at reaching the desired water quality, instead of treating the problems as isolated points. When a regional focus is employed, a great variety of alternative strategies becomes available, normally leading to lower costs and greater safety. An adequate organisational and institutional structure is essential.

Among the main control measures, there are:

- *wastewater treatment*

- *flow regularisation in the water body*
- *aeration of the water body*
- *aeration of the treated wastewater*
- *allocation of other uses for the water body*

### a)   Wastewater treatment

Individual or collective sewage treatment before discharge is usually the main and often the only control strategy. However, its possible combination with some of the other presented strategies should be analysed, aiming at obtaining a technically favourable solution at the lowest cost. Wastewater treatment is the main alternative analysed in the present book.

### b)   Flow regularisation of the water body

This alternative generally consists of building an upstream dam, in order to augment the low flow under critical conditions. The most attractive option is to include multiple uses for the dam, such as irrigation, hydroelectric power generation, recreation, water supply and others.

Another positive aspect is that the effluent from the dams can contain higher levels of dissolved oxygen because of the aeration at the effluent weir.

It must be remembered that the implementation of dams is a delicate topic from an environmental point of view. Also, if the upstream catchment area is not properly protected, the reservoir can become a point of localised pollution and risks of eutrophication.

### c)   Aeration of the water body

Another possibility is to provide aeration in the water body at a point downstream of the discharge, maintaining the DO concentrations above the minimum allowable.

An advantage of this alternative resides in the fact that the assimilative capacity of the water course can be totally used in periods of high flow and the aeration can be limited to dry periods only. This is a form of collective treatment and involves the distribution of the costs between the various beneficiaries.

The following aeration methods can be employed:

- diffused-air aeration
- surface (mechanical) aeration
- aeration at weirs
- turbine aeration
- injection by pressure

Besides this, natural waterfalls can contribute significantly to the DO increase (see Section 3.2.5.2, equations 3.11 and 3.12).

### d)   Aeration of the treated wastewater

At the effluent weir of the WWTP, after satisfaction of the oxygen demand, the effluent can suffer a simple aeration, usually by means of weirs. These devices can increase the DO concentration in the order of some milligrams per litre (1 to 3 mg/L) contributing to the fact that, already at the mixing point, a slightly higher DO concentration is achieved. In anaerobic effluents, however, aeration must be avoided because it causes the release of hydrogen sulphide, which causes problems of bad odours.

### e)   *Allocation of other uses for the water body*

In case it is not possible (mainly for economic reasons) to control the polluting discharges in order to preserve the water quality as a function of the intended uses of the water body, these uses could be re-evaluated in the river or in selected reaches.

Therefore, it could be necessary to attribute less noble uses for a certain reach of the river, due to the unfeasibility of implementing the control measure at the desired level. The allocation of uses for the water body should be carried out in such a way as to optimise regional water resources, aiming at their various uses (Arceivala, 1981).

---

**Example 3.2**

The city and the industry from the general example in Chapter 2 (Section 2.2.7) discharge together their effluents into a water course. The catchment area upstream does not present any other significant discharges. Downstream of the discharge point the stream travels a distance of 50 km until it reaches the main river. In this downstream reach there are no other significant discharges.

Main data:

- Wastewater characteristics (values obtained from the mentioned example, for year 20 of operation):
    - Average wastewater flow: 0.114 m³/s
    - BOD concentration: 341 mg/L
- Catchment area characteristics:
    - Drainage area upstream of the discharge point: 355 km²
    - Specific discharge of the water body (minimum flow per unit area of the basin) 2.0 L/s.km²
- Water body characteristics:
    - Altitude: 1,000 m
    - Water temperature: 25 °C
    - Average depth: 1.0 m
    - Average velocity: 0.35 m/s

Assume all other necessary data.

Required items:

- Calculate and plot the DO sag curve until the stream joins the main river
- Present wastewater treatment alternatives for the pollution control of the water body
- Calculate and plot the DO sag curves for the alternatives analysed

---

---

### *Example 3.2 (Continued)*

**Solution:**

**Determination of the input data (raw wastewater)**

a) River flow ($Q_r$)

Minimum specific discharge: $Q_{r_{spec}} = 2.0$ L/s.km$^2$
Drainage basin area: $A = 355$ km$^2$

$$Q_r = Q_{r_{spec}} \cdot A = 2.0\,\text{L/s.km}^2 \times 355\,\text{km}^2 = 710\,\text{L/s} = 0.710\,\text{m}^3/\text{s}$$

b) Wastewater flow ($Q_w$)

$$Q_w = 0.114\,\text{m}^3/\text{s (stated in the problem)}$$

c) Dissolved oxygen in the river ($DO_r$)

Considering that the water body does not present significant discharges, adopt the DO concentration upstream of the discharge as 90% of the saturation value. Saturation concentration: $C_s = 7.5$ mg/L ($25\,°C$, 1,000 m of altitude) (see item j below)

$$DO_r = 0.9 \times C_s = 0.9 \times 7.5\,\text{mg/L} = 6.8\,\text{mg/L}$$

d) Dissolved oxygen in the sewage ($DO_w$)

$$DO_w = 0.0\,\text{mg/L (adopted)}$$

e) Biochemical oxygen demand in the river ($BOD_r$)

From Table 3.6, for a clean river:

$$BOD_r = 2.0\,\text{mg/L}$$

f) Biochemical oxygen demand of the wastewater ($BOD_w$)

$$BOD_w = 341\,\text{mg/L (stated in the problem)}$$

g) Deoxygenation coefficient ($K_1$)

As laboratory tests were not possible, $K_1$ is adopted as an average value from the literature (raw sewage – see Table 3.3):

$$K_1 = 0.38\,\text{d}^{-1}(20\,°C,\text{ base e})$$

Correction of $K_1$ for a temperature of $25\,°C$ (Equation 3.8):

$$K_{1_T} = K_{1_{20C}} \cdot \theta^{(T-20)} = 0.38 \times 1.047^{(25-20)} = 0.48\text{d}^{-1}$$

---

### *Example 3.2 (Continued)*

h) Reaeration coefficient ($K_2$)

Depth of the water body: H $=$ 1.0 m
Velocity of the water body: v $=$ 0.35 m/s

Formula to be used, according with the applicability range (see Table 3.5 and Figure 3.11): O'Connor and Dobbins formula:

$$K_2 = 3.73 \cdot \frac{v^{0.5}}{H^{1.5}} = 3.73 \cdot \frac{(0.35 \text{ m/s})^{0.5}}{(1.0 \text{ m})^{1.5}} = 2.21 \text{d}^{-1} (20\,^\circ\text{C, base e})$$

Correction for the temperature of 25 °C (Equation 3.13):

$$K_{2_T} = K_{2_{20C}} \theta^{(T-20)} = 2.21 \times 1.024^{(25-20)} = 2.49 \text{ d}^{-1}$$

i) Travel time

Velocity of the water body: v $=$ 0.35 m/s
Travel distance: d $=$ 50,000 m

The travel time to arrive at the confluence with the main river is (Equation 3.29):

$$t = \frac{d}{v.86,\,400} = \frac{50,\,000 \text{ m}}{0.35 \text{ m/s} . 86,400 \text{ s/d}} = 1.65 \text{ d}$$

j) Saturation concentration of dissolved oxygen ($C_s$)

Water temperature: T $=$ 25 °C
Altitude: 1,000 m

From Table 3.6:

$$C_s = 7.5 \text{ mg/L}$$

l) Minimum allowable dissolved oxygen ($DO_{min}$)

$$DO_{min} = 5.0 \text{ mg/L (adopted)}$$

### Example 3.2 (Continued)

*Summary:*

**INPUT DATA**

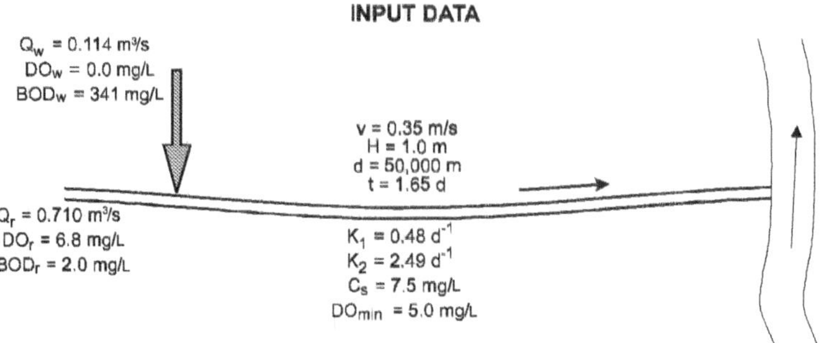


Fig Ex 3.2d Input data for the example. Raw wastewater.

**Calculation of the output data – raw wastewater**

a) Oxygen concentration at the mixing point ($C_0$)

From Equation 3.19:

$$C_0 = \frac{Q_r.DO_r + Q_w.DO_w}{Q_r + Q_w} = \frac{0.710 \times 6.8 + 0.114 \times 0.0}{0.710 + 0.144} = 5.9 \text{ mg/L}$$

The dissolved oxygen deficit is (see Equation 3.20):

$$D_0 = C_s - C_0 = 7.5 - 5.9 = 1.6 \text{ mg/L}$$

b) Ultimate BOD concentration at the mixing point ($L_0$)

The transformation factor $BOD_5$ to BOD ultimate is given by Equation 3.23:

$$K_T = \frac{BOD_u}{BOD_5} = \frac{1}{1 - \exp(-5.K_1)} = \frac{1}{1 - \exp(-5 \times 0.48)} = 1.10$$

The $BOD_5$ at the mixing point is obtained from Equation 3.21:

$$BOD5_0 = \frac{(Q_r.BOD_r + Q_w.BOD_w)}{Q_r + Q_w}$$

$$= \frac{(0.710 \times 2.0 + 0.114 \times 341)}{0.710 + 0.114} = 49 \text{ mg/L}$$

The ultimate BOD at the mixing point is obtained from Equation 3.22:

$$L_0 = BOD5_0.K_T = 49 \times 1.10 = 54 \text{ mg/L}$$

***Example 3.2 (Continued)***

c) Critical time ($t_c$)

From Equation 3.25:

$$t_c = \frac{1}{K_2 - K_1}.\ln\left\{\frac{K_2}{K_1}.\left[1 - \frac{D_0.(K_2 - K_1)}{L_0.K_1}\right]\right\}$$

$$= \frac{1}{2.49 - 0.48}.\ln\left\{\frac{2.49}{0.48}.\left[1 - \frac{1.6.(2.49 - 0.48)}{54 \times 0.48}\right]\right\} = 0.75d$$

The critical distance is obtained from the critical time and the velocity:

$$d_c = t.v.86{,}400 = 0.75 \times 0.35 \times 86{,}400 = 22{,}680 \text{ m} = 22.7 \text{ km}$$

d) Critical concentration of the dissolved oxygen ($DO_c$)

The critical deficit is given by Equation 3.26:

$$D_c = \frac{K_1}{K_2}.L_0.e^{-K_1.t_c} = \frac{0.48}{2.49}.54.e^{-0.48 \times 0.75} = 7.2 mg/L$$

The critical concentration is given by Equation 3.27:

$$DO_c = C_s - D_c = 7.5 - 7.2 = 0.3 \text{ mg/L}$$

Environmental control measures need to be adopted, since there are DO concentrations lower than the minimum allowable ($DO_{min} = 5.0$ mg/L).

In case a negative value of DO concentration had been calculated, one should always keep in mind that negative concentrations have no physical meaning. The Streeter–Phelps model would be no longer valid under these conditions (from the moment when DO = 0 mg/L), and the calculation and the graph must be interrupted at this point. However, even in this case the model played an important role, since the requirement for control measures was identified.

e) DO sag curve

Along the water course, in the absence of specific data, it is assumed that the dilution by natural contributions (direct drainage) is counterbalanced by the BOD load occasionally distributed along the reach (diffuse pollution).

In case there were significant tributaries or sewage discharges downstream, the water body should be subdivided into new reaches. It is an essential condition of the Streeter–Phelps model that each reach is homogeneous.

From Equation 3.24:

$$C_t = C_s - \left[\frac{K_1.L_0}{K_2 - K_1}.(e^{-K_1.t} - e^{-K_2.t}) + (C_s - C_0).e^{-K_2.t}\right]$$

$$= 7.5 - \left[\frac{0.48 \times 54}{2.49 - 0.48}.(e^{-0.48 \times t} - e^{-2.49 \times t}) + (7.5 - 5.9).e^{-2.49 \times t}\right]$$

---

### *Example 3.2 (Continued)*

For various values of t:

| d (km) | t (d) | $C_t$ (mg/L) |
|--------|-------|--------------|
| 0.0    | 0.00  | 5.9          |
| 5.0    | 0.17  | 3.1          |
| 10.0   | 0.33  | 1.5          |
| 15.0   | 0.50  | 0.6          |
| 20.0   | 0.66  | 0.3          |
| 25.0   | 0.83  | 0.3          |
| 30.0   | 0.99  | 0.5          |
| 35.0   | 1.16  | 0.8          |
| 40.0   | 1.32  | 1.1          |
| 45.0   | 1.49  | 1.5          |
| 50.0   | 1.65  | 1.9          |

It can be observed that in practically all the distance, the DO is below the minimum allowable level of 5.0 mg/L. The DO profile can be visualised in Figure 3.16.

If a negative DO concentration had occurred, the model would stop being used at the point when DO became less than zero, and the negative values should not be reported or plotted.

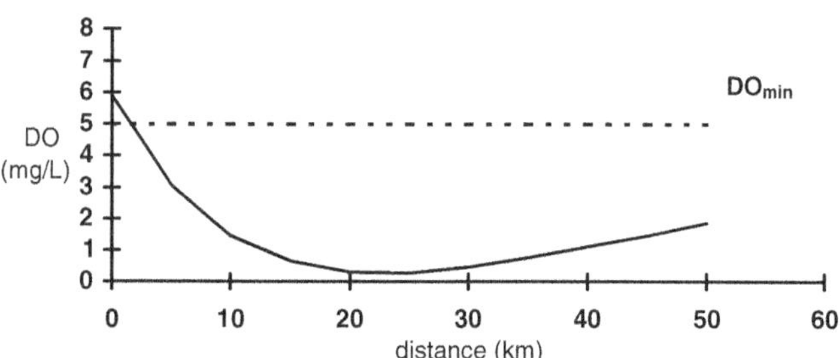


DO profile in the river. Raw sewage.

## Calculation of the output data – treated wastewater

After the confirmation of the need for wastewater treatment, the different alternatives of BOD removal efficiencies must be investigated. The concept of treatment level (primary, secondary) used in this example is covered in Chapter 4.

For the sake of simplicity, in this example it is assumed that the domestic and industrial wastewaters are mixed and treated together, at the same plant and, therefore, with the same removal efficiency. Other approaches are possible, involving different plants and treatment efficiencies if the domestic and industrial effluents are separated.

### *Example 3.2 (Continued)*

a) *Alternative 1*: Primary treatment – Efficiency 35%

From Equation 3.28, the BOD of the treated wastewater is:

$$BOD_{tw} = BOD_{rw} \cdot \left(1 - \frac{E}{100}\right) = 341 \cdot \left(1 - \frac{35}{100}\right) = 222 \text{ mg/L}$$

The new coefficient $K_1$ (treated wastewater at primary level) can be obtained from Table 3.3, and is adopted in this example as:

$$K_1 = 0.35 \text{ d}^{-1}(T = 20\,^\circ C)$$

$$K_1 = 0.44 \text{ d}^{-1}(\text{after correction for } T = 25\,^\circ C \text{ using Equation 3.8})$$

The remaining input data are the same. The calculation sequence is also the same.

The calculated and plotted DO values are shown in item d.

The critical DO concentration (2.8 mg/L) occurs at a distance of 22.1 km. The minimum allowable value (5.0 mg/L) continues not to be complied with in a large part of the river reach. The efficiency of the proposed treatment is insufficient. Therefore a higher efficiency must be adopted, associated with secondary treatment level.

b) *Alternative 2*: Secondary treatment – Efficiency 65%

All secondary-level sewage treatment processes reach a BOD removal efficiency of at least 65%, even the simplest ones. In this item, sewage treatment by UASB (Upflow Anaerobic Sludge Blanket) reactors is analysed.

The effluent BOD from the treatment plant is:

$$BOD_{tw} = 341 \cdot \left(1 - \frac{65}{100}\right) = 119 \text{ mg/L}$$

The new coefficient $K_1$ (treated wastewater at secondary level) can be obtained from Table 3.3, and is adopted in this example as:

$$K_1 = 0.18 \text{ d}^{-1}(T = 20\,^\circ C)$$

$$K_1 = 0.23 \text{ d}^{-1}(T = 25\,^\circ C)$$

It was assumed that the effluent DO from the treatment plant is zero (0.0 mg/L), since the effluent is anaerobic. If a different treatment process is adopted and the effluent has higher levels of DO in the effluent, this must be taken into consideration. Naturally, if only anaerobic reactors are adopted, aeration of the effluent must not be practised since hydrogen sulphide may be released into the atmosphere.

The calculated DO values and the graph of the DO profile are presented in item d.

---

### *Example 3.2 (Continued)*

Along the whole length of the water course the DO values are above the minimum allowable concentration (the critical DO is 5.4 mg/L, greater than the minimum allowable of 5.0 mg/L). In this way, from the viewpoint of the receiving body, this alternative is *satisfactory.* Existing BOD discharge standards are not analysed here. These standards vary from country to country or from region to region and they can be taken into consideration when applicable. In the present case, the BOD of the discharge is 119 mg/L. In the case that the environmental agency establishes discharge standards for BOD of, say, 25 mg/L, these standards will not be satisfied in this alternative. Under certain conditions, environmental agencies relax the discharge standard, provided that the standard for the receiving body standard is satisfied.

Assuming that the environmental agency has accepted this alternative of 65% BOD removal efficiency, which has been shown to be sufficient in terms of DO, there is no need to investigate other alternatives of greater removal efficiencies, which probably would have higher costs. The most economic situation is usually that in which the critical DO is only marginally greater than the minimum allowable DO. This aspect is of great importance for developing countries. Similarly, there is no need to analyse efficiencies lower than 65%, since this is already on the lower boundary of typical efficiencies for secondary treatment level.

*If the efficiency of 65% had been unsatisfactory, new efficiencies should be tried in a sequentially increasing way, until the receiving body standard is reached.*

c) Summary

The alternative to be adopted is alternative 2 – sewage treatment at a secondary level, with a BOD removal efficiency of 65%.

The DO concentrations in the water body for the three scenarios are presented below.

| | | DO concentration (mg/L) | | |
|---|---|---|---|---|
| d (km) | t (d) | E = 0% | E = 35% | E = 65% |
| 0.0 | 0.00 | 5.9 | 5.9 | 5.9 |
| 5.0 | 0.17 | 3.1 | 4.3 | 5.6 |
| 10.0 | 0.33 | 1.5 | 3.5 | 5.5 |
| 15.0 | 0.50 | 0.6 | 3.0 | 5.4 |
| 20.0 | 0.66 | 0.3 | 2.8 | 5.4 |
| 25.0 | 0.83 | 0.3 | 2.8 | 5.4 |
| 30.0 | 0.99 | 0.5 | 3.0 | 5.4 |
| 35.0 | 1.16 | 0.8 | 3.1 | 5.5 |
| 40.0 | 1.32 | 1.1 | 3.4 | 5.5 |
| 45.0 | 1.49 | 1.5 | 3.6 | 5.6 |
| 50.0 | 1.65 | 1.9 | 3.8 | 5.7 |

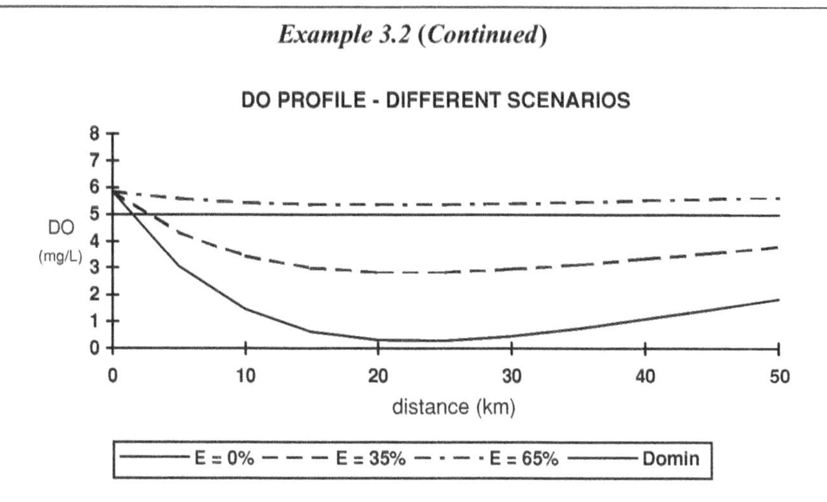


DO profiles for three different BOD removal efficiencies in the wastewater treatment.

The values above were obtained using a spreadsheet. Small differences in decimals may occur, depending on the criteria employed for rounding the values of the intermediate calculations, especially if they are performed using calculators.

## 3.3 CONTAMINATION BY PATHOGENIC MICROORGANISMS

### 3.3.1 Introduction

One of the most important aspects of water pollution is that related with public health, associated with water-borne diseases. This topic, including the main pathogens of interest and the concept of indicator organisms of faecal contamination, is discussed in Chapter 2.

A water body receiving the discharge of sewage may incorporate into itself a wide range of pathogenic organisms. This fact may not generate a direct impact on the aquatic organisms themselves, but may affect some of prevailing uses of the water, such as potable water supply, irrigation and bathing.

Therefore, it is very important to know the behaviour of the pathogenic organisms in the water body, starting from the discharge point until places where water is likely to be used. It is known that most of these agents have optimal conditions for their growth and reproduction in the human intestinal tract. Once submitted to the adverse conditions that prevail in the water body, they tend to decrease in number, characterising the so-called **decay**.

In Chapter 2 it was seen that the bacteria of the coliform group are used as **indicators of faecal contamination**; that is, they indicate if the water has been contaminated by faeces and, as a result, if it presents a potential for having pathogens

Table 3.8. Important factors that contribute to bacterial decay in water bodies

| Physical factors | Physical-chemical factors | Biological and biochemical factors |
|---|---|---|
| • solar light (ultraviolet radiation)<br>• temperature (values in water are usually lower than those in human bodies<br>• adsorption<br>• flocculation<br>• sedimentation | • osmotic effects (salinity)<br>• pH<br>• chemical toxicity<br>• redox potential | • lack of nutrients<br>• predation<br>• competition |

and therefore transmitting diseases. The present item covers the qualitative and quantitative relations associated with the decay of the coliforms in water bodies. It is assumed that this decay represents, with a certain safety, an indication of the behaviour of the pathogens (especially bacteria) discharged into the water body.

## 3.3.2 Bacterial decay kinetics

### 3.3.2.1 Intervening factors

Coliforms and other microorganisms of intestinal origin present a natural mortality when exposed to environmental conditions that are different from the previously preponderant conditions inside the human system, which were ideal for their development and reproduction. Table 3.8 lists important factors that contribute to the bacterial decay in water bodies (Arceivala, 1981; EPA, 1985; Thomann and Mueller, 1987). These factors may act simultaneously and with different degrees of importance.

### 3.3.2.2 Kinetics of bacterial decay

The bacterial mortality rate is generally estimated by Chick's law, according to which, the higher the concentration of bacteria, the higher is the decay rate (first-order reaction):

$$\frac{dN}{dt} = -K_b.N \tag{3.33}$$

where:

$N$ = number of coliforms (organisms /100 ml)
$K_b$ = coefficient of bacterial decay (d$^{-1}$)
$t$ = time (d)

Table 3.9. Formulas for the calculation of the coliform concentrations in water bodies

| Hydraulic regime | Scheme | Coliform concentration N (organisms/100ml) |
|---|---|---|
| Plug flow *(e.g.: rivers)* | | $N = N_0 . e^{-K_b . t}$ |
| Completely mixed *(e.g.: lakes)* | | $N = \dfrac{N_0}{1 + K_b . t}$ |

$N_0$ = number of coliforms in the influent (organisms/100 ml). In plug-flow reactors, coliforms at time $t = 0$
$N$ = number of coliforms after time t (organisms/100 ml)
$K_b$ = coefficient of bacterial decay (d$^{-1}$)
$t$ = time (d)

The formula to calculate the coliform concentration after a time t depends on the hydraulic regime of the water body. Rivers are usually represented as plug-flow reactors, while reservoirs are frequently represented as completely-mixed reactors. Depending on the characteristics of the water body, the formulas shown in Table 3 can be used.

For completely-mixed reactors, the time t corresponds to the detention time, given by: $t = V/Q$. The concentration of the coliforms at any point in the reactor is the same, coinciding with the effluent concentration.

### 3.3.2.3 Bacterial decay coefficient

Values of $K_b$ obtained in various studies in fresh water vary within a wide range. Typical values, however, are close to (Arceivala, 1981; EPA, 1985; Thomann and Mueller, 1987):

$$K_b = 0.5 \text{ to } 1.5 \, d^{-1} \text{ (base e, } 20\,°C) \quad \text{Typical value} \approx 1.0 \, d^{-1}$$

The effect of temperature on the decay coefficient can be formulated as:

$$K_{b_T} = K_{b_{20}} . \theta^{(T-20)} \tag{3.34}$$

where:
   $\theta$ = temperature coefficient (−)

A typical value for $\theta$ can be **1.07** (Castagnino, 1977; Thomann and Mueller, 1987), though there is a great variation in the data presented in the literature.

Table 3.10. Main processes for the removal of
pathogenic organisms in wastewater treatment

| Type | Process |
| --- | --- |
| Natural | Maturation ponds<br>Land infiltration |
| Artificial | Chlorination<br>Ozonisation<br>Ultraviolet radiation<br>Membranes |

*Note:* for a description of the process – see Chapter 4

### 3.3.3  Control of the contamination by pathogenic organisms

The best measure to control contamination of a water body by pathogenic organisms from sewage is through their removal at the wastewater treatment stage. However, this approach is not practised throughout the world. In various countries there is systematic disinfection of the sewage treatment effluent, while in others disinfection is only carried out in the potable water treatment. However, in any case, approaches that preserve the defined uses of the water body should be adopted.

The wastewater treatment processes usually applied are very efficient in the removal of suspended solids and organic matter, but generally insufficient for the removal of pathogenic microorganisms. In spite of the great importance of this item in developing countries, it has not yet received due consideration. Table 4.9 in Chapter 4 lists the coliform removal efficiencies obtained in the main wastewater treatment systems. It should be always remembered that the coliforms are not a direct indication of the presence of pathogens, and they may represent only those organisms that have similar decay (or removal) mechanisms and similar (or higher) mortality rates. Protozoan cysts and helminth eggs are removed by different mechanisms (e.g. sedimentation) and are not well represented by coliforms.

Even though removal efficiencies of 90% shown in Table 4.9 may seem high, it should be borne in mind that, when dealing with coliforms, much higher efficiencies are generally necessary in order to have low concentrations in the water body, as a result of the very high concentrations in the raw sewage. High coliform removal efficiencies can be obtained by the processes listed in Table 3.10, which are further detailed in Chapter 4.

The processes listed above are capable of reaching coliform removal efficiencies of 99.99% or more. Frequently the coliform removal efficiency is expressed in a logarithmic scale, according to:

| Removal efficiencies | |
| --- | --- |
| Log units | Percentage (%) |
| 1 | 90 |
| 2 | 99 |
| 3 | 99.9 |
| 4 | 99.99 |

For instance, a coliform concentration, which is reduced from $10^7$ organisms/100 ml to $10^4$ organisms/100 ml, is reduced in 3 orders of magnitude, or 99.9%. If the logarithms of the concentrations are calculated, the reduction is from 7 to 4 units, in other words, 3 log units. Coliform concentrations are frequently represented in terms of the order of magnitude (powers of 10) or in their logarithms, considering their great variability and the uncertainty in more precise numerical values, and because coliform data usually tend to follow a log-normal distribution. The following formulas relate the efficiency expressed as percentage removal with log units removed.

$$\boxed{\text{Efficiency } (\%) = (N_0 - N)/N_0 = 100 \times (1 - 10^{-\text{log.units removed}})} \quad (3.35)$$

$$\boxed{\text{Log units removed} = -\log_{10}[1 - (\text{Efficiency } (\%)/100)]} \quad (3.36)$$

Not all countries or regions have coliform standards for the water body. When existent, they vary as a function of the water use and a number of local aspects. Values are usually situated around $10^2$ to $10^3$ faecal (thermotolerant) coliforms per 100 ml.

---

### Example 3.3

Calculate the concentration profile of faecal (thermotolerant) coliforms in the river of Example 3.2. Calculate the coliform removal efficiency necessary in the wastewater treatment, so that the river presents a coliform concentration lower than $10^3$ CF/100 ml.

Data:
- river flow: $Q_r = 0.710$ m$^3$/s
- wastewater flow: $Q_w = 0.114$ m$^3$/s
- water temperature: $T = 25\,°C$
- travel distance: $d = 50$ km
- velocity of the water: $v = 0.35$ m/s

**Solution:**

a) Faecal coliform concentration in the raw sewage

Adopt a faecal coliform concentration of $N_{rw} = 1 \times 10^7$ org/100 mL in the raw wastewater (see Chapter 2).

b) Faecal coliform concentration in the wastewater–river mixture, after the discharge

Assume that the river is clean upstream of the discharge, with a negligible concentration of coliforms ($N_r = 0$ organisms/100 mL).

***Example 3.3 (Continued)***

The concentration in the mixing point is calculated by a weighted average with the flows:

$$N_0 = \frac{Q_r.N_r + Q_w.N_{rw}}{Q_r + Q_w} = \frac{0.710 \times 0 + 0.114 \times 1 \times 10^7}{0.710 + 0.114}$$

$$= 1.38 \times 10^6 \, org/100 \, mL$$

c) Concentration profile along the distance

The faecal coliform concentration is calculated by the equation for plug flow (rivers), presented in Table 3.9. Adopting $K_b = 1.0 \, d^{-1}$ (20 °C), the value for the temperature of 25 °C is obtained:

$$K_{b_T} = K_{b_{20}}.^{(T-20)} = 1.0 \times 1.07^{(25-20)} = 1.40 \, d^{-1}$$

The concentrations as a function of time are calculated from:

$$N = N_0.e^{-K_b.t} = 1.38 \times 10^6.e^{-1.4.t}$$

Varying t, the values of $N_t$ are obtained. The correspondence between distance and time is given by:

$$d = v.t = (0.35 \, m/s \times 86,400 \, s/d).t/(1000 \, m/l_2 m$$

The table and graph below present $N_t$ for various values of t and d:

| d (km) | t (d) | $N_t$ (organisms/100 mL) |
|---|---|---|
| 0.0 | 0.00 | $1.38 \times 10^6$ |
| 5.0 | 0.17 | $1.09 \times 10^6$ |
| 10.0 | 0.33 | $8.69 \times 10^5$ |
| 15.0 | 0.50 | $6.89 \times 10^5$ |
| 20.0 | 0.66 | $5.47 \times 10^5$ |
| 25.0 | 0.83 | $4.34 \times 10^5$ |
| 30.0 | 0.99 | $3.44 \times 10^5$ |
| 35.0 | 1.16 | $2.73 \times 10^5$ |
| 40.0 | 1.32 | $2.17 \times 10^5$ |
| 45.0 | 1.49 | $1.72 \times 10^5$ |
| 50.0 | 1.65 | $1.36 \times 10^5$ |

---

### *Example 3.3 (Continued)*

**FAECAL COLIFORM PROFILE - RAW WASTEWATER**

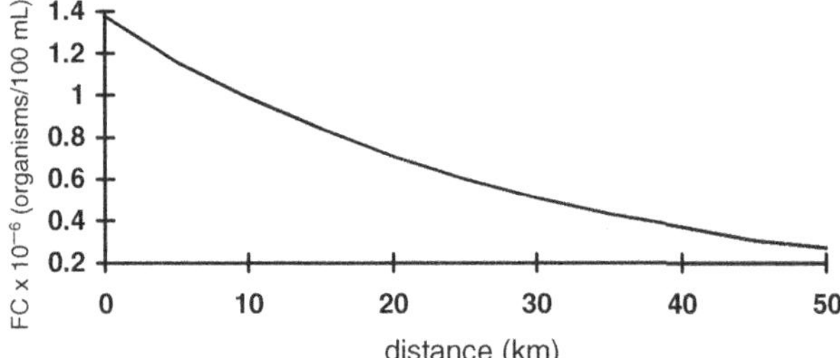


In spite of the considerable decrease along the travel distance, the concentrations are still very high and far greater than the desired value of $10^3$ organisms/100 mL.

d) Maximum allowable concentration in the wastewater

At the discharge point, the faecal coliform concentration needs to be less than 1,000 organisms/100 mL. Using the equation for the concentration in the mixing point, the maximum desirable concentration in the treated wastewater is obtained.

$$N_0 = \frac{Q_r.N_r + Q_w.N_{tw}}{Q_r + Q_w} = 1,000 = \frac{0.710 \times 0 + 0.114 \times N_{tw}}{0.710 + 0.114}$$

$$N_{tw} = 7,228 \text{ organisms/100 mL}$$

e) Required removal efficiency of faecal coliforms in the wastewater treatment

The required efficiency is:

$$E = \frac{1.0 \times 10^7 - 7,228}{1.0 \times 10^7} = 0.9993 = 99.93\%$$

In log units, the removal efficiency is:

$$\text{Log units removed} = -\log_{10}[1 - (E\,(\%)/100)] = -\log_{10}[1 - 0.9993]$$
$$= 3.15 \text{ log units}$$

Therefore, the high efficiency of 99.93% (3.15 log units) for the removal of faecal coliforms in the wastewater treatment will be required. Such a high efficiency is not usually reached in the conventional treatment processes, requiring a specific stage for coliform removal (see Table 3.10).

---

### Example 3.4

Calculate the concentration of the faecal coliforms in a *reservoir* with a volume of 5,000,000 $m^3$. The reservoir receives, together, a river and a sewage discharge, both with the same characteristics as in Example 3.3. Calculate the necessary coliform removal efficiency in the wastewater treatment, so that the reservoir has faecal (thermotolerant) coliform concentrations less than or equal to 1000 FC/100 ml.

Data:
- river flow: $Q_r = 0.710$ $m^3/s$
- wastewater flow: $Q_w = 0.114$ $m^3/s$
- water temperature: $T = 25\,^\circ C$

**Solution:**

a) Faecal coliform concentration in the raw sewage

$N_{rw} = 1 \times 10^7$ organisms/100 mL (same as in Example 3.3).

b) Faecal coliform concentration in the wastewater–river mixture

$N_0 = 1.38 \times 10^6$ organisms/100 mL (same as in Example 3.3)

c) Detention time in the reservoir

Total influent flow to the reservoir:

$$Q = Q_r + Q_w = 0.710 + 0.114 = 0.824\,m^3/s$$

$$t = \frac{V}{Q} = \frac{5,000,000\,m^3}{(0.824\,m^3/s) \times (86.400\,s/d)} = 70.2\,d$$

d) Coliform concentration in the reservoir

Assuming a complete-mix model and a $K_b$ value of 1.4 $d^{-1}$ (equal to Example 3.3, for $T = 25\,^\circ C$), the concentrations of coliforms in the reservoir and in the reservoir effluent are given by (see equation in Table 3.9):

$$N = \frac{N_0}{1 + K_b.t} = \frac{1.38 \times 10^6}{1 + 1.4 \times 70.2} = 13,900\,\text{organisms}/100\,mL$$

$$= 1.39 \times 10^4\,\text{organisms}/100\,mL$$

This value is above the desired standard of 1,000 organisms/100 mL.

---

**Example 3.4 (Continued)**

e) Maximum allowable concentration in the wastewater

Using the same equation for completely-mixed reactors:

$$N = \frac{N_0}{1 + K_b.t} = 1,000 = \frac{N_0}{1 + 1.4 \times 70.2}$$

$$N_0 = 99,280 \text{ organisms}/100\,\text{mL} = 9.93 \times 10^4 \text{ organisms}/100\,\text{mL}$$

At the sewage–river mixing point, the concentration must be 99.280 organisms/100 mL. Using the equation for the concentration in the mixture (weighted averages), the maximum desirable concentration in the sewage is obtained.

$$N_0 = \frac{Q_r.N_r + Q_w.N_w}{Q_r + Q_w} = 99.280 = \frac{0.710 \times 0 + 0.114 \times N_w}{0.710 + 0.114}$$

$$N_w = 717.603 \text{ organisms}/100\,\text{ml} = 7.18 \times 10^5 \text{organisms}/100\,\text{ml}$$

f) Required efficiency for coliform removal in the wastewater treatment

$$E = \frac{1.0 \times 10^7 - 7.18 \times 10^5}{1.0 \times 10^7} = 0.928 = 92.8\%$$

This efficiency is lower than the efficiency required in Example 3.3 but this is due to the high detention time in the reservoir (70.2 days) compared with the reduced time in the river (1.65 days). If both systems had the same detention time, the plug-flow reactor (river) would have been more efficient compared with the completely-mixed reactor (reservoir).

---

## 3.4 EUTROPHICATION OF LAKES AND RESERVOIRS

### 3.4.1 The eutrophication process

**Eutrophication** is the excessive growth of aquatic plants, either planktonic, attached or rooted, at such levels as to cause interference with the desired uses of the water body (Thomann and Mueller, 1987). As discussed in this chapter, the main stimulating factor is an excessive level of **nutrients** in the water body, principally nitrogen and phosphorus.

In this chapter, the water bodies under consideration are lakes and reservoirs. The process of eutrophication can also occur in rivers, though this is less frequent, owing to the environmental conditions being less favourable for the growth of algae and other plants, because of factors such as turbidity and high velocities.

The following description illustrates the possible sequence of the eutrophication process in a water body, such as a lake or reservoir (see Figure 3.16). The level of eutrophication is usually associated with the predominant land use and occupation in the catchment area.

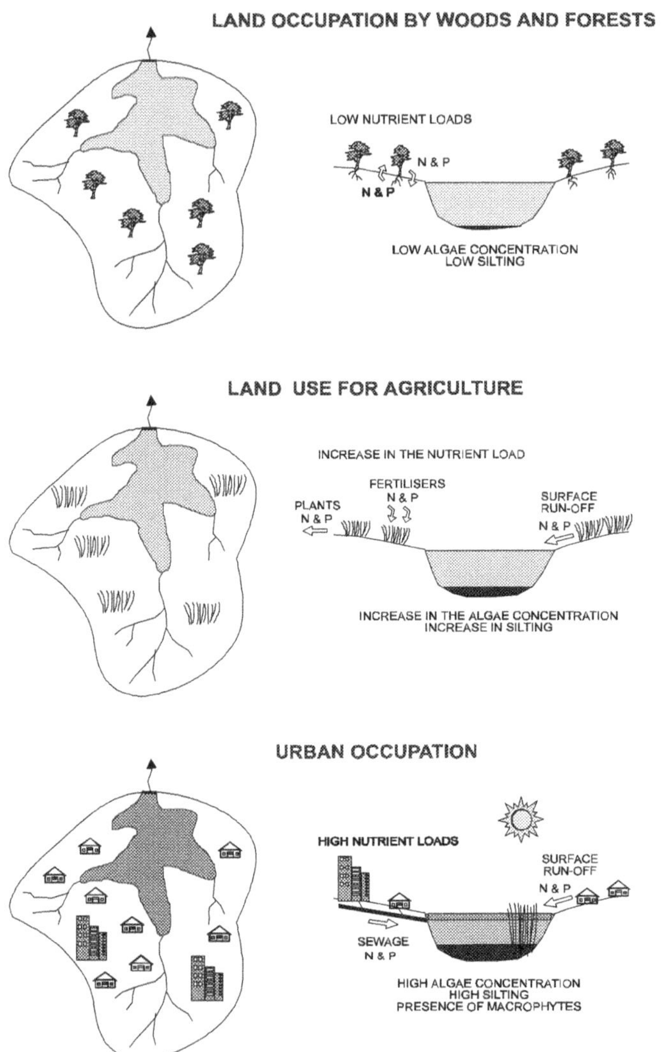


Figure 3.16.   Sequence of the eutrophication process in a lake or reservoir. Association between land use and eutrophication.

## a)   Occupation by woods and forests

A lake situated in a catchment area occupied by woods and forests usually presents a low productivity; that is to say, there is little biological activity of production (synthesis) in it. Even in these natural conditions and in the absence of human interference, the lake tends to accumulate solids that settle, which form a layer of sludge at the bottom. With the decomposition of the settled material, there is a certain increase, still incipient, of the level of nutrients in the liquid mass. As a result, there is a progressive increase in the population of aquatic plants and, in consequence, of the other organisms situated at a higher level in the food chain.

In the catchment area, the larger part of the nutrients is retained within a nearly closed cycle. The plants die and, in the soil, are decomposed, releasing nutrients. In a region of woods and forests, the infiltration capacity of the rainwater into the soil is high. The nutrients then percolate into the soil, where they are absorbed by the roots of the plants, making part again of their composition and closing the cycle. The input of nutrients to the water body is small.

The water body still presents a low trophic level.

## b)   Agricultural occupation

The removal of natural vegetation from the catchment area for agricultural use generally leads to an intermediate stage in the deterioration process of the water body. The crops planted in the basin are harvested and transported for human consumption, probably outside the catchment area. With this, there is a removal of nutrients that is not naturally compensated, causing a break in the internal cycle. To compensate this removal and to make the agriculture more intensive, fertilisers containing high levels of nitrogen and phosphorus are added artificially. The farmers aim at guaranteeing a high production and thus add high quantities of N and P, frequently greater than the assimilative capacity of the plants.

The substitution of woods by plants for agricultural purposes can also cause a reduction in the infiltration capacity of the soil. Therefore the nutrients, already added in excess, are less retained and run off the soil until they eventually reach a lake or reservoir.

The increase in the nutrient level in the water body causes a certain increase in the number of algae and, in consequence, of the other organisms located at higher levels in the food chain, culminating with the fish. This relative increase in the productivity of the water body can even be welcome, depending on its intended uses, as would be the case, for instance, in aquaculture. The balance between the positive and negative aspects will depend, to a large extent, on the nutrient assimilative capacity of the water body.

## c)   Urban occupation

If an agricultural or forest area in a catchment area is substituted by urban occupation, a series of consequences should take place, this time at a faster rate.

- *Silting*. The implementation of housing developments implies land movement for the works. Urbanisation also reduces the water infiltration capacity into the soil. The soil particles tend to be transported to the lower parts of the catchment area until they reach the lake or reservoir. In these water bodies, they tend to settle, owing to the low horizontal velocities and turbulence. The sedimentation of the soil particles causes silting and reduces the net volume of the water body. The settled material also serves as a support medium for the growth of rooted plants of larger dimensions (macrophytes) near the shores. In spite of some ecological advantages (e.g. physical retention of pollutants, reduction of sediment resuspension, refuge for fishes and macroinvertebrates), these plants cause an evident deterioration in the visual aspect of the water body.

- *Urban stormwater drainage*. Urban drainage transports a far greater load of nutrients in comparison with the other types of occupation of the catchment area. This nutrient input contributes to a rise in the level of algae in the reservoir.
- *Sewage*. The greatest deterioration factor, however, is associated with wastewater originating from urban activities. The wastewater contains nitrogen and phosphorus present in faeces and urine, food remains, detergents and other by-products of human activity. The N and P contribution from sewage is much higher than the contribution originating from urban drainage.

Therefore, there is a great increase in the input of N and P onto the lake or reservoir, bringing as a result an elevation in the population of algae and other plants. Depending on the assimilative capacity of the water body, the algal population can reach very high values, bringing about a series of problems, which are described in the subsequent item. In a period with high sunshine (light energy for photosynthesis), the algae can reach superpopulations and be present at massive concentrations at the surface layer. This surface layer hinders the penetration of light energy for the lower layers in the water body, causing the death of algae situated in these regions. The death of these algae brings in itself a series of other problems. These events of superpopulation of algae are called **algal blooms**.

## 3.4.2 Problems of eutrophication

The following are the main undesired effects of eutrophication (Arceivala, 1981; Thomann and Mueller, 1987; von Sperling, 1994):

- *Recreational and aesthetic problems*. Reduction of the use of water for recreation, bathing and as a general tourist attraction because of the:
  - frequent algal blooms
  - excessive vegetation growth
  - disturbances with mosquitoes and insects
  - occasional bad odours
  - occasional fish mortality
- *Anaerobic conditions in the bottom of the water body*. The increase in productivity of the water body causes a rise in the concentration of heterotrophic bacteria, which feed on the organic matter from algae and other dead microorganisms, consuming dissolved oxygen from the liquid medium. In the bottom of the water body there are predominantly anaerobic conditions, owing to the sedimentation of organic matter and the small penetration of oxygen, together with the absence of photosynthesis (absence

of light). With the anaerobiosis, reducing conditions prevail, leading to compounds and elements being present in a reduced state:
* iron and manganese are found in a soluble form, which may bring problems with the water supply;
* phosphate is also found in a soluble form, and may represent an internal source of phosphorus for algae;
* hydrogen sulphide may also causes problems of toxicity and bad odours.
* *Occasional anaerobic conditions in the water body as a whole.* Depending on the degree of bacterial growth, during periods of total mixing of the liquid mass (thermal inversion) or in the absence of photosynthesis (night time), fish mortality and the reintroduction of reduced compounds from the bottom onto the whole liquid mass could occur, leading to a large deterioration in the water quality.
* *Occasional fish mortality.* Fish mortality could occur as a result of:
  * anaerobiosis (mentioned above)
  * ammonia toxicity. Under conditions of high pH (frequent during periods of high photosynthetic activity), ammonia may be present in significant amounts in its free form ($NH_3$), toxic for the fish, instead of the ionised form ($NH_4^+$), which is non-toxic.
* *Greater difficulty and increase in the costs of water treatment.* The excessive presence of algae substantially affects the treatment of water abstracted from a lake or reservoir, due to the necessity of:
  * removal of the algae themselves
  * colour removal
  * taste and odour removal
  * higher consumption of chemical products
  * more frequent filter backwashings
* *Problems with industrial water supply.* Elevation in the costs of industrial water supply due to reasons similar to those already mentioned, and also to the presence of algae deposits in cooling waters.
* *Water toxicity.* Impairment of water for human and animal supply because of the presence of toxic secretions from cyanobacteria (cyanotoxins).
* *Alteration in the quality and quantity of commercial fish.*
* *Reduction in navigation and transport capacity.* The excessive growth of rooted macrophytes interferes with navigation, aeration and transport capacity of the water body.
* *Negative interference in the equipment for energy generation* (macrophytes in turbines).
* *Gradual disappearance of the lake.* As a result of eutrophication and silting, there is an increase in the accumulation of material and vegetation, and the lake becomes progressively shallower until it disappears. This tendency of disappearing (conversion to swamps or marshes) is irreversible, although usually extremely slow. With human interference, the process can accelerate abruptly. In case there is no control in the source or the dredging of the sediments, the water body could disappear relatively quickly.

Table 3.11. Trophic characterisation of lakes and reservoirs

| Item | Trophic level | | | | |
| --- | --- | --- | --- | --- | --- |
| | Ultraoligotrophic | Oligotrophic | Mesotrophic | Eutrophic | Hypereutrophic |
| Biomass | Very low | Low | Intermediate | High | Very high |
| Fraction of green algae and/or cyanobacteria | Low | Low | Variable | High | Very high |
| Macrophytes | Low or absent | Low | Variable | High or low | Low |
| Production dynamics | Very low | Low | Intermediate | High | High, unstable |
| Oxygen dynamics at the upper layer | Normally saturated | Normally saturated | Variable around supersaturation | Frequently supersaturated | Very unstable, from supersaturation to absence |
| Oxygen dynamics of the lower layer | Normally saturated | Normally saturated | Variable below saturation | Below saturation to complete absence | Very unstable, from supersaturation to absence |
| Impairment of multiple uses | Low | Low | Variable | High | Very high |

Adapted from Vollenweider (cited by Salas and Martino, 1991)

### 3.4.3  Trophic levels

With the objective of characterising the stage of eutrophication of the water body, allowing the undertaking of preventative and/or corrective measures, it is interesting to adopt a classification system. In a simplified way, there are the following trophic levels:

- *oligotrophic* (clear lakes with a low productivity)
- *mesotrophic* (lakes with an intermediate productivity)
- *eutrophic* (lakes with a high productivity, compared with the natural basic level)

This classification can be further detailed, with the inclusion of other trophic levels, such as: ultraoligotrophic, oligotrophic, oligomesotrophic, mesotrophic, mesoeutrophic, eutrophic, eupolitrophic and hypereutrophic (from lowest to highest productivity).

A qualitative classification between the main trophic levels may be as presented in Table 3.11.

The quantification of the trophic level is, however, more difficult, especially for tropical lakes. Von Sperling (1994) presents a collection of various references, in terms of total phosphorus concentration, chlorophyll $a$ and transparency, which shows the large amplitude in the ranges proposed by various authors. Besides this, the cited reference presents other possible indices to be used, always with the safeguard of the difficulty in generalising the data from one water body to another. It should be kept in mind that tropical water bodies present a larger capacity of phosphorus assimilation in comparison with water bodies in temperate climates. An interpretation of the synthesis reported by von Sperling may be as presented in Table 3.12 in terms of the total phosphorus concentration.

The establishment of the trophic levels based only on phosphorus is due to a mathematical modelling convenience. In the same way that in the other water pollution topics covered in the book, representative variables, such as dissolved oxygen (pollution by organic matter) and coliforms (contamination by pathogens),

Table 3.12. Approximate range of values of total phosphorus for the main trophic levels

| Trophic level | Total phosphorus concentration in the reservoir (mgP/m$^3$) |
|---|---|
| Ultraoligotrophic | <5 |
| Oligotrophic | <10–20 |
| Mesotrophic | 10–50 |
| Eutrophic | 25–100 |
| Hypereutrophic | >100 |

*Source:* table constructed using the data presented by von Sperling (1994)
*Note:* the overlapping of values between two ranges indicates the difficulty in establishing rigid ranges

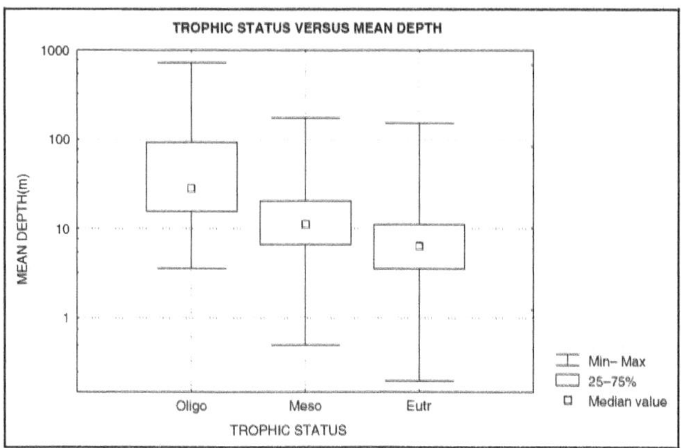


Figure 3.17.   Box-and-whisker plot of mean depth for three groups of trophic levels. *Source:* von Sperling et al (2002) – 269 lakes and reservoirs. Note: mean depth: lake volume / area of the lake

phosphorus is adopted as a representative of the trophic level in this chapter. However, as seen in Section 3.4.5, there are some situations in which nitrogen controls the eutrophication process.

The researcher should always be open to include other variables in the analysis, in order to get a picture as close as possible to the actual behaviour of the water body under study. Von Sperling et al (2002), analysing data from more than 1,500 lakes and reservoirs around the world, investigated the correlation between morphometric variables and trophic level (using a subset of 269 water bodies, which had information on trophic level). From the statistical analysis it was clear that depth (mean depth, maximum depth and relative depth) is the morphometric variable most closely related with the trophic status: the shallower the water body, the greater the tendency for having a higher trophic level, mainly because of a higher light penetration over the full water body. Figure 3.17 shows a resulting box-and-whisker plot for the mean depth, where this association is clearly seen.

The association between trophic levels and water uses is shown in Table 3.13.

### 3.4.4 Dynamics of lakes and reservoirs

The vertical temperature profile in lakes and reservoirs varies with the seasons of the year. This temperature variation affects the density of the water and as a result the mixing and stratification capacity of the water body

During warm periods, the temperature at the surface layer is much higher than the temperature at the bottom, because of solar radiation. Owing to this fact, the density of the water at the surface is lower than the density of the bottom layer,

Table 3.13. Association between trophic levels and water uses in a water body

| Use | Trophic level | | | | | |
| --- | --- | --- | --- | --- | --- | --- |
| | Ultraoligotrophic | Oligotrophic | Mesotrophic | Mesoeutrophic | Eutrophic | Hypereutrophic |
| *Drinking water supply* | | Desirable | Acceptable | | | |
| *Process water supply* | | | Desirable | Acceptable | | |
| *Cooling water supply* | | | | | Acceptable | |
| *Primary contact recreation* | | | Desirable | Acceptable | | |
| *Secondary contact recreation* | | | Desirable | | Acceptable | |
| *Landscaping* | | | | Acceptable | | |
| *Fish culture (sensitive species)* | | | Desirable | Acceptable | | |
| *Fish culture (tolerant species)* | | | | | Acceptable | |
| *Irrigation* | | | | | | Acceptable |
| *Energy production* | | | | | | Acceptable |

*Source:* adapted from Thornton e Rast (1994)

which causes the existence of distinct layers in the water body:

- *epilimnion*: upper, warmer, less dense, with higher circulation
- *thermocline*: transition layer
- *hypolimnion*: bottom layer, colder, denser, with greater stagnation

The difference in densities can be such as to cause a complete **stratification** in the water body, with the three layers not mixing with each other. This stratification has a great influence on the water quality. Depending on the trophic level of the water body, a complete absence of dissolved oxygen in the hypolimnion can occur. As a result, in this layer there will be a predominance of iron, manganese and other compounds in their reduced forms.

With the arrival of a cold period, the upper layer in the lake cools, causing a certain homogenisation in the temperature along the depth. With the homogeni-sation of the temperature, there is a greater similarity in the water densities. The upper layer, suddenly cooled, tends to go to the bottom of the lake, dislodging the bottom layer and causing a complete turn over in the lake. This phenomenon is called thermal inversion or **turn over**. In lakes that present a high concentration of reduced compounds in the hypolimnion, the reintroduction of these compounds into the whole liquid mass of the lake can cause a large deterioration in the water quality. The decrease in dissolved oxygen due to the reintroduced demand by the reduced organic and inorganic compounds, together with the resuspension of the anaerobic bottom layer, can cause fish mortality.

Figure 3.18 presents typical temperature and DO profiles under conditions of stratification and turn over.

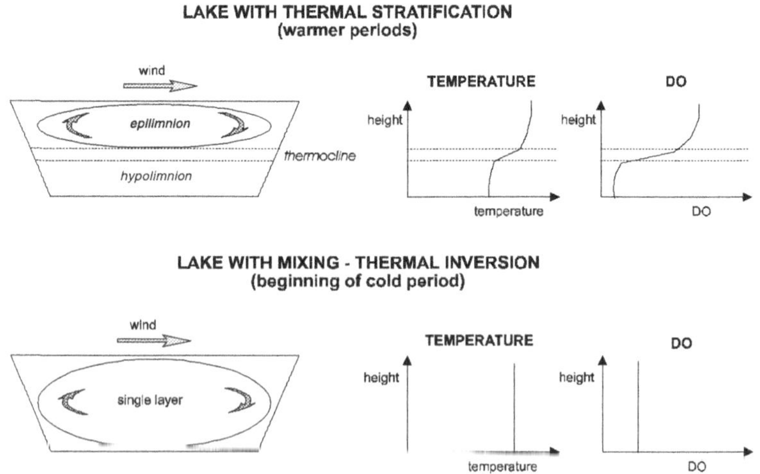


Figure 3.18.   Typical temperature and DO profiles in a lake under stratification and turn-over conditions.

Lakes and reservoirs may present different mixing patterns and frequencies (von Sperling, 1999; Dantas, 2000):

a) **Holomictic lakes**. *Complete circulation over the whole water column.* Depending on the number of circulations per year, they can be classified as:
  • **Monomictic**. *One circulation per year.* Usually located where there are clear seasonal variations. Two types:
    − **Warm monomictic**. *One circulation in winter.* Located in temperate regions, high-altitude subtropical regions and in tropical regions.
    − **Cold monomictic**. *One circulation in summer.* Located in subpolar regions and high-altitude regions in temperate climates.
  • **Dimictic**. *Two circulations per year, one in spring, one in autumn.* Located in temperate climates. Warm months: stratification; autumn: cooling of upper layer and mixing. Cold months: ice cover; spring: ice melting, wind-induced mixing.
  • **Oligomictic**. *Few circulations per year.* Usually deep lakes in wet tropics, where there is little seasonal variation. Warm water along the water column.
  • **Polimictic**. *Many circulations per year.* Usually shallow lakes with daily circulations, unprotected from wind action, and located in warmer regions. Influence from daily temperature variations. Day hours: stratification. Night hours: cooling of upper layer and mixing.
b) **Meromictic lakes**. *Circulation does not occur at the whole water column.* Bottom layer (monimolimnion): stagnated due to high concentration of dissolved substances. Little influence from temperature.
c) **Amictic lakes**. *No circulation.* Usually ice-covered lakes at very high altitudes in equatorial regions or high latitudes.

For warm regions (the main focus of this book), the prevailing mixing patterns are either *warm monomictic* or *polimictic*. The variables that most significantly affect the mixing pattern are those related with depth (mean depth, maximum depth, relative depth) (von Sperling et al, 2002).


## 3.4.5 Limiting nutrient

The **limiting nutrient** is the one that, being essential for a certain population, limits its growth. According to Liebig's law, a limiting nutrient is the one whose concentration is closest to the minimum related to the organism's demand. With low concentrations of the limiting nutrient, the population growth is low. With an increase in the limiting nutrient concentration, the population growth also increases. This situation persists until the point in which the concentration of this nutrient starts to be so high in the system, that another nutrient starts to be the new limiting factor, since it is not present at concentrations sufficiently high for the requirements of the large population. This nutrient is now the new limiting nutrient, because there is no impact in increasing the concentration of the first nutrient, since the population will not rise, because it will be limited by the insufficiency of the new limiting nutrient.

Thomann and Mueller (1987) suggested the following criterion, based on the ratio between the nitrogen and phosphorus concentrations (N/P), in order to make a preliminary estimate of whether the algal growth is being controlled by phosphorus or nitrogen.

- large lakes, with a predominance of diffuse sources: **N/P $\gg$ 10**: *limited by phosphorus*
- small lakes, with a predominance of point sources: **N/P $\ll$ 10**: *limited by nitrogen*

According to Salas and Martino (1991), most of the tropical lakes in Latin America are limited by phosphorus. Another aspect is that, even if the external input of nitrogen is controlled, there are organisms (cyanobacteria) which are capable of fixing atmospheric nitrogen. These organisms would be not reduced in numbers with the decrease in the influent load of nitrogen. Because of this, *usually a larger priority is given to the control of the phosphorus sources when the eutrophication of a lake or reservoir needs to be controlled*. The present text follows this approach.

## 3.4.6 Estimation of the phosphorus load into a lake or reservoir

The main sources of phosphorus to a lake or reservoir are, in increasing order of importance:

- *Stormwater drainage*
  - Areas with woods and forests
  - Agricultural areas
  - Urban areas
- *Wastewater*

Stormwater drainage from areas with ample vegetation coverage, such as woods and forests, transports a lower quantity of phosphorus. In these areas, phosphorus is not in excess in the environment, since the ecosystem is close to an equilibrium, without having large excesses or scarcities of the main elements.

Drainage from agricultural areas leads to higher and more variable P loads, depending on the soil retention capacity, irrigation, type of fertilisers and climatic conditions.

Urban drainage is associated with the highest loads. Domestic sewage transported by waterborne sewerage systems is usually the greatest source of phosphorus. Phosphorus can be found in human wastes, household detergents and other by-products of human activities. Regarding industrial wastewater, the generalisation of its contribution is difficult because of the variability of the various industrial wastewaters, even within the same industrial processing activity.

Table 3.14. Typical values of unit phosphorus contributions

| Source | Type | Typical values | Unit |
|---|---|---|---|
| Drainage | Areas of woods and forests | 10 | kgP/km$^2$.year |
|  | Agricultural areas | 50 | kgP/km$^2$.year |
|  | Urban areas | 100 | kgP/km$^2$.year |
| Domestic sewage | Domestic | 0.5 | kgP/inhab.year |

*Note:* values may vary widely from place to place; data presented are only references of orders of magnitude

Table 3.14 presents typical values of the unit phosphorus contribution, compiled from various references (von Sperling, 1985). The unit of time adopted is "year", convenient for modelling of P in lakes. Naturally the values can vary widely, from place to place. However, the values presented aim only to show an order of magnitude of the typical values.

## 3.4.7 Estimation of the phosphorus concentration in the water body

Literature presents a series of simplified empirical models to estimate the phosphorus concentration in a water body, as a function of influent load, detention time and geometric characteristics. The empirical models can be applied in any of the following applications:

- *Estimation of the trophic level.* Once the phosphorus concentration in the water body has been estimated, the trophic level of the lake can be evaluated, based on the considerations of Section 3.4.3.
- *Estimation of the maximum allowable load.* The maximum allowable P load into the lake can be estimated, such that the resulting P concentration is lower than a maximum desired value (e.g. a concentration that characterises eutrophic conditions).

The empirical approach has been more applied for planning than conceptual and more sophisticated models, because of the difficulty in structuring and obtaining the coefficients and input data necessary for these models.

The most widely known empirical model is that proposed by Vollenweider (1976), which has been developed, however, for temperate climatic conditions. The model, presented in a convenient form for the present text, is:

$$\boxed{P = L.10^3 / \left[ V.\left( \tfrac{1}{t} + K_s \right) \right]} \tag{3.37}$$

where:
    P = phosphorus concentration in the water body (gP/m$^3$)
    L = influent phosphorus load (kgP/year)

V = volume of the reservoir (m$^3$)

t = hydraulic detention time (year)

$K_s$ = loss coefficient of P by sedimentation (1/year)

Vollenweider obtained the value for $K_s$ by regression analysis with the detention time in some reservoirs. The obtained value was:

$$K_s = 1/\sqrt{t} \qquad (3.38)$$

Castagnino (1982), theoretically analysing the P loss by sedimentation in **tropical lakes,** found a value of $K_s$ equal to 2.5 times the value of Vollenweider. This magnifying factor of 2.5 is composed by a factor of 1.3 for the faster sedimentation at higher temperatures and a factor of 1.9 for the faster phytoplankton growth rate (1.3 × 1.9 = 2.5). According to Castagnino, the corrected $K_s$ value for tropical conditions is:

$$K_s = 2.5/\sqrt{t} \qquad (3.39)$$

Salas and Martino (1991), analysing experimental data from 40 lakes and reservoirs in Latin America and the Caribbean obtained, by regression analysis, the following relation for $K_s$:

$$\boxed{K_s = 2/\sqrt{t}} \qquad (3.40)$$

With the values obtained by Salas and Martino (1991), the *P concentration* in the reservoir becomes:

*P concentration in the reservoir:*

$$\boxed{P = L.10^3 \left/ \left[ V.\left( \frac{1}{t} + \frac{2}{\sqrt{t}} \right) \right] \right.} \qquad (3.41)$$

Equation 3.41 can be arranged to lead to the maximum allowable *P load* into a lake or reservoir, so as not to surpass a maximum desired phosphorus concentration in the water body.

*Maximum allowable P load:* $\qquad \boxed{L = P.V.\left( \frac{1}{t} + \frac{2}{\sqrt{t}} \right) \left/ 10^3 \right.} \qquad (3.42)$

To use Equation 3.42, L must be estimated so that P is below the limit for eutrophic conditions. From Table 3.12, the P concentration range in an eutrophic water body is 25 to 100 mgP/m$^3$or 0.025 to 0.100 gP/m$^3$. The establishment of a more relaxed or restrictive value for P must be done for each case, analysing the multiple uses of the reservoir and its level of importance.

Because of the fact that it was developed based on tropical water bodies, the empirical model proposed by Salas and Martino (1991) is probably the most adequate to be used for planning and management of lakes and reservoirs in warm-climate regions. Naturally, the critical view and experience of the researcher must always be present, to avoid distortions, given the specificity of each reservoir and lake under study.

## 3.4.8  Control of eutrophication

The control strategies usually adopted can be classified into two broad categories (Thomann and Mueller, 1987; von Sperling, 1995a):

- *Preventive measures* (action in the catchment area)
    - Reduction of external sources
- *Corrective methods* (action in the lake or reservoir)
    - Mechanical processes
    - Chemical processes
    - Biological processes

### a)  Preventive methods

Preventive methods, which comprise the reduction of the phosphorus input to the lake by acting on external sources, can include strategies related to the wastewater or to the stormwater drainage. The control strategies for wastewater are illustrated in Figure 3.19.

*Control of wastewater*

- Wastewater treatment with nutrient removal (tertiary treatment)
- Sewage diversion to downstream of the reservoir, associated with conventional (secondary) wastewater treatment
- Exportation of the wastewater to another catchment area without lakes or reservoirs, followed by conventional wastewater treatment
- Land infiltration of the wastewater

*Control of stormwater drainage*

- Control of the land use and occupation in the catchment area
- Protective green belts along the reservoir and its tributaries
- Construction of P retention reservoirs upstream of the main reservoir

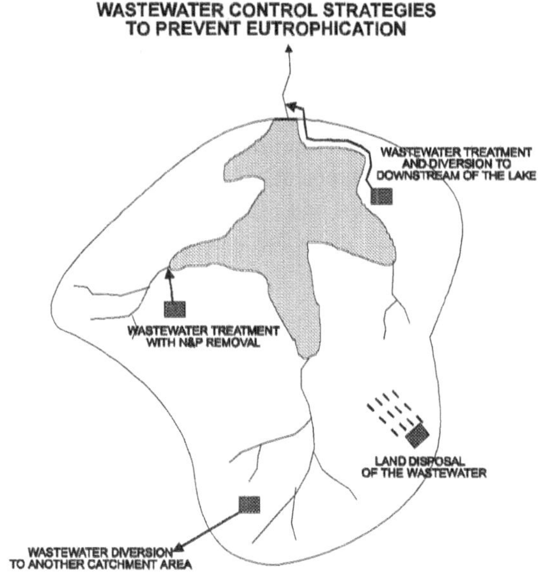


Figure 3.19.    Wastewater control strategies aiming at the prevention of nutrient inputs into the reservoir.

Regarding wastewater treatment with phosphorus removal, this can be undertaken by biological and/or physical–chemical processes.

*Biological phosphorus removal* from the wastewater is currently consolidated and undertaken in many countries. The process is based on alternating between aerobic and anaerobic conditions, a situation which makes a certain group of microorganisms (phosphate accumulating organisms) assimilate a higher quantity of phosphorus than would be required in their usual metabolic processes. When removing these bacteria from the system in the biological excess sludge, the phosphorus absorbed by them is also removed. With biological P removal, effluents with concentrations of 0.5 mgP/L can be reached, although it is more appropriate to consider a more conservative value of 1.0 mgP/L. Chapters 35 and 36 cover in detail the process of biological P removal.

Phosphorus removal by *physical–chemical processes* is based on the precipitation of phosphorus after the addition of aluminium sulphate, ferric chloride or lime. The consumption of chemical products and the sludge generation are high. Physical–chemical polishing after biological P removal can generate effluents with concentrations in the order of 0.1 mgP/L.

### b)    Corrective methods

Corrective methods that can be adopted can include one or more of the strategies listed in Table 3.15 (von Sperling, 1995a).

Whenever possible, greater emphasis should be given to preventive measures, usually cheaper and more effective.

Table 3.15. Corrective methods for the recovery of lakes and reservoirs

| Processes | Technique | Characteristics |
|---|---|---|
| Mechanical | Hypolimnetic aeration | • Injection of compressed air or oxygen into the bottom layers of the lake, promoting the stabilisation of the organic matter accumulated at the bottom and avoiding the release of nutrients from the sediments.<br>• Presents high operational costs and requires the acquisition of specialised equipment, but is a widely applied and highly efficient technique. |
| | Destratification | • Consists of the injection of compressed air or oxygen into the bottom layers of the lake, leading to the circulation of the whole water body.<br>• Use of simpler equipment.<br>• Presents as inconvenient the transportation of reducing compounds to the upper layer, leading to the fertilisation of the epilimnion. |
| | Removal of deep waters | • Aims at the removal and substitution of deep waters by upper-layer waters, richer in oxygen, reducing the accumulation of nutrients in the hypolimnion.<br>• The volume of liquid removed through hydrostatic pressure or pumping can be used for irrigation or directed to a WWTP. |
| | Addition of water with a higher quality | • Dilution technique that reduces the nutrient concentration in the water body.<br>• Its application limits the formation of hydrogen sulphide in the hypolimnion, thus avoiding fish mortality. |
| | Sediment removal | • Upper layers of the sediment are removed by dredging, favouring the exposure of layers with a lower polluting potential.<br>• The sludge removed, after treatment, can be used as a soil conditioner. |
| | Covering of the sediment | • Corrective method to avoid the release of nutrients in the deeper layers.<br>• The sediment is isolated from the rest of the water body by covering it with plastic material or finely particulated substances.<br>• Expensive method which presents difficulties in its installation. |
| | Removal of the aquatic macrophytes | • Aquatic macrophytes, which, when present in excessive numbers may interfere with various water uses, can be removed by manual or mechanical processes. |
| | Removal of the planktonic biomass | • The planktonic biomass, which presents a large pollutant storage capacity, may be removed by centrifuging or by the use of microsieves. |
| | Shading | • Acts against the excessive vegetation growth, by reducing the level of solar radiation received, by means of:<br>– tree planting in the shores of small water bodies<br>– installation of panels in the shores<br>– application of supernatant material or light dyes at the surface layer |

(Continued)

Table 3.15 (*Continued*)

| Processes | Technique | Characteristics |
|---|---|---|
| Chemical | Chemical precipitation of phosphorus | • Recommended in the case of diffused sources of phosphorus, in which the removal of nutrients is impractical. |
| | Oxidation of the sediment with nitrate | • Efficient for the reduction of the internal fertilisation problem.<br>• Avoids the excessive decrease of the oxygen concentration in deep waters. |
| | Application of herbicides | • Avoids excessive vegetation growth.<br>• Associated with problems of toxicity, taste and odour and bioaccumulation. |
| | Application of lime | • Used for the sediment disinfection and to eliminate algae and submerged plants in small water bodies, and also for the neutralisation of the water in acidic lakes. |
| Biological | Use of fish that feed on the plants | • Reduces the plant community because of the activity of herbivorous fish. |
| | Use of cyanophages | • Reduces the density of cyanobacteria, by the attack of specific viruses.<br>• Little employed. |
| | Manipulation of the food chain | • Reduces the phytoplanktonic community by the increase of the zooplanktonic population |

*Source:* von Sperling (1995a)

---

**Example 3.5**

Estimate the trophic level of a reservoir based on the phosphorus concentration. In case eutrophic conditions are identified, estimate the maximum allowable P load so that eutrophic conditions are avoided.

Data:
- Reservoir volume: $10 \times 10^6$ m$^3$
- Average influent flow (tributaries + wastewater): $50 \times 10^6$ m$^3$/year
- Drainage area: 60 km$^2$
  - Woods and forests: 40 km$^2$
  - Agriculture: 10 km$^2$
  - Urban area: 10 km$^2$
- Contributing population (connected to the sewerage system): 16,000 inhabitants
- Wastewater characteristics: raw domestic sewage

**Solution:**

a) Estimation of the influent P load into the reservoir

Adopting the unit load values proposed in Table 3.14, the influent loads are:
- Raw domestic sewage: 16,000 inhab. $\times$ 0.5 kgP/inhab.year = 8,000 kgP/year

---

### *Example 3.5 (Continued)*

- Drainage from the wooded areas: $40 \text{ km}^2 \times 10 \text{ kgP/km}^2.\text{year} = 400 \text{ kgP/year}$
- Drainage from the agricultural areas: $10 \text{ km}^2 \times 50 \text{ kgP/km}^2.\text{year} = 500 \text{ kgP/year}$
- Drainage from the urban areas: $10 \text{ km}^2 \times 100 \text{ kgP/km}^2.\text{year} = 1{,}000 \text{ kgP/year}$

Total influent load into the reservoir: $8{,}000 + 400 + 500 + 1{,}000 = 9{,}900 \text{ kgP/year}$

b) Estimation of the hydraulic detention time

The hydraulic detention time is given by:

$$t = \frac{V}{Q} = \frac{10 \times 10^6 \text{ m}^3}{50 \times 10^6 \text{ m}^3/\text{year}} = 0.20 \text{ years}$$

c) Estimation of the phosphorus concentration in the reservoir

Adopting the model of Salas and Martino (1991), Equation 3.41:

$$P = \frac{L.10^3}{V.\left(\frac{1}{t} + \frac{2}{\sqrt{t}}\right)} = \frac{9{,}900 \times 10^3}{10 \times 10^6.\left(\frac{1}{0.20} + \frac{2}{\sqrt{0.20}}\right)} = 0.105 \text{ gP/m}^3 = 105 \text{ mgP/m}^3$$

d) Evaluation of the trophic level of the reservoir

Based on the P concentration of 105 mgP/m$^3$ and on the interpretation of Table 3.12, the reservoir is in the borderline between eutrophy and hypereutrophy. Therefore, control methods are necessary, so that the lake does not present eutrophic conditions.

e) Reduction of the influent phosphorus load

Through the adoption of preventive methods of wastewater and stormwater control, the influent phosphorus load into the reservoir can be drastically reduced. The influent load must be reduced down to a value that is below the limit for eutrophic conditions. Using Table 3.12, a not very conservative value of 50 mgP/m$^3$ can be used as the limit between mesotrophy and eutrophy. Under these conditions, the maximum allowable phosphorus load into the reservoir is given by Equation 3.42:

$$L = P.V.\left(\frac{1}{t} + \frac{2}{\sqrt{t}}\right)/10^3 = 0.050 \times 10 \times 10^6.\left(\frac{1}{0.20} + \frac{2}{\sqrt{0.20}}\right)/10^3$$
$$= 4{,}736 \text{ kgP/year}$$

The influent load needs to be reduced from 9,900 kgP/year to 4,736 kgP/year. Integrated action between wastewater and stormwater control can reach this reduction without difficulty.

## 3.5 QUALITY STANDARDS FOR WASTEWATER DISCHARGES AND WATERBODIES

### 3.5.1 Introduction

This section presents a discussion on the establishment of quality standards. In the perspective of this book, these standards are an important topic in the prevention and control of the impacts of the discharges of wastewater, which are the main issue of this chapter. This section, based on von Sperling and Fattal (2001) and von Sperling & Chernicharo (2002), analyses the practical implementation of standards, with a special focus on developing countries.

The impact of the discharge of urban wastewater into rivers, lakes, estuaries and the sea is a matter of great concern in most countries. An important point in this scenario is the establishment of an adequate legislation for the protection of the quality of water resources, this being a crucial point in the environmental and public health development of all countries. Most *developed nations* have already surpassed the basic stages of water pollution problems, and are currently fine-tuning the control of micro-pollutants, the impacts of pollutants in sensitive areas or the pollution caused by drainage of stormwater. However, *developing nations* are under constant pressure, from one side observing or attempting to follow the inter-national trends of frequently lowering the limit concentrations of the standards, and from the other side being unable to reverse the continuous trend of environmental degradation. The increase in the sanitary infrastructure can barely cope with the net population growth in many countries. The implementation of sanitation and sewage treatment depends largely on political will and, even when this is present, financial constraints are the final barrier to undermine the necessary steps towards environmental restoration and public health maintenance. Time passes, and the distance between desirable and achievable, between laws and reality, continues to enlarge.

Figure 3.20 presents a comparison between the current status of developed and developing countries in terms of actual effluent concentrations of a particular pollutant and its associated discharge standard. In most developed countries, compliance occurs for most of the time, and the main concern relates to occasional episodes of non-compliance, at which most of the current effort is concentrated. However, in most developing nations the concentrations of pollutants discharged into the water bodies are still very high, and efforts are directed towards reducing the distance to the discharge standards, and eventually achieving compliance.

An adequate legislation for the protection of public health and the quality of water resources is an essential tool in the environmental development of all countries. The transfer of written codes from paper into really practicable standards, which are used not merely for enforcement, but mainly as an integral part of the public health and environmental protection policy, has been a challenge for most countries.

Besides the water quality *requirements* (see Section 1.3) that represent in a generalised and conceptual way the desired quality for a water, there is the need to establish quality standards, supported by a legal framework. *Standards* must be

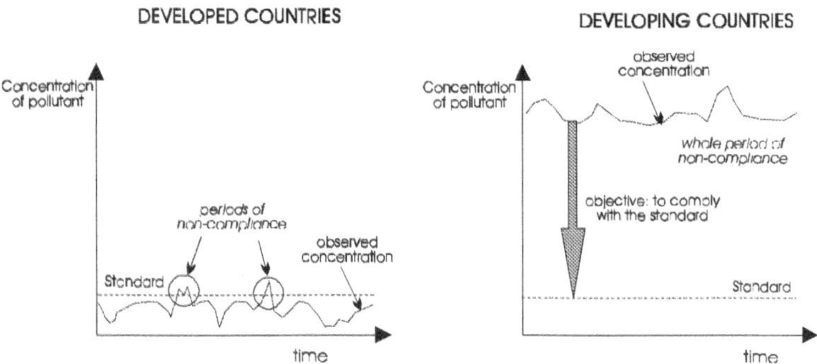


Figure 3.20.   Comparison between developed and developing countries in terms of compliance to discharge standards. Source: von Sperling & Fattal (2001), von Sperling & Chernicharo (2002).

complied with, through enforcement of the legislation, by the entities involved the discharges and use of the water.

The following concepts are also important. *National standards* are defined by each country, have legal status and are based on the specific conditions of the country itself. Depending on the political structure of the country, *regional standards* may also be developed, for each state or other form of political division. Usually, regional standards are at least equal to national standards and are often more stringent or complete. *Guidelines* or recommendations are proposed by entities of wide acceptance (e.g. World Health Organization (WHO)), are generic by nature and usually aim at the protection of public health and environment in worldwide terms.

Economic, social and cultural aspects, prevailing diseases, acceptable risks and technological development are all particular to each country or region, and are better taken into account by the country or region itself, when converting guidelines into national/regional standards. This adaptation is crucial, and adequate consideration of the guidelines prior to the adoption of standards may be an invaluable tool in the health and environmental development of a country, whereas inadequate consideration may lead to discredit, frustration, unnecessary monetary expenditure, unsustainable systems and other problems. The setting of standards should be based on sound, logical, scientific grounds and should be aimed at achieving a measured or estimated benefit or minimising a given risk for a known cost (Johnstone and Horan, 1994).

In practical terms, there are the following types of standards or guidelines of direct interest to the topics of this book:

- *Discharge* (emission, effluent) standards
- Quality standards for the *receiving water body*
- Standards or quality guidelines for a certain *use of the treated effluent* (e.g.: irrigation)

### 3.5.2  Considerations about the development of discharge standards in developing countries

#### 3.5.2.1  *Typical problems with setting up and implementing standards in developing countries*

Table 3.16 presents a list of common problems associated with setting up and implementing standards, especially in developing countries.

Table 3.16. Common problems associated with setting up and implementing standards, especially in developing countries

| Problem | How it should be | How it frequently is |
|---|---|---|
| Guidelines are directly taken as national standards | Guidelines are general worldwide values. Each country should adapt the guidelines, based on local conditions, and derive the corresponding national standards. | In many cases the adaptation is not done in developing countries, and the worldwide guidelines are directly taken as national standards, without recognising the country's singularities. |
| Guideline values are treated as absolute values, and not as target values | Guideline values should be treated as target values, to be attained on a short, medium or local term, depending on the country's technological, institutional or financial conditions. | Guideline values are treated as absolute rigid values, leading to simple "pass" or "fail" interpretations, without recognising the current difficulty of many countries to comply with them. |
| Protection measures that do not lead to immediate compliance with the standards do not obtain licensing or financing | Environmental agencies should license and banks should fund control measures (e.g. wastewater treatment plants) which allow for a stepwise improvement of water quality, even though standards are not immediately achieved. | The environmental agencies or financial institutions do not support control measures which, based on their design, do not prove to lead to compliance with the standards. Without licensing or financing, intermediate measures are not implemented. The ideal solution, even though approved, is also not implemented, because of lack of funds. As a result, no control measures are implemented. |
| Standards are frequently copied from developed countries | National standards should be based on the country's specific economical, institutional, technological and climatic conditions. | National standards are frequently directly copied from developed countries' standards, either because of lack of confidence on their own capacity, desirability to achieve developed countries' status, lack of knowledge or poor knowledge transfer from international consulting companies. Cost implications are not taken into account. The standards become purely theoretical and are not implemented or enforced. |

Table 3.16 (*Continued*)

| Problem | How it should be | How it frequently is |
| --- | --- | --- |
| Developed countries sometimes attempt to reach developed countries' status too quickly | If the guidelines and even the standards are treated as target values, time would be necessary to lead to compliance. Each country, based on the economic and technological capacity, should take the time that is reasonably necessary to achieve compliance. Developing countries are naturally likely to take more time than developed countries. Developing countries should understand that current standards in developed countries result from a long period of investment in infrastructure, during which standards progressively improved. | The desire to achieve developed countries' status too quickly can lead to the use of inappropriate technology, thus creating unsustainable systems. |
| Some standards are excessively stringent or excessively relaxed | Standards should reflect water quality criteria and objectives, based on the intended water uses. | In most cases, standards are excessively stringent, more than would be necessary to guarantee the safe use of water. In this case, they are frequently not achieved. Designers may also want to use additional safety factors in the design, thus increasing the costs. In other cases, standards are too relaxed, and do not guarantee the safe intended uses of the water. |
| There is no affordable technology to lead to compliance of standards | Control technologies should be within the countries financial conditions. The use of appropriate technology should be always pursued. | Existing technologies are in many cases too expensive for developing countries. Either because the technology is inappropriate, or because there is no political will or the countries' priorities are different, control measures are not implemented. |
| Compliance with standards are at a lower level of priority compared to other basic environmental sanitation needs | Each country, based on the knowledge of its basic conditions and needs, should set priorities to be achieved. If standards are well set up, they will naturally be integrated with the environmental control measures. | Basic water supply and sanitation needs are so acute in some countries, that standards are seen as an unnecessary sophistication. |

(*Continued*)

Table 3.16 (*Continued*)

| Problem | How it should be | How it frequently is |
|---|---|---|
| Standards are not actually enforced | Standards should be enforceable and actually enforced. Standard values should be achievable and allow for enforcement, based on existing and affordable control measures. Environmental agencies should be institutionally well developed in order to enforce standards. | Standards are not enforced, leading to a discredit in their usefulness and application, and creating the culture that standards are to remain on paper only. |
| Discharge standards are not compatible with water quality standards | In terms of pollution control, the true objective is the preservation of the quality of the water bodies. Discharge standards exist only by practical (and justifiable) reasons. However, discharge standards should be compatible with water quality standards, assuming a certain dilution or assimilation capacity of the water bodies. | Even if water quality standards are well set up, based on water quality objectives, discharge standards may not be compatible with them. Some parameters in the discharge standards may be too stringent and others too relaxed. In this case, different assimilation capacities of the water bodies are implicit. The aim of protecting the water bodies is thus not guaranteed. |
| Number of parameters are frequently inadequate (too many or too few) | The list of parameters covered by the national standards should reflect the desired protection of the intended water uses, without excesses or limitations. | In some countries, the standards include an excessively large list of parameters, many of which have no actual regional importance, are very costly to monitor or are not supported by satisfactory laboratory capabilities. In other situations, standards cover only a limited list of parameters, which are not sufficient to safeguard the intended water uses. |
| Monitoring requirements are undefined or inadequate | Monitoring requirements and frequency of sampling should be defined, in order to allow proper statistical interpretation of results. The cost implications for monitoring need to be taken into account in the overall regulatory framework. | In many cases, monitoring requirements are not specified, leading to difficulty in the interpretation of the results. In other cases, monitoring requirements are excessive and thus unnecessarily costly. Still in other cases, monitoring requirements are very relaxed, not allowing interpretation of results with confidence. |

Table 3.16 (*Continued*)

| Problem | How it should be | How it frequently is |
|---|---|---|
| Required percentage of compliance is not defined | It should be clear how to interpret the monitoring results and the related compliance with the standards (e.g. mean values, maximum values, absolute values, percentiles or other criteria). | The non specification of how to treat the monitoring results may lead to different interpretations, which may result in diverging positions as to whether compliance has been achieved or not. |
| Low standard values are sometimes below laboratory detection limits | If standards are treated as target values and are well linked with the water quality objectives, they should not be limited by current laboratory detection limits. In due time, laboratory techniques will improve and be consistent with the standard values. | Standards which are below the detection limit are sometimes seen as unjustifiable, which may be true in some cases, but not in many other cases. |
| There is no institutional development which could support and regulate the implementation of standards | The efficient implementation of standards requires an adequate infrastructure and institutional capacity to license, guide and control polluting activities and enforce standards. | In many countries the health and environmental agencies are not adequately structured or sufficiently equipped, leading to a poor control of the various activities associated with the implementation of standards. |
| Reduction of health or environmental risks due to compliance with the standards is not immediately perceived by decision makers or the population. | Decision makers and the population at large should be well informed about the benefits and costs associated with the maintenance of good water quality, as specified by the standards. | Decision makers are frequently more sensitive to costs than to benefits resulting from the implementation of control measures. The population is not well informed, and does not drive politicians and decision makers in order to invest in health and environmental protection. |
| Excessive expenditure on unjustifiably high standards may lead to population disagreement with really worthwhile standards | Standards should really reflect the water quality objectives, and these objectives should result from a consensus from the various segments of the society, directly involved in the catchment area. | Representatives of the society frequently do not participate in the decision-making process. High costs, which are not seen as bringing correspondingly high benefits, may lead to discredit and disagreement when aiming at implementing standards that are really important for the involved community. |

*Source:* von Sperling & Fattal (2001), von Sperling & Chernicharo (2002).

### 3.5.2.2 *Stepwise implementation of standards*

Usually the stepwise implementation of a wastewater treatment plant is through the *physical expansion* of the size or number of units. A plant can have, for instance,

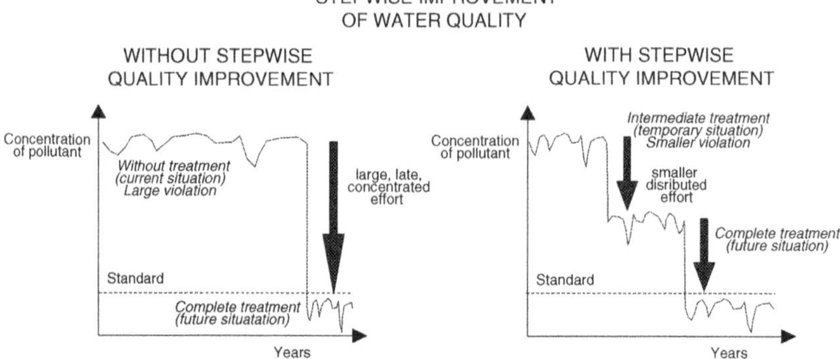


Figure 3.21.   Concept of the stepwise improvement of water quality. Source: von Sperling & Fattal (2001), von Sperling & Chernicharo (2002).

two tanks built in the first stage, and another tank built in the second stage, after it has been verified that the influent load has increased, frequently due to the population growth. This stepwise implementation is essential, in order to allow reduction in present value construction costs.

However, another concept of stepwise implementation, which should be put in practice, especially in developing countries, is the *gradual improvement of the water or wastewater quality*. It should be possible, in a large number of situations, to implement in the first stage a less efficient process, or a process that removes less pollutants, transferring to a second stage the improvement towards a system more efficient or more wide-reaching in terms of pollutants. If the planning is well structured, the environmental agency could make allowances in the sense of permitting a temporary small violation in the standards in the first stage. Naturally a great deal of care must be exercised in not allowing that a temporary situation becomes permanent, which is a very common occurrence in developing countries. This alternative of stepwise development of water or wastewater quality is undoubtedly much more desirable than a large violation of the standards, whose solution is often unpredictable over time.

Figure 3.21 presents a typical situation concerning the implementation of wastewater treatment. If a country decides to implement treatment plants that can potentially lead to an immediate compliance with the standards, this will require a large and concentrated effort, since the current water quality is probably very poor, especially in developing countries. This large effort is naturally associated with a large cost. In most instances, the country cannot afford this large cost, and the plant construction is postponed and eventually never put into effect. On the other hand, if the country decides to implement only a partial treatment, financial resources may be available. A certain improvement in the water quality is obtained and health and environmental risks are reduced, even though the standards have not been satisfied. In this case, the standards are treated as target values, to be achieved whenever possible. The environmental agency is a partner in the solution of the problem, and establishes a programme of future improvements. After some time, there are additional funds for expanding the efficiency of the treatment plant, and the standards are finally satisfied. In this case, compliance with the standards is likely to be obtained before the alternative without stepwise implementation.

Not only wastewater systems should expand on a stepwise basis on developing countries, but also the standards for water quality. There should be a knowledge about the targets that are desired to be achieved over time, and these targets could eventually be the same as the general guidelines. However, with the standards the approach should be different, and the numeric values of the limit concentrations should progress stepwisely towards stringency. The standards should be adapted periodically, eventually reaching the same values as those in the guidelines.

The advantages of a stepwise implementation of standards and sanitary infrastructure are listed in Table 3.17.

An important issue in the stepwise approach is how to guarantee that the second and subsequent stages of improvement will be implemented, and not interrupted in the first stage. Because of financial restrictions, there is always the risk that the subsequent stages will be indefinitely postponed, under the argument that the priority has now shifted to systems that have not yet implemented the first stage. Even though this might well be a justifiable argument, it cannot be converted into a commonly used excuse. The environmental agency must set up scenarios of intervention targets with the entity responsible for the sanitary system. The scenarios should include the minimum intervention, associated with the first stage, and subsequent prospective scenarios, including required measures, benefits, costs and timetable. The formalisation of the commitment also helps in ensuring the continuation of the water quality improvement.

### 3.5.2.3 *The principle of equity*

The principle of equity means that all peoples, irrespective of race, culture, religion, geographic position or economic status are entitled to the same life expectancy and quality of life. Broadly speaking, the reasons for a lower quality of life are associated with environmental conditions, and if these improve, life quality is expected to rise accordingly. On this basis, there is no justification for accepting different environmental guideline values between developed and developing countries.

If guideline values (e.g. WHO guidelines) are treated as absolute values, than only developed countries are more likely to achieve them, and developing nations possibly will not be able to afford the required investments. However, if guideline values are treated as **targets**, than hopefully all countries will eventually be able to achieve them, some on a short, some on a medium and others only on a long term.

Figure 3.22 illustrates this point, for three different countries. For all of them, the guideline values are the same. The very developed country has been already compliant, and presents a better water quality than actually required. The developed country requires only a small effort and achieves compliance in a short term. The developing country requires a stepwise approach and achieves compliance only on a long term. However, in the end all countries will hopefully be compliant with the guidelines.

Table 3.17. Advantages of a stepwise implementation of standards and sanitary infrastructure

| Advantage | Comment |
| --- | --- |
| Polluters are more likely to afford gradual investment for control measures | Polluters and/or water authorities will find it much more feasible to divide investments in different steps, than to make a large and in many cases unaffordable investment |
| The present value of construction costs is reduced | The division of construction costs into different stages leads to a lower present value than a single, large, initial cost. This aspect is more relevant in countries in which, due to inflation problems, interest rates are high. |
| The cost-benefit of the first stage is likely to be more favourable than in the subsequent stages | In the first stage, when environmental conditions are poor, usually a large benefit is achieved with a comparatively low cost. This means that already in the first stage a significant benefit is likely to be achieved, with only a fraction of the overall costs. In the subsequent stages, the increase of the benefit is not so substantial, but the associated costs are high. The cost-benefit is then less favourable. |
| There is more time and better conditions to know the wastewater characteristics | The operation of the system will involve monitoring, which, on its course, will allow a good knowledge of the wastewater characteristics. The design of the second or subsequent stages will be based on the actual characteristics, and not on generic values taken from the literature. |
| There is the opportunity to optimise operation, without necessarily making physical expansion | The experience in the operation of the system will lead to a good knowledge of its behaviour. This will allow, in some cases, the optimisation of the process (improvement of efficiency or capacity), without necessarily requiring the physical expansion of the system. The first stage will be analogous to a pilot plant. |
| There is time and opportunity to implement, in the second stage, new techniques or better developed processes | The availability of new or more efficient processes for wastewater treatment is always increasing with time. Process development is continuous and fast. The second or subsequent steps can make use of better and/or cheaper technologies, than it would be possible with a single step. |
| The country has more time to develop its own standards | As time passes, the experience in operating the system and evaluating its positive and negative implications in terms of water quality, health status and environmental conditions will lead to the establishment of standards that are really appropriate for the local conditions. |
| The country has more time and better conditions to develop a suitable regulatory framework and institutional capacity | Experience obtained in the operation of the system and in setting up the required infrastructure and institutional capacity for regulation and enforcement will also improve progressively, as the system expands on the second and subsequent stages. |

*Source:* von Sperling & Fattal (2001), von Sperling & Chernicharo (2002).

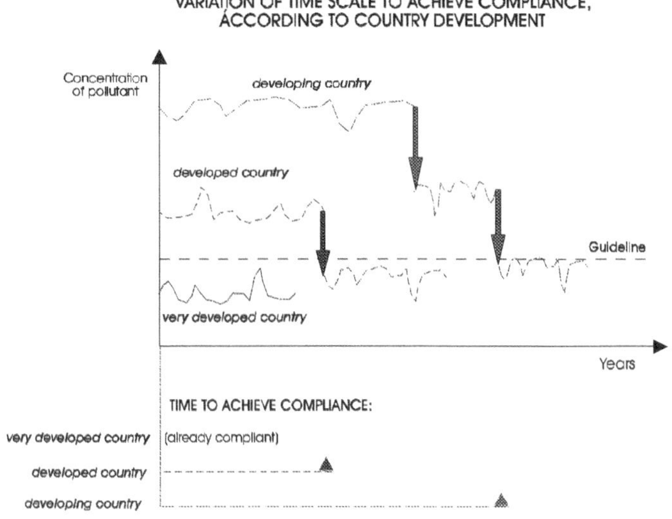


Figure 3.22.   Variation of time scale to achieve guideline compliance, for a very developed country, a developed country and a developing country (in all cases, the guideline value is the same). Source: von Sperling & Fattal (2001).

## 3.5.2.4 Institutional development

An efficient implementation of standards must go in parallel with the development of the institutional framework necessary for monitoring, controlling, regulating and enforcing the standards. This topic is well discussed by Johnstone and Horan (1996) and some of the points are summarised below.

Institutional development takes time and the models cannot be directly copied from developed countries. Even though lessons should be learned from other countries that have already passed the basic steps of institutional development, an adaptation is also required in order to accommodate the countries' specific economic, cultural and social conditions. However, experience from other countries can help in structuring the organisations, especially when they are introduced for the first time. It must be recognised that institutional development is a continuous process, building on the experience of prior organisations.

Another important point is the need to separate the duties and responsibilities of regulating quality with those of achieving standards. This is especially true when private sector operators have to comply with standards.

The main points to be emphasised for developing countries are (Johnstone and Horan, 1996): (a) consider the process of institutional development and technical improvements to be long term; (b) build on past experiences; (c) separate regulatory and operational duties and responsibilities; (d) develop regulatory systems and procedures needed to enforce standards; (e) ensure that sufficient legal powers are in force; (f) recognise the costs of regulation and legal enforcement.

Table 3.18. European Community requirements for discharges from urban wastewater treatment plants

| Parameter | Concentration | Minimum percentage of reduction [1] | Notes |
|---|---|---|---|
| $BOD_5$ [2][3] | 25 mg/L $O_2$ | 70–90 % | – |
| COD [3] | 125 mg/L $O_2$ | 75 % | – |
| Total suspended solids | 35 mg/L [4]<br>60 mg/L<br>150 mg/L | 90 %<br>70%<br>– | P.E. greater than 10,000 inhab.<br>P.E. between 2,000 and 10,000 inhab.<br>For ponds effluents |
| Total nitrogen [5] [6] | 10 mg/L (⁻)<br><br>15 mg/L | 70–80 | P.E. greater than 100,000 inhab.<br><br>P.E. between 10,000 and 100,000 inhab. |
| Total phosphorus [5] | 1 mg/L<br><br>2 mg/L | 80 | P.E. greater than 100,000 inhab.<br><br>P.E. between 10,000 and 100,000 inhab. |

*Source:* Official Journal of the European Communities No. L 135/40 (Council of the European Communities, 1991).
P.E. = Population Equivalent (inhabitants)
BOD, COD and SS: the maximum number of samples that are allowed to fail the requirements is specified in the Directive
(1) Removal in relation to the load of the influent.
(2) The parameter can be replaced by another parameter: Total Organic Carbon (TOC) or Total Oxygen Demand (TOD) if a relationship can be established between $BOD_5$ and the substitute parameter.
(3) Analyses concerning discharges from ponds (lagoons) shall be carried out in filtered samples. However, the concentration of total suspended solids in unfiltered samples shall not exceed 150 mg/L.
(4) This requirement is optional.
(5) Total N and Total P: requirement for discharge in sensitive water bodies only (one or both parameters may be applied, depending on the local situation). Values are annual means.
(6) Alternatively, the daily average of Total Nitrogen must not exceed 20 mg/L (for water temperature of 12 °C or more).

## 3.5.3 Examples of standards and guidelines

### 3.5.3.1 Introduction

As commented, discharge standards vary from country to country and, in many cases, from state to state, reflecting their specificities, development stage, economical level, commitment of environmental protection and various other factors.

In any case, much more important than the discharge standards are the *quality standards for the water body*, because the quality in the water body is the one really associated with its uses. *Discharge standards* exist because of a practical aspect: it is easier for the environmental agencies to control, monitor and enforce point discharges, whose responsible agent is known. In a water body receiving multiple discharges, the occasional detection of non-compliance of the standards in the water body is not a trivial matter in terms of assigning those responsible. Therefore, discharge standards play an important role in most countries in the world.

Table 3.19. WHO recommended microbiological quality guidelines for treated wastewater reuse in agricultural irrigation

| Category | Reuse conditions | Exposed group | Intestinal nematodes (eggs/L)[b] (arithmetic mean) | Faecal coliforms (FC/100 mL)[c] (geometric mean) |
|---|---|---|---|---|
| A | Irrigation of crops likely to be eaten uncooked, sports fields, public parks[d] | Workers, consumers, public | $\leq 1$ | $\leq 1000$[d] |
| B | Irrigation of cereal crops, industrial crops, fodder crops, pasture and trees | Workers | $\leq 1$ | No standard recom-mended |
| C | Localised irrigation of crops in category B if exposure to workers and the public does not occur | None | Not applicable | Not applicable |

*Source:* WHO (1989).

(a)  In specific cases, local epidemiological, sociocultural and environmental factors should be taken into account and the guidelines modified accordingly.
(b)  *Ascaris* and *Trichuris* species and hookworms.
(c)  During the irrigation period.
(d)  A more stringent guideline limit ($\leq 200$ faecal coliforms/100 mL) is appropriate for public lawns, such as hotel lawns, with which the public may come into direct contact.
(e)  In the case of fruit trees, irrigation should cease two weeks before fruit is picked, and no fruit should be picked off the ground. Sprinkler irrigation should not be used.

This section presents some important standards and guidelines and suggestions to be applied for domestic sewage:

- European Community Directive concerning urban wastewater treatment
- WHO guidelines for treated-wastewater reuse in agricultural irrigation
- Possible discharge standards to be applied for domestic sewage

Various other important standards and guidelines exist, but their compilation is beyond the scope of this book.

## 3.5.3.2 *European Community Directive for urban wastewater treatment*

This section summarises the main requirements for urban (domestic + non-domestic) wastewater treatment plants in Europe – Council Directive 91/271/EEC, 21/05/1991 (Council of the European Communities, 1991). This directive speci-fies the minimum removal efficiencies and limit concentrations of $BOD_5$, COD, SS, N and P. Note that the values for N and P apply only when the discharge is to sensitive water bodies. The criteria for classifying a sensitive water body are presented in the legislation, but these are typically lakes, reservoirs, estuaries, bays and coastal waters, subject to certain conditions. The values presented in Table 3.18 for concentration or removal efficiency must apply.

Table 3.20. Possible discharge standards, according to different levels of stringency, for the main pollutants in domestic sewage

| Parameter | Discharge to | Discharge standard (mg/L) | | |
|---|---|---|---|---|
| | | Less stringent | Stringent | Very stringent |
| BOD | Any water body | 60 | 20–30 | 10 |
| COD | Any water body | 200 | 100–150 | 50 |
| SS | Any water body | 60 | 20–30 | 10 |
| Total N | Sensitive water body | – | 10–15 | 10 |
| Total P | Sensitive water body | – | 1–2 | 1 |

The directive also specifies (a) the minimum number of annual samples and (b) the maximum permitted number of samples that could fail to conform (for BOD, COD and SS).

### 3.5.3.3 *WHO guidelines for the reuse of treated wastewater in agricultural irrigation*

The WHO (1989) established microbiological quality guidelines for treated wastewater reuse in agricultural irrigation (Table 3.19). In these guidelines, two types or indicator organisms are applied: faecal coliforms and helminth (nematode) eggs, depending on the type of irrigation and on the group exposed. The guidelines also present suggestions for treatment processes to be applied and other relevant information.

### 3.5.3.4 *Possible discharge standards for domestic sewage*

Table 3.20 presents a simplified synthesis of possible discharge standards, according to different restriction levels, for the main pollutants of interest in domestic sewage. Depending on each country, region or situation, less stringent, stringent or very stringent standards may be adopted (or a combination of them, depending on the relative degree of importance of each parameter.

# 4

# Overview of wastewater treatment systems

## 4.1 WASTEWATER TREATMENT LEVELS

In planning studies for the implementation of the wastewater treatment, the following points must be clearly addressed:

* *environmental impact studies on the receiving body*
* *treatment objectives*
* *treatment level and removal efficiencies*

The environmental impact studies that are necessary for the evaluation of the compliance with the receiving body standards were detailed in Chapter 3. The requirements to be reached for the effluent are also a function of the specific legislation that defines the quality standards for the effluent and for the receiving body. The legislation was also covered in Chapter 3.

The removal of pollutants during treatment in order to reach a desired quality or required discharge standard is associated with the concepts of *treatment level* and *treatment efficiency*.

Wastewater treatment is usually classified according to the following levels (see Tables 4.1 and 4.2):

* Preliminary
* Primary

Table 4.1. Wastewater treatment levels

| Level | | Removal |
|---|---|---|
| Preliminary | • | Coarse suspended solids (larger material and sand) |
| Primary | • | Settleable suspended solids |
| | • | Particulate (suspended) BOD (associated to the organic matter component of the settleable suspended solids) |
| Secondary | • | Particulate (suspended) BOD (associated to the particulate organic matter present in the raw sewage, or to the non settleable particulate organic matter, not removed in the possibly existing primary treatment) |
| | • | Soluble BOD (associated to the organic matter in the form of dissolved solids) |
| Tertiary | • | Nutrients |
| | • | Pathogenic organisms |
| | • | Non-biodegradable compounds |
| | • | Metals |
| | • | Inorganic dissolved solids |
| | • | Remaining suspended solids |

*Note:* depending on the treatment process adopted, the removal of nutrients (by biological processes) and pathogens can be considered an integral part of secondary treatment.

- Secondary
- Tertiary

The objective of **preliminary treatment** is only the removal of coarse solids, while **primary treatment** aims at removing settleable solids and part of the organic matter. *Physical* pollutant removal mechanisms are predominant in both levels. In **secondary treatment** the aim is the removal of organic matter and possibly nutrients (nitrogen and phosphorus) by predominantly *biological* mechanisms. The objective of **tertiary treatment** is the removal of specific pollutants (usually toxic or non-biodegradable compounds) or the complementary removal of pollutants that were not sufficiently removed in the secondary treatment. Tertiary treatment is rare in developing countries.

The *removal efficiency* of a pollutant in the treatment or in a treatment stage is given by the formula:

$$E = \frac{C_o - C_e}{C_o}.100 \qquad (4.1)$$

where:

$E$ = removal efficiency (%)
$C_o$ = influent concentration of the pollutant (mg/L)
$C_e$ = effluent concentration of the pollutant (mg/L)

Table 4.2. Characteristics of the main wastewater treatment levels

| Item | Treatment level[1] | | |
|---|---|---|---|
|  | Preliminary | Primary | Secondary |
| Pollutants removed | • Coarse solids | • Settleable solids<br>• Particulate BOD | • Non-settleable solids<br>• Fine particulate BOD<br>• Soluble BOD<br>• Nutrients[4]<br>• Pathogens[4] |
| Removal efficiencies | – | • SS: 60–70%<br>• BOD: 25–40%<br>• Coliforms: 30–40% | • SS: 65–95%<br>• BOD: 60–99%<br>• Coliforms: 60–99%[3] |
| Predominant treatment mechanism | • Physical | • Physical | • Biological |
| Complies with usual discharge standards?[2] | • No | • No | • Usually yes |
| Application | • Upstream of pumping stations<br>• Initial treatment stage | • Partial treatment<br>• Intermediate stage of a more complete treatment | • More complete treatment (for organic matter) |

*Notes:*

(1)   A secondary level WWTP usually has preliminary treatment, but may or may not have primary treatment (depends on the process).

(2)   Discharge standard as stated in the legislation. The environmental agency may authorise other values, if environmental studies demonstrate that the receiving body is able to assimilate a higher loading.

(3)   The coliform removal efficiency can be higher if a specific removal stage is included.

(4)   Depending on the treatment process, nutrients and pathogens may be removed in the secondary stage.


## 4.2 WASTEWATER TREATMENT OPERATIONS, PROCESSES AND SYSTEMS

The treatment methods are composed by unit operations and processes, and their integration makes up the treatment systems.

The concepts of unit operations and unit process are frequently used interchangeably, because they can occur simultaneously in the same treatment unit. In general, the following definitions can be adopted (Metcalf & Eddy, 1991):

- **Physical unit operations:** treatment methods in which *physical forces* are predominant (e.g. screening, mixing, flocculation, sedimentation, flotation, filtration)

- **Chemical unit processes:** treatment methods in which the removal or the conversion of the contaminants occurs by the addition of *chemical products* or due to *chemical reactions* (e.g. precipitation, adsorption, disinfection).
- **Biological unit processes:** treatment methods in which the removal of the contaminants occurs by means of *biological activity* (e.g. carbonaceous organic matter removal, nitrification, denitrification)

Various mechanisms can act separately or simultaneously in the removal of the pollutants, depending on the process being used. The main mechanisms are listed in Table 4.3.

Table 4.4 lists the main processes, operations, and treatment systems frequently used in the treatment of *domestic sewage*, as a function of the pollutant to be removed. These methods are employed in the **liquid phase** (or liquid lines), which corresponds to the main flow of the liquid (sewage) in sewage treatment works. On the other hand, the **solid phase** (covered in Section 5) is associated with the solid by-products generated in the treatment, notably sludge. The present text concentrates on the *biological treatment* of wastewater, which is the reason why physical – chemical treatment systems are not covered (these depend on the addition of chemical products and are used more frequently for the treatment of industrial wastewaters).

Table 4.5 presents a summary of the main secondary level domestic sewage treatment systems. The technology of wastewater treatment has various other processes and variants, but the present book addresses only the most frequently used systems in warm-climate countries. The flowsheets of the systems described in this table are presented in Figures 4.1a–f. The integration between the various operations and processes listed in Table 4.5 can be seen in the flowsheets. In all flowsheets, besides going to the receiving water body, the effluent may be reused (agricultural / industrial / other) if conditions so permit.

In order to allow a better understanding of the main wastewater treatment systems, the remainder of the chapter is devoted to a preliminary description of them. Further details may be found in various chapters throughout this book.

Table 4.3. Main mechanisms for the removal of pollutants in wastewater treatment

| Pollutant | Subdivision | | Main removal mechanisms |
|---|---|---|---|
| Solids | Coarse solids (> ~1 cm) | *Screening* | Retention of the solids with dimensions greater than the spacing between the bars |
| | Suspended solids (> ~1 μm) | *Sedimentation* | Separation of the particles with a density greater than the sewage |
| | Dissolved solids (< ~1 μm) | *Adsorption* | Retention on the surface of biomass flocs or biofilms |
| Organic matter | BOD in suspension (particulate BOD) (> ~1 μm) | *Sedimentation* | Separation of the particles with a density greater than the sewage |
| | | *Adsorption* | Retention on the surface of biomass flocs or biofilms |
| | | *Hydrolysis* | Conversion of the BOD in suspension into soluble BOD by means of enzymes, allowing its stabilisation |
| | | *Stabilisation* | Utilisation by biomass as food, with conversion into gases, water and other inert compounds. |
| | Soluble BOD (< ~1 μm) | *Adsorption* | Retention on the surface of biomass flocs or biofilms |
| | | *Stabilisation* | Utilisation by biomass as food, with conversion into gases, water and other inert compounds. |
| Pathogens | Larger dimensions and/or with protective layer (protozoan cysts and helminth eggs) | *Sedimentation* | Separation of pathogens with larger dimensions and density greater than the sewage |
| | | *Filtration* | Retention of pathogens in a filter medium with adequate pore size |
| | Lower dimensions (bacteria and viruses) | *Adverse environmental conditions* | Temperature, pH, lack of food, competition with other species, predation |
| | | *Ultraviolet radiation* | Radiation from the sun or artificial |
| | | *Disinfection* | Addition of a disinfecting agent, such as chlorine |

*(Continued)*

Table 4.3  (*Continued*)

| Pollutant | Subdivision | | Main removal mechanisms |
|---|---|---|---|
| Nitrogen | Organic nitrogen | *Ammonification* | Conversion of organic nitrogen into ammonia |
| | Ammonia | *Nitrification* | Conversion of ammonia into nitrite, and the nitrite into nitrate, by means of nitrifying bacteria |
| | | *Bacterial assimilation* | Incorporation of ammonia into the composition of bacterial cells |
| | | *Stripping* | Release of free ammonia ($NH_3$) into the atmosphere, under high pH conditions |
| | | *Break-point chlorination* | Conversion of ammonia into chloramines, through the addition of chlorine |
| | Nitrate | *Denitrification* | Conversion of nitrate into molecular nitrogen ($N_2$), which escapes into the atmosphere, under anoxic conditions |
| Phosphorus | Phosphate | *Bacterial assimilation* | Assimilation in excess of the phosphate from the liquid by phosphate accumulating organisms, which takes place when aerobic and anaerobic conditions are alternated |
| | | *Precipitation* | Phosphorus precipitation under conditions of high pH, or through the addition of metallic salts |
| | | *Filtration* | Retention of phosphorus-rich biomass, after stage of biological excessive P assimilation |

Table 4.4.  Treatment operations, processes and systems frequently used for the removal of pollutants from domestic sewage

| Pollutant | Operation, process or treatment system |
|---|---|
| Suspended solids | • Screening<br>• Grit removal<br>• Sedimentation<br>• Land disposal |
| Biodegradable organic matter | • Stabilisation ponds and variants<br>• Land disposal<br>• Anaerobic reactors<br>• Activated sludge and variants<br>• Aerobic biofilm reactors |
| Pathogenic organisms | • Maturation ponds<br>• Land disposal<br>• Disinfection with chemical products<br>• Disinfection with ultraviolet radiation<br>• Membranes |

Table 4.4 (*Continued*)

| Pollutant | Operation, process or treatment system |
| --- | --- |
| Nitrogen | • Nitrification and biological<br>• denitrification<br>• Maturation and high-rate ponds<br>• Land disposal<br>• Physical–chemical processes |
| Phosphorus | • Biological removal<br>• Maturation and high-rate ponds<br>• Physical chemical processes |

Table 4.5. Summary description of the main biological wastewater treatment systems

| STABILISATION PONDS | |
| --- | --- |
| *Facultative pond* | Wastewater flows continuously through a pond especially constructed for wastewater treatment. The wastewater remains in the ponds for many days. The soluble and fine particulate BOD is aerobically stabilised by bacteria which grow dispersed in the liquid medium, while the BOD in suspension tends to settle, being converted anaerobically by bacteria at the bottom of the pond. The oxygen required by the aerobic bacteria is supplied by algae through photosynthesis. The land requirements are high. |
| *Anaerobic pond – facultative pond* | Around 50 to 65% of the BOD is converted in the anaerobic pond (deeper and with a smaller volume), while the remaining BOD is removed in the facultative pond. The system occupies an area smaller than that of a single facultative pond. |
| *Facultative aerated lagoon* | The BOD removal mechanisms are similar to those of a facultative pond. However, oxygen is supplied by mechanical aerators instead of through photosynthesis. The aeration is not enough to keep the solids in suspension, and a large part of the sewage solids and biomass settles, being decomposed anaerobically at the bottom. |
| *Completely mixed aerated lagoon – sedimentation pond* | The energy introduced per unit volume of the pond is high, what makes the solids (principally the biomass) remain dispersed in the liquid medium, in complete mixing. The resulting higher biomass concentration in the liquid medium increases the BOD removal efficiency, which allows this pond to have a volume smaller than a facultative aerated lagoon. However, the effluent contains high levels of solids (bacteria) that need to be removed before being discharged into the receiving body. The sedimentation pond downstream provides conditions for this removal. The sludge of the sedimentation pond must be removed every few years. |
| *High rate ponds* | High rate ponds are conceived in order to maximise algal production, in a totally aerobic environment. To accomplish this, lower depths are employed, allowing light penetration throughout the liquid mass. Therefore, photosynthetic activity is high, leading to high dissolved oxygen concentrations and pH levels. These factors contribute to the increase of the pathogens die-off and to the removal of nutrients. High rate ponds usually receive a high organic load per unit surface area. Usually a moderate agitation in the liquid is introduced, caused by a low-power mechanical equipment. |

*(Continued)*

Table 4.5 (*Continued*)

| | |
|---|---|
| *Maturation ponds* | The main objective of maturation ponds is the removal of pathogenic organisms. In maturation ponds prevail environmental conditions which are adverse to these organisms, such as ultraviolet radiation, high pH, high DO, lower temperature (compared with the human intestinal tract), lack of nutrients and predation by other organisms. Maturation ponds are a post-treatment stage for BOD-removal processes, being usually designed as a series of ponds or a single-baffled pond. The coliform removal efficiency is very high. |
| **LAND DISPOSAL** | |
| *Slow rate system* | The objectives may be for (a) wastewater treatment or (b) water reuse through crop production or landscape irrigation. In each case, design criteria are different. Wastewater is applied to the soil, supplying water and nutrients necessary for plant growth. Part of the liquid evaporates, part percolates into the soil, and the largest fraction is absorbed by the plants The surface application rates are very low. The liquid can be applied by sprinkling, graded-border, furrow and drip irrigation. |
| *Rapid infiltration* | Wastewater is applied in shallow basins. The liquid passes through the porous bottom and percolates into the soil. The evaporation loss is lower in view of the higher application rates. Vegetation may or may not be used. The application is intermittent, which provides a rest period for the soil. The most common types are: application for groundwater recharge, recovery using underdrains and recovery using wells. |
| *Subsurface infiltration* | Pre-settled sewage (usually from septic tanks) is applied below the soil surface. The infiltration trenches or chambers are filled with a porous medium, which provides transportation, storage and partial treatment, followed by the infiltration itself. |
| *Overland flow* | Wastewater is distributed in the upper part of vegetated slopes, flows over the slopes and is collected by ditches at the lower part. Treatment occurs in the root-soil system. The application is intermittent. Distribution of wastewater may be by high-pressure sprinklers, low-pressure sprays and gated or perforated pipes or channels. |
| *Constructed wetlands* | While the former systems are land-based systems, these are aquatic-based systems. The systems are composed by shallow basins or channels in which aquatic plants grow. The system can be of free-water surface (water level above ground level) or subsurface flow (water level below ground level). Biological, chemical and physical mechanisms act on the root–soil system. |
| **ANAEROBIC SYSTEMS** | |
| *Upflow anaerobic sludge blanket reactor (UASB)* | BOD is converted anaerobically by bacteria dispersed in the reactor. The liquid flow is upwards. The upper part of the reactor is divided into settling and gas collection zones. The settling zone allows the exit of the clarified effluent in the upper part and the return of the solids (biomass) by gravity to the system, increasing its concentration in the reactor. Amongst the gases formed is methane. The system has no primary sedimentation tank. The sludge production is low, and the excess sludge wasted is already thickened and stabilised. |

Table 4.5  (*Continued*)

| | |
|---|---|
| *Anaerobic filter* | BOD is converted anaerobically by bacteria that grow attached to a support medium (usually stones) in the reactor. The tank works submerged and the flow is upwards. The system requires a primary sedimentation tank (frequently septic tanks). The sludge production is low and the excess sludge is already stabilised. |
| *Anaerobic reactor – post-treatment* | UASB reactors produce an effluent that has difficulty in complying with most existing discharge standards. Therefore, some form of post treatment is frequently necessary. The post treatment may be biological (aerobic or anaerobic) or physical-chemical (with the addition of coagulants). Practically all wastewater treatment processes may be used as a post treatment of the anaerobic reactors. The global efficiency of the system is usually similar to the one that would be obtained if the process were being applied for raw wastewater. However, land, volume and energy requirements are lower. Sludge production is also lower. |

ACTIVATED SLUDGE

| | |
|---|---|
| *Conventional activated sludge* | The biological stage comprises two units: aeration tank (reactor) and secondary sedimentation tank. The biomass concentration in the reactor is very high, due to the recirculation of the settled solids (bacteria) from the bottom of the secondary sedimentation tank. The biomass remains in the system longer than the liquid, which guarantees a high BOD removal efficiency. It is necessary to remove a quantity of the sludge (biomass) that is equivalent to what is produced. This excess sludge removed needs to be stabilised in the sludge treatment stage. The oxygen supply is done by mechanical aerators or by diffused air. Upstream of the reactor there is a primary sedimentation tank to remove the settleable solids from the raw sewage. |
| *Activated sludge (extended aeration)* | Similar to the previous system, but with the difference that the biomass stays longer in the system (the aeration tanks are bigger). With this, there is less substrate (BOD) available for the bacteria, which makes them use their own cellular material as organic matter for their maintenance. Consequently, the removed excess sludge (bacteria) is already stabilised. Primary sedimentation tanks are usually not included. |
| *Intermittently operated activated sludge (sequencing batch reactors)* | The operation of the system is intermittent. In this way, the reaction (aerators on) and settling (aerators off) stages occur in different phases in the same tank. When the aerators are turned off, the solids settle, which allows the removal of the clarified effluent (supernatant). When the aerators are turned on again, the settled solids return to the liquid mass, with no need of sludge recirculation pumps. There are no secondary sedimentation tanks. It can be in the conventional or extended aeration modes. |
| *Activated sludge with biological nitrogen removal* | The biological reactor incorporates an anoxic zone (absence of oxygen, but presence of nitrates). The anoxic zone can be upstream and/or downstream of the aerated zone. The nitrates formed in the nitrification process that takes place in the aerobic zone are used in the respiration of facultative microorganisms in the anoxic zones, being reduced to gaseous molecular nitrogen, which escapes to the atmosphere. |

(Continued)

Table 4.5 (*Continued*)

| | |
|---|---|
| *Activated sludge with biological nitrogen and phosphorus removal* | Besides the aerobic and anoxic zones, the biological reactor also incorporates an anaerobic zone, situated at the upper end of the tank. Internal recirculations make the biomass to be successively exposed to anaerobic and aerobic conditions. With this alternation, a certain group of microorganisms absorbs phosphorus from the liquid medium, in quantities that are much higher than those which would be normally necessary for their metabolism. The withdrawal of these organisms in the excess sludge results in the removal of phosphorus from the biological reactor. |
| **AEROBIC BIOFILM REACTORS** | |
| *Low rate trickling filter* | BOD is stabilised aerobically by bacteria that grow attached to a support medium (commonly stones or plastic material). The sewage is applied on the surface of the tank through rotating distributors. The liquid percolates through the tank and leaves from the bottom, while the organic matter is retained and then further removed by the bacteria. The free spaces permit the circulation of air. In the low rate system there is a low availability of substrate (BOD) for the bacteria, which makes them undergo self-digestion and leave the system stabilised. Sludge that is detached from the support medium is removed in the secondary sedimentation tank. The system requires primary sedimentation. |
| *High rate trickling filter* | Similar to the previous system but with the difference that a higher BOD load is applied. The bacteria (excess sludge) need to be stabilised within the sludge treatment. The effluent from the secondary sedimentation tank is recirculated to the filter in order to dilute the influent and to guarantee a homogeneous hydraulic load. |
| *Submerged aerated biofilter* | The submerged aerated biofilter is composed by a tank filled with a porous material (usually submerged), through which sewage and air flow permanently. The air flow is always upwards, while the liquid flow can be downward or upward. The biofilters with granular material undertake, in the same reactor, the removal of soluble organic compounds and particulate matter. Besides being a support medium for biomass growth, the granular material acts also as a filter medium. Periodic backwashings are necessary to eliminate the excess biomass accumulated, reducing the head loss through the medium. |
| *Rotating biological contactor (biodisc)* | The biomass grows adhered to a support medium, which is usually composed by a series of discs. The discs, partially immersed in the liquid, rotate, exposing their surface alternately to liquid and air. |

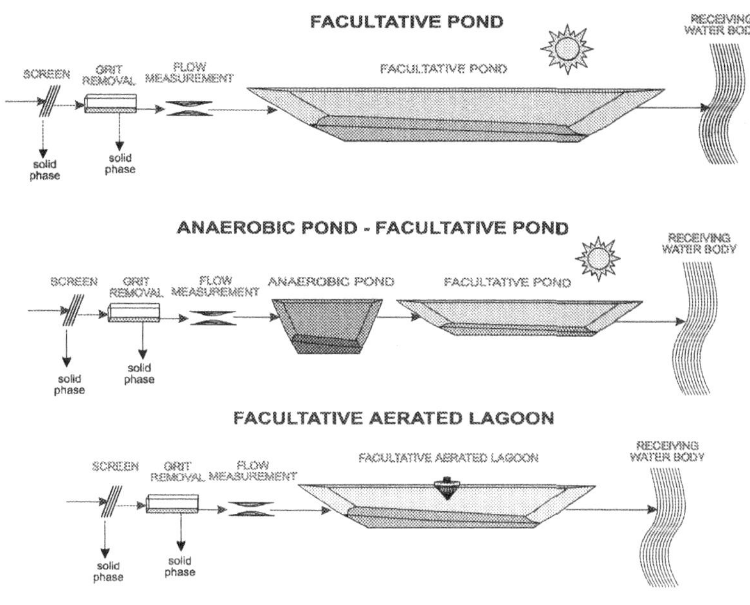


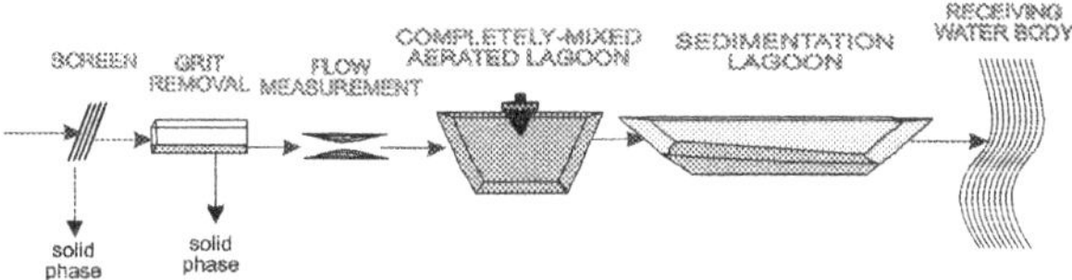


Figure 4.1a.   Flowsheet of stabilisation pond systems (liquid phase only).

## LAND DISPOSAL SYSTEMS

### LOW-RATE INFILTRATION

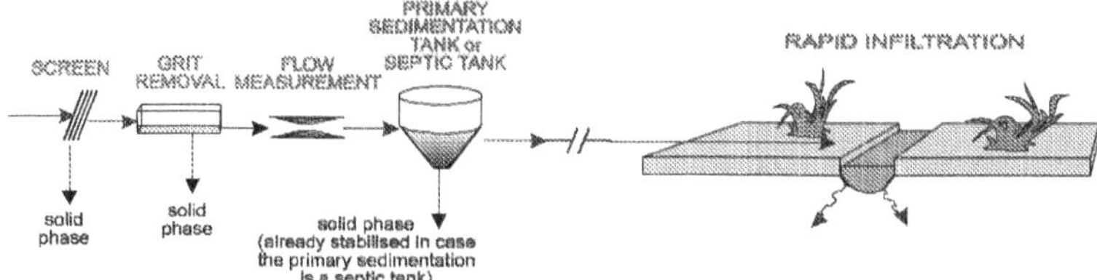


### RAPID INFILTRATION

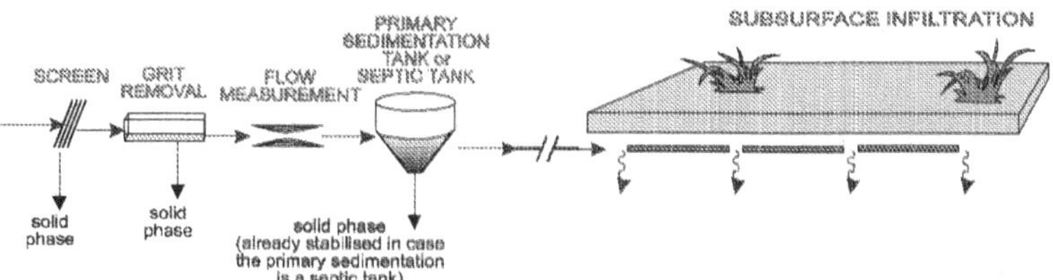


### SUBSURFACE INFILTRATION

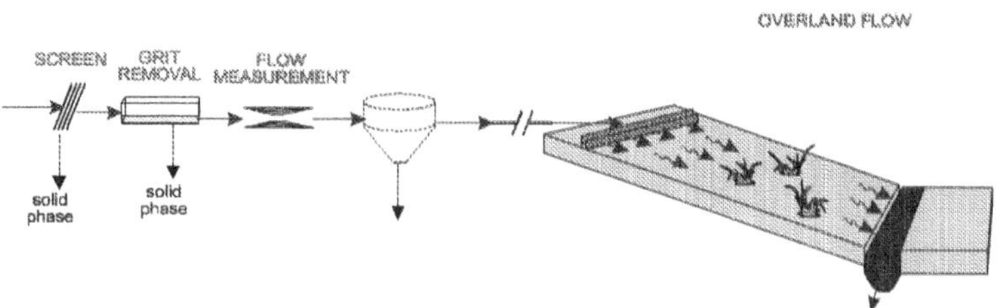


### OVERLAND FLOW



Figure 4.1b. Flowsheet of soil-based land treatment systems (liquid phase only).

# CONSTRUCTED WETLANDS

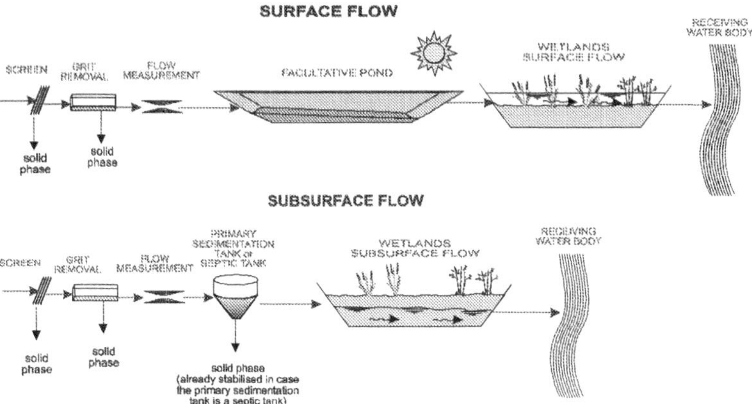



Figure 4.1c.    Flowsheet of aquatic-based land treatment systems (liquid phase only).

# ANAEROBIC SYSTEMS

## UPFLOW ANAEROBIC SLUDGE BLANKET REACTOR

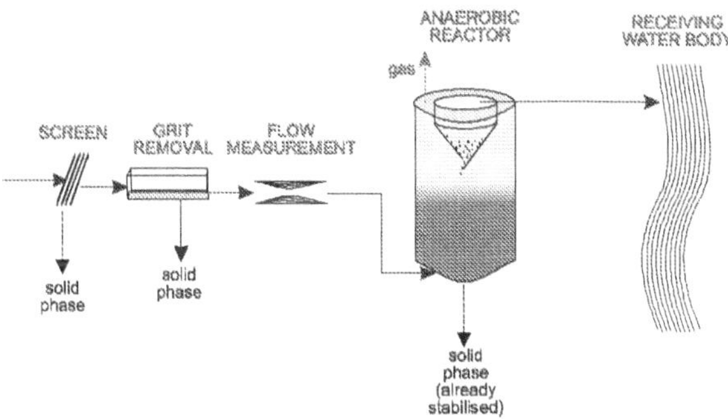



## SEPTIC TANK - ANAEROBIC FILTER

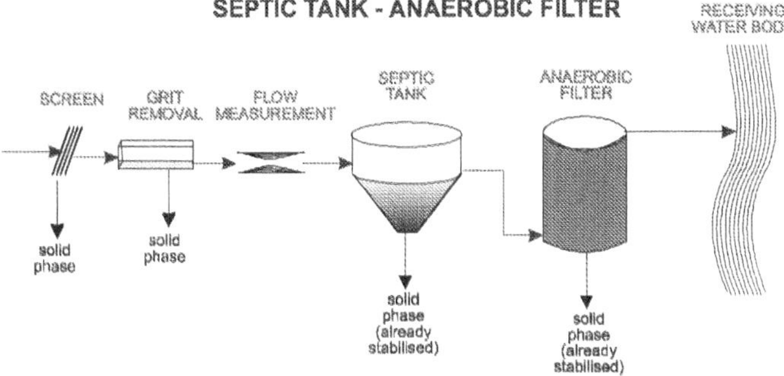



Figure 4.1d.    Flowsheet of anaerobic reactors (liquid phase only).

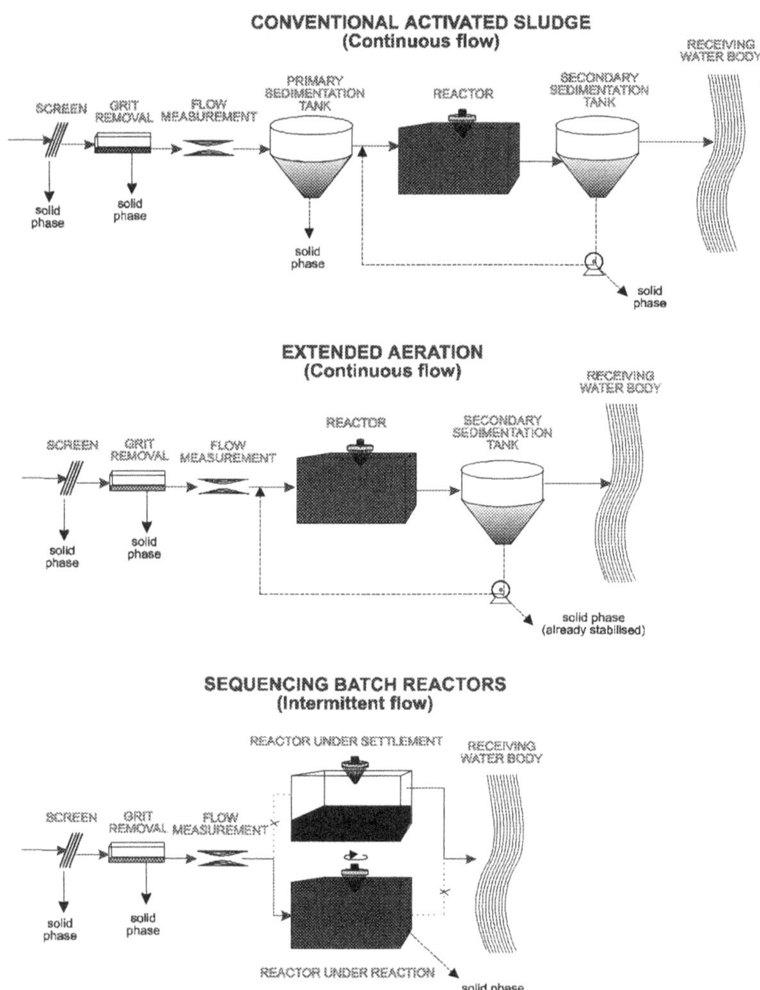


Figure 4.1e.   Flowsheet of activated sludge systems (liquid phase only).

## AEROBIC BIOFILM REACTORS

### LOW RATE TRICKLING FILTER

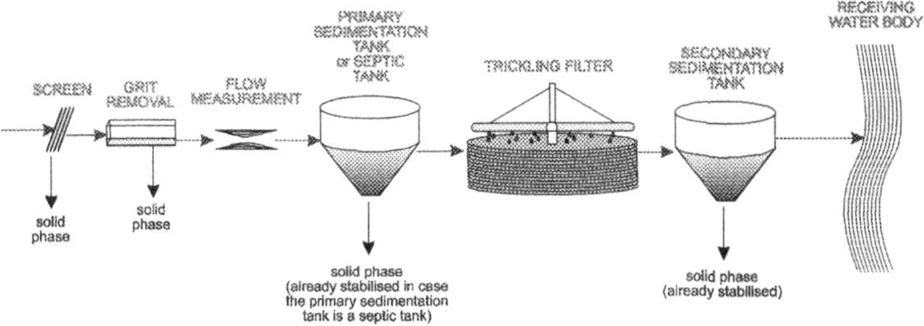


### HIGH RATE TRICKLING FILTER



### SUBMERGED AERATED BIOFILTER

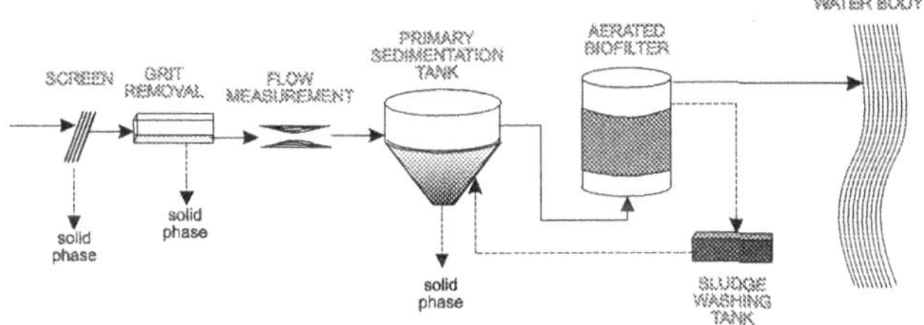


### ROTATING BIOLOGICAL CONTACTOR

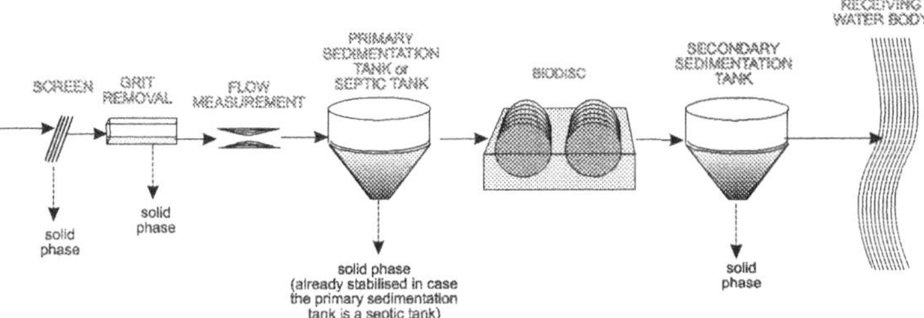


Figure 4.1f.   Flowsheet of aerobic biofilm reactors (liquid phase only).

## 4.3 PRELIMINARY TREATMENT

Preliminary treatment is mainly intended for the removal of:

- *Coarse solids*
- *Grit*

The basic removal mechanisms are of a *physical* order. Besides the coarse solids removal units, there is also a *flow measurement* unit. This usually consists of a standardised flume (e.g. Parshall flume), where the measured liquid level can be correlated with the flow. Weirs (rectangular or triangular) and closed-pipe measurement mechanisms can also be adopted. Figure 4.2 presents a typical flowsheet of the preliminary treatment.

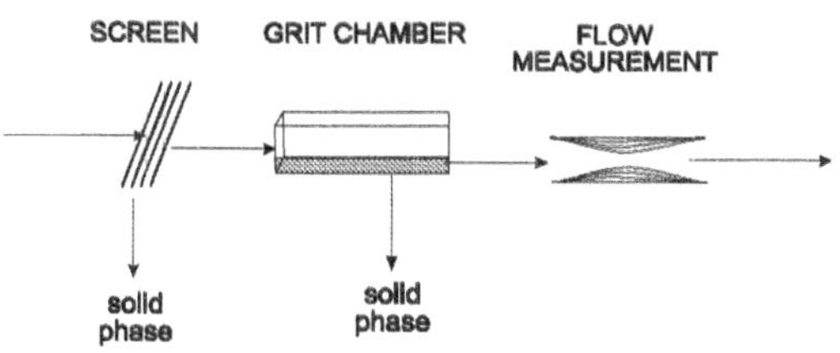


Figure 4.2.   Typical flowsheet of the preliminary treatment

The removal of *coarse solids* is frequently done by *screens* or racks, but static or rotating screens and comminutors can also be used. In the screening, material with dimensions larger than the spaces between the bars is removed (see Figure 4.3). There are coarse, medium, and fine screens, depending on the spacing between the bars. The removal of the retained material can be manual or mechanised.

The main objectives of the removal of coarse solids are:

- *protection of the wastewater transport devices (pumps and piping)*
- *protection of the subsequent treatment units*
- *protection of the receiving bodies*

The removal of *sand* contained in the sewage is done through special units called *grit chambers* (see Figure 4.4). The sand removal mechanism is simply by sedimentation: the sand grains go to the bottom of the tank due to their larger dimensions and density, while the organic matter, which settles much slower, stays in suspension and goes on to the downstream units.

There are many processes, from manual to completely mechanised units, for the removal and transportation of the settled grit. The basic purposes of grit removal are:

- *to avoid abrasion of the equipment and piping*
- *to eliminate or reduce the possibility of obstructions in piping, tanks, orifices, siphons, etc*
- *to facilitate the transportation of the liquid, principally the transfer of the sludge in its various phases*

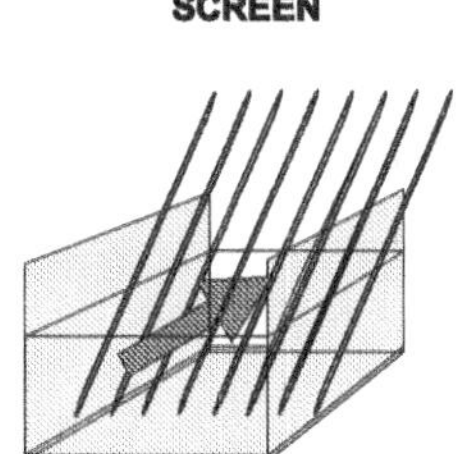



Figure 4.3.   Schematics of a screen

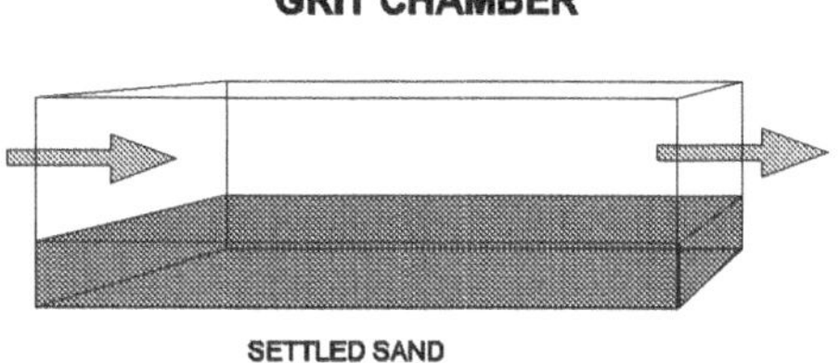



Figure 4.4.   Diagram of a grit chamber

## 4.4 PRIMARY TREATMENT

Primary treatment aims at the removal of:

- *settleable suspended solids*
- *floating solids*

After passing the preliminary treatment units, sewage still contains non-coarse suspended solids, which can be partially removed in sedimentation units. A significant part of these suspended solids is comprised of organic matter in suspension. In this way, its removal by simple processes such as sedimentation implies a reduction in the BOD load directed to the secondary treatment, where its removal is more expensive.

The sedimentation tanks can be circular (Figure 4.5) or rectangular. Sewage flows slowly through the sedimentation tanks, allowing the suspended solids with a greater density than the surrounding liquid to slowly settle to the bottom. The mass of solids accumulated in the bottom is called raw *primary sludge*. This sludge is removed through a single pipe in small sized tanks or through mechanical scrapers and pumps in larger tanks. Floating material, such as grease and oil, tends to have a lower density than the surrounding liquid and rise to the surface of the sedimentation tanks, where they are collected and removed from the tank for subsequent treatment.

The efficiency of primary treatment in the removal of suspended solids, and, as result, BOD, may be enhanced by the addition of coagulants. This is called *advanced primary treatment* or *chemically enhanced primary treatment* (CEPT).

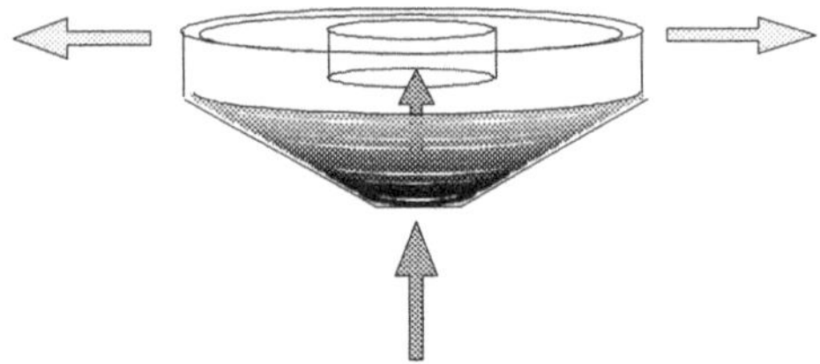


Figure 4.5.   Schematics of a circular primary sedimentation tank

Coagulants may be aluminium sulphate, ferric chloride or other, aided or not by a polymer. Phosphorus may be also removed by precipitation. More sludge is formed, resulting from the higher amount of solids removed from the liquid and from the chemical products added. The primary sludge may be digested by conventional digesters, but in some cases it may also be stabilised by lime (simplifying the flowsheet, but further increasing the amount of sludge to be disposed of).

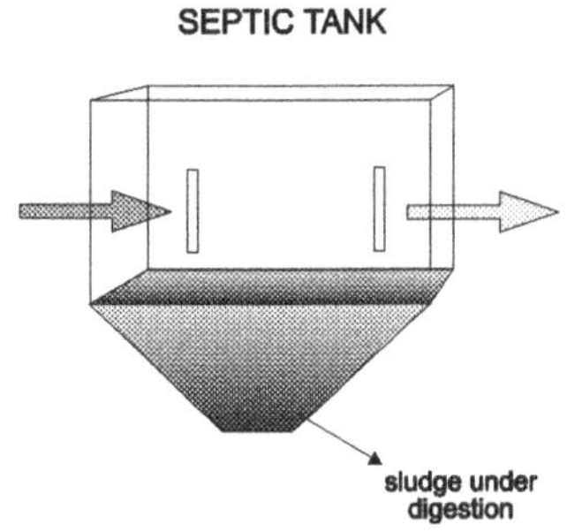


Figure 4.6.   Schematics of a single-chamber septic tank

*Septic tanks* are also a form of primary treatment (Figure 4.6). The septic tanks and their variants, such as Imhoff tanks, are basically sedimentation tanks, where the settleable solids are removed to the bottom. These solids (sludge) remain at the bottom of the tanks for a long period of time (various months) which is sufficient for their digestion. This stabilisation occurs under anaerobic conditions.

## 4.5 SECONDARY TREATMENT

### 4.5.1 Introduction

The main objective of secondary treatment is the removal of *organic matter*. Organic matter is present in the following forms:

- *dissolved organic matter* (**soluble** or **filtered BOD**) that is not removed by merely physical operations, such as the sedimentation that occurs in primary treatment;
- *organic matter in suspension* (**suspended** or **particulate BOD**), which is largely removed in the occasionally existing primary treatment, but whose solids with slower settleability (finer solids) remain in the liquid mass.

The secondary treatment processes are conceived in such a way as to accelerate the decomposition mechanisms that naturally occur in the receiving bodies. Thus, the decomposition of the degradable organic pollutants is achieved under controlled conditions, and at smaller time intervals than in the natural systems.

The essence of secondary treatment of domestic sewage is the inclusion of a *biological stage*. While preliminary and primary treatments have predominantly physical mechanisms, the removal of the organic matter in the secondary stage is carried out through biochemical reactions, undertaken by microorganisms.

A great variety of microorganisms take part in the process: bacteria, protozoa, fungi and others. The basis of the whole biological process is the effective contact between these organisms and the organic matter contained in the sewage, in such a way that it can be used as food for the microorganisms. The microorganisms convert the organic matter into carbon dioxide, water and cellular material (growth and reproduction of the microorganisms) (see Figure 4.7). This biological decomposition of the organic matter requires the presence of oxygen as a fundamental component of the aerobic processes, besides the maintenance of other favourable environmental conditions, such as temperature, pH, contact time, etc.

Figure 4.7.   Simplified diagram of bacterial metabolism

Secondary treatment generally includes preliminary treatment units, but may or may not include primary treatment units. There exists a large variety of secondary treatment processes, and the most common ones are:

- *Stabilisation ponds*
- *Land disposal systems*
- *Anaerobic reactors*
- *Activated sludge systems*
- *Aerobic biofilm reactors*

These processes have been summarised in Table 4.5, and a simplified description is presented below. In Section 4.7.2, there is a general comparison between all the processes described, including basic data (efficiencies, land requirements, power requirements, costs, sludge production, etc.), together with qualitative comparisons and a list of advantages and disadvantages.

## 4.5.2 Stabilisation ponds

The following variants of stabilisation ponds are described briefly in this section:

- Facultative ponds
- Anaerobic pond – facultative ponds systems
- Facultative aerated lagoons
- Complete-mix aerated lagoon – sedimentation pond systems
- High rate ponds
- Maturation ponds

### a) Facultative ponds

Stabilisation ponds are units specially designed and built with the purpose of treating sewage. However, the construction is simple and is principally based on earth movement for digging, filling and embankment preparation.

When facultative ponds receive raw sewage, they are also called *primary ponds* (a secondary pond would be the one which would receive its influent from a previous treatment unit, such as anaerobic ponds – see item *b* in this section).

Amongst the stabilisation ponds systems, the process of facultative ponds is the simplest, relying only on natural phenomenon. The influent enters continuously in one end of the pond and leaves in the opposite end. During this time, which is of the order of many days, a series of events contribute to the purification of the sewage.

Part of the organic matter in suspension (*particulate BOD*) tends to settle, constituting the bottom sludge. This sludge undergoes a decomposition process by anaerobic microorganisms and is converted into carbon dioxide, methane and other compounds. The inert fraction (non-biodegradable) stays in this bottom layer.

The dissolved organic matter (*soluble BOD*), together with the small-dimension organic matter in suspension (*fine particulate BOD*), does not settle and stays dispersed in the liquid mass. Its decomposition is through **facultative** bacteria that have the capacity to survive, either in the presence or in the absence of free oxygen (but presence of nitrate), hence the designation of facultative, which also defines the name of the pond. These bacteria use the organic matter as energy source, which is released through respiration. The presence of oxygen is necessary in aerobic respiration, and it is supplied to the medium by the photosynthesis carried out by the algae. There is an equilibrium between consumption and the production of oxygen and carbon dioxide (see Figure 4.8).

> **Bacteria** ➜ *respiration*:
>
> - oxygen consumption
> - carbon dioxide production
>
> **Algae** ➜ *photosynthesis*:
>
> - oxygen production
> - carbon dioxide consumption

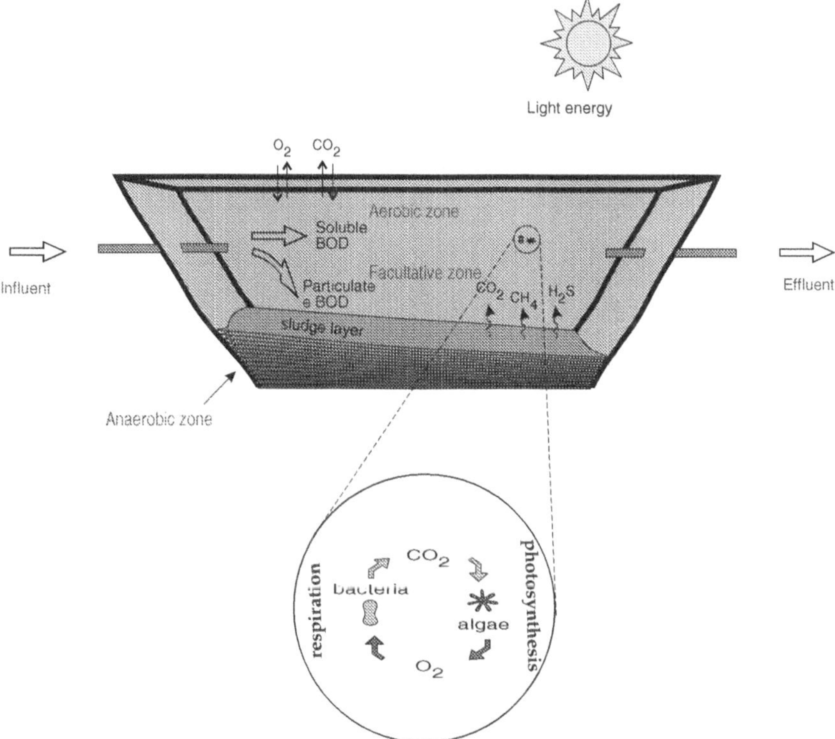


Figure 4.8.   Simplified diagram of a facultative pond

A light energy source, in this case represented by the sun, is necessary for photosynthesis to occur. For this reason, locations with high solar radiation and low cloudiness are favourable for the implementation of facultative ponds.

Photosynthesis is higher near the water surface, as it depends on solar energy. Typical pond depths are between 1.5 and 2.0 m. When deep regions in the pond are reached, the light penetration is low, what causes the predominance of oxygen consumption (*respiration*) over its production (*photosynthesis*), with the possible absence of dissolved oxygen at a certain depth. Besides, photosynthesis only occurs during the day, and during the night the absence of oxygen can prevail. Owing to these facts, it is essential that the main bacteria responsible for the stabilisation of the organic matter are facultative, so that they can survive and proliferate, either in the presence or in the absence of oxygen (but only under anoxic, and not strict anaerobic conditions).

The process of facultative ponds is essentially natural, as it does not require any equipment. For this reason, the stabilisation of the organic matter takes place at lower rates, implying the need of a long detention time in the pond (usually greater than 20 days). To be effective, photosynthesis needs a large exposure surface area to make the most of the solar energy by the algae, also implying the need of large units. As a result, the total area required by facultative ponds is the largest within all the wastewater treatment processes (excluding the land disposal processes). On the other hand, because the process is totally natural, it is associated to a high operational simplicity, which is a factor of fundamental importance in developing countries.

Figure 4.9 presents a typical flowsheet of a facultative pond system.

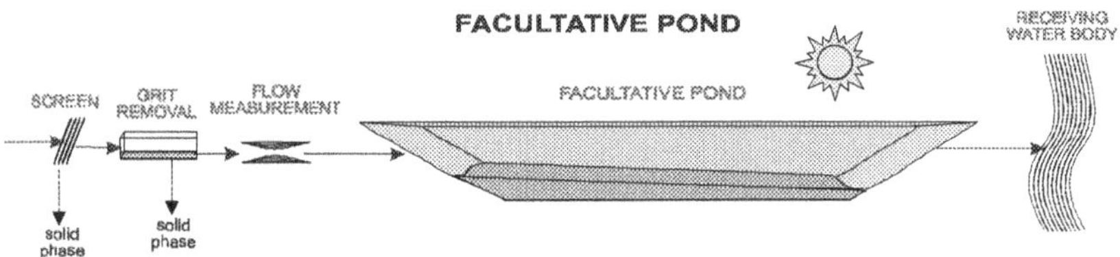


Figure 4.9.   Typical flowsheet of a facultative pond system

## b) Anaerobic pond – facultative ponds systems

The process of facultative ponds, in spite of having a satisfactory efficiency, requires a large area that is often not available in the locality in question. There is therefore, the need to find solutions that could imply the reduction of the total area required. One of these solutions is the system of anaerobic ponds followed by facultative ponds. In this case, the facultative pond is also called a *secondary* pond, since it receives the influent from an upstream treatment unit, and not the raw sewage.

The raw sewage enters a pond that has smaller dimensions and is deeper (around 4 to 5 m). Owing to the smaller dimensions of this pond, photosynthesis practically does not occur. In the balance between oxygen consumption and production, consumption is much higher. Therefore, anaerobic conditions predominate in this first pond, which is consequently called an *anaerobic pond*.

Anaerobic bacteria have a slower metabolic and reproduction rate than the aerobic bacteria. For a detention time of only 2 to 5 days in the anaerobic pond, there is only partial decomposition of the organic matter. However, the BOD removal of the order of 50 to 70%, even if insufficient, represents a large contribution, substantially reducing the load to the facultative pond that is situated downstream.

The facultative pond receives a load of only 30 to 50% of the raw sewage load, which therefore allows it to have smaller dimensions. The overall area requirement (anaerobic + facultative pond) is such, that land savings in the order of 1/3 are obtained, compared with just a single facultative pond.

The working principles of this facultative pond are exactly as described in item a. Figure 4.10 shows a typical flowsheet of a system of anaerobic ponds followed by facultative ponds.

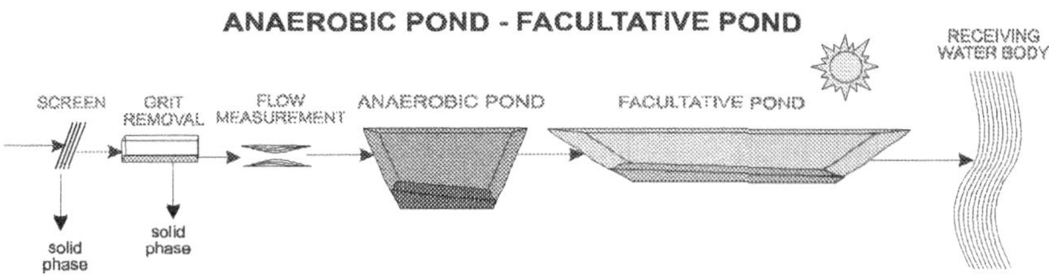


Figure 4.10.   Typical flowsheet of a system of anaerobic ponds followed by facultative ponds

The efficiency of the system is similar or only slightly higher than that of a single facultative pond. The system is also conceptually simple and easy to operate. However, the existence of an anaerobic stage in an open unit is always a cause for concern, due to the possibility of the release of malodours. If the system is well balanced, then the generation of bad smells should not occur. However, occasional operational problems could lead to the release of hydrogen sulphide, responsible for a bad smell. For this reason, this system should, whenever possible, be located far from residences.

## c) Facultative aerated lagoon

If a predominantly aerobic system is desired, with even smaller dimensions, a *facultative aerated lagoon* can be used. The main difference in relation to a conventional facultative pond is regarding the form of the oxygen supply. While in facultative ponds the oxygen is mainly obtained from photosynthesis, in the case of facultative aerated lagoons the oxygen is supplied by mechanical equipment called *aerators*.

The most commonly used mechanical aerators in aerated ponds are those with a vertical axis that rotates at a high speed, causing great turbulence in the water. This turbulence favours the penetration of atmospheric oxygen into the liquid mass, where it is then dissolved. A greater oxygen introduction is achieved, in comparison with the conventional facultative pond, which leads to a faster decomposition of the organic matter. Because of this, the detention time of the wastewater in the pond can be less (in the order of 5 to 10 days for domestic sewage). Consequently, the land requirements are much smaller.

The pond is called *facultative* by the fact that the level of energy introduced by the aerators is only sufficient for the oxygenation, but not to maintain the solids (biomass and wastewater solids) in suspension in the liquid mass. In this way, the solids tend to settle and constitute a sludge layer at the bottom of the pond, to be decomposed anaerobically. Only the soluble and fine particulate BOD remains in the liquid mass, undergoing aerobic decomposition. Therefore, the pond behaves like a conventional facultative pond (see Figure 4.11).

## FACULTATIVE AERATED LAGOON

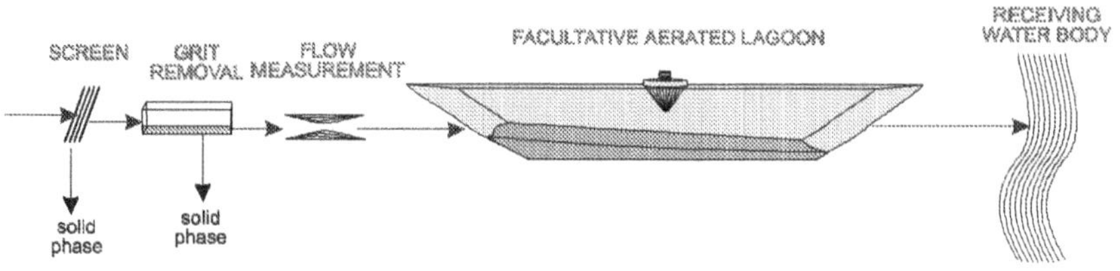


Figure 4.11.　Typical flowsheet of a system of facultative aerated lagoons

Aerated lagoons are less simple in terms of operation and maintenance, when compared with the conventional facultative ponds, owing to the introduction of mechanisation. Therefore, the reduction of the reduction of the land requirements is achieved with a certain rising in the operational level, along with the introduction of electricity consumption.

### d) Complete-mix aerated lagoon – sedimentation pond systems

A way of reducing the aerated pond volume even further is to increase the aeration level per unit volume of the lagoon, thus creating a turbulence that, besides guaranteeing oxygenation, allows all the solids to be maintained in suspension in the liquid medium. The denomination of *complete mix* is because of the high degree of energy per unit volume, which is responsible for the total mixing of all the constituents in the pond. Amongst the solids maintained in suspension and in complete mixing are the biomass, besides the organic matter of the raw sewage. There is, therefore, a larger concentration of bacteria in the liquid medium as well as a larger organic matter – biomass contact. Consequently, the efficiency of the system increases and allows the volume of the aerated pond to be greatly reduced. The typical detention time in an aerated pond is in the order of 2 to 4 days.

However, in spite of the high efficiency of this lagoon in the removal of the organic matter originally present in the sewage, a new problem is created. The biomass stays in suspension in all the volume and thus leaves with the pond effluent. This biomass is, in a way, also organic matter (particulate BOD), even if of a different nature of the BOD of the raw sewage. If this new organic matter were discharged into the receiving body, it would also exert an oxygen demand and cause a deterioration in the water quality.

Therefore, it is important that there is a unit downstream in which the suspended solids (predominantly the biomass) can settle and be separated from the liquid (final effluent). This unit is a *sedimentation pond*, with the main purpose of permitting the settling and accumulation of the solids.

The sedimentation pond is designed with short detention times, around 2 days. In this period, the solids will go to the bottom where they will undergo digestion and be stored for a period of some years, after which they will be removed. There are also sedimentation ponds with continuous removal of the bottom sludge, using, for instance, pumps mounted on rafts.

The land required for this pond system is the smallest within the pond systems. The energy requirements are similar to or only slightly higher than those in the facultative aerated lagoons. However, the aspects related to sludge handling can be more complicated, due to the fact that there is a smaller storage period in the pond compared with the other systems. If the sludge is removed periodically, this will take place with an approximate frequency of around 2 to 5 years. The removal of the sludge is a laborious and expensive task.

Figure 4.12 illustrates the flowsheet of the system.

**COMPLETE-MIX AERATED LAGOON - SEDIMENTATION POND**

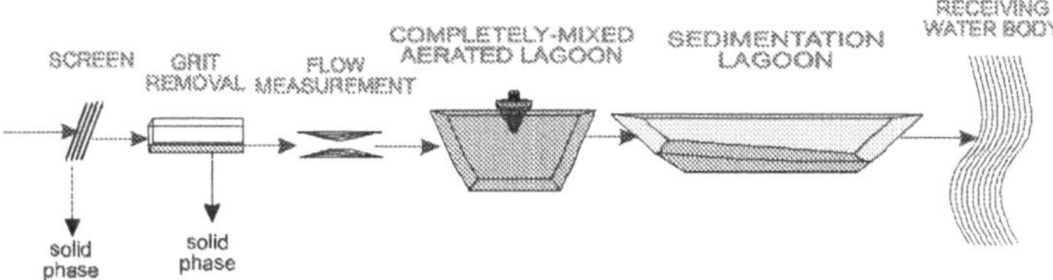


Figure 4.12.   Typical flowsheet of a system of complete-mix aerated lagoons – sedimentation ponds

## e) High rate ponds

High rate algal ponds are conceived to maximise the production of algae in a totally aerobic environment. To accomplish this, the ponds are shallow (less than 1.0 m depth), thus guaranteeing the penetration of light in all the liquid mass. Consequently, the photosynthetic activity is high, leading to high dissolved oxygen concentrations and an increase in pH (consumption of carbonic acid in the photosynthesis). These factors contribute to the increase in the death rate of the pathogenic microorganisms and the removal of nutrients, which is the main objective of the high rate ponds.

Ammonia removal occurs by stripping of the free ammonia ($NH_3$), since in high pH conditions the ammonia equilibrium shifts in the direction of free ammonia (with a reduction in the concentration of the ammonium ion $NH_4^+$). The increase in the $NH_3$ concentration leads to its release to the atmosphere.

Phosphorus removal also occurs as a result of the high pH, which causes the precipitation of the phosphates into the form of hydroxyapatite or struvite.

The high rate ponds receive a high organic load per unit surface area. There is usually the introduction of moderate agitation in the pond, which is achieved by means of a horizontal-axis rotor or equivalent equipment. Its function is not to aerate, but to gently move the liquid mass. This agitation improves the contact of the influent with the bacteria and algae, reduces dead zones and facilitates the exposure of a larger quantity of algae to sun light.

The configuration of the pond can be in the form of a carrousel, similar to an oxidation ditch (Figure 4.13). The high rate ponds can come after facultative ponds, in which a large part of the BOD is removed, leaving the polishing in terms of pathogen and nutrient removal for the high rate ponds.

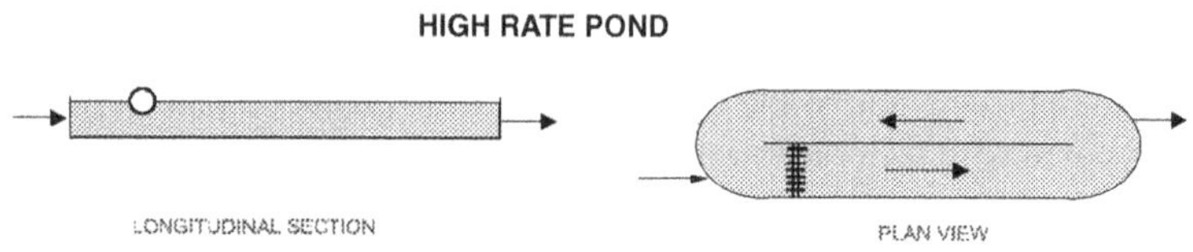


Figure 4.13.    Schematics of a high rate pond

## f) Maturation ponds

Maturation ponds aim at polishing the effluent from any stabilisation pond system previously described or, in broader terms, from any sewage treatment system. The main objective of maturation ponds is the *removal of pathogenic organisms* and not an additional BOD removal. Maturation ponds are an economic alternative for the disinfection of the effluent, in comparison to more conventional methods, such as chlorination.

The ideal environment for pathogenic microorganisms is the human intestinal tract. Outside it, whether in the sewerage system, in the sewage treatment or in the receiving body, the pathogenic organisms tend to die. Various factors contribute to this, such as temperature, solar radiation, pH, food shortage, predator organisms, competition, toxic compounds, etc. The maturation pond is designed in such a way as to optimise some of these mechanisms. Many of these mechanisms become more effective with lower pond depths, which justifies the fact that maturation ponds are shallower than the other types of ponds. Within the mechanisms associated with the pond depth, the following can be mentioned (van Haandel and Lettinga, 1994; van Buuren et al, 1995):

- Solar radiation (ultraviolet radiation)
- High pH (pH > 8.5)
- High DO concentration (favouring an aerobic community, which is more efficient in the elimination of coliforms)

Maturation ponds should reach extremely high coliform removal efficiencies (E > 99.9 or 99.99% ), so that the effluent can comply with usual standards or guidelines for direct use (e.g. for irrigation) or for the maintenance of the various uses of the receiving body (e.g. bathing). The ponds also usually reach total removal of helminth eggs.

In order to maximise the coliform removal efficiency, the maturation ponds are designed with one of the following two configurations: (a) three or four ponds in series (see Figure 4.14) or (b) a single pond with baffles.

**ANAEROBIC POND – FACULTATIVE POND – MATURATION PONDS**

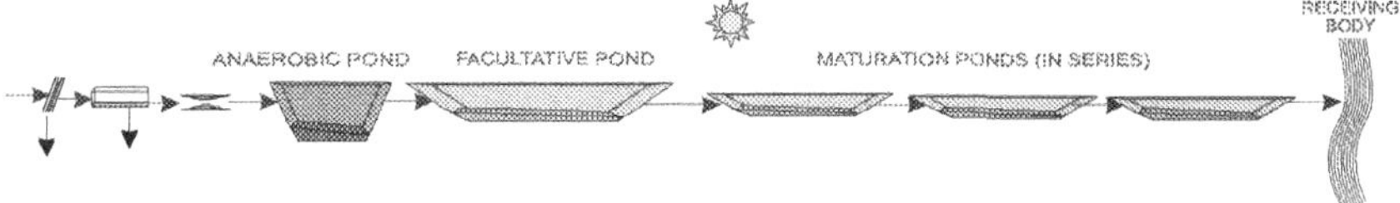


Figure 4.14.   Typical flowsheet of a system of stabilisation ponds followed by maturation ponds in series.

## 4.5.3  Land disposal

The most common destinations for the final disposal of treated liquid effluents are water courses and the sea. However, land disposal is also a viable process, applied in various locations around the world.

The land application of wastewater can be considered as a form of *final disposal*, of *treatment* (primary, secondary or tertiary level) or both. Land application of wastewater leads to groundwater recharge and/or to evapotranspiration. Sewage supplies the plants with water and nutrients.

In the soil, a pollutant has basically four possible destinations:

* *retention in the soil matrix*
* *retention by the plants*
* *appearance in the underground water*
* *collection by underdrains*

Various mechanisms in the soil act in the removal of the pollutants:

* *physical* (settling, filtration, radiation, volatilisation, dehydration)
* *chemical* (oxidation and chemical reactions, precipitation, adsorption, ion exchange, complexation, photochemical breakdown)
* *biological* (biodegradation and predation)

The capacity of the soil to assimilate complex organic compounds depends on its properties and on climatic conditions. Infiltration rates and types of vegetation are important factors in the use of soil as a medium for the degradation of organic compounds. These reactions require good soil aeration, which is inversely related to the humidity of the soil. Insufficient aeration leads to a lower assimilation capacity of the organic compounds by the soil. Almost all soil types are efficient in the removal of organic matter. The removal results from the filtering action of the soil, followed by the biological oxidation of the organic material. Soils with clay (fine texture) or soils with a considerable quantity of organic matter will also retain wastewater constituents through mechanisms of adsorption, precipitation and ion exchange.

The most common types of land application are:

*Soil-based systems:*

- slow-rate systems
- rapid infiltration
- subsurface infiltration
- overland flow

*Aquatic-based systems:*

- constructed wetlands

Aquatic-based systems (constructed wetlands) are included in this section for didactic reasons, although they could have been presented in the section of stabilisation ponds, which are also aquatic-based systems.

The selection of the treatment method is a function of various factors, including required efficiency, climatic conditions, depth to ground water, soil permeability, slope etc. The application of wastewater can be done by methods such as sprinklers, furrows, graded border, drip irrigation and others.

### a) Slow-rate systems

Depending on the design objective, slow-rate systems can be classified according to two types (WPCF, 1990):

- **Slow infiltration systems**. Main objective: *wastewater treatment*. The amount of wastewater applied is not controlled by crop requirements. For municipal wastewaters, loading rates are controlled either by nitrogen loading or soil permeability. The systems are designed to maximise the amount of wastewater applied per unit land area.
- **Crop irrigation systems**. Main objective: *water reuse for crop production* (wastewater treatment is an additional objective). The systems are designed to apply sufficient wastewater to meet crop irrigation requirements. Loading rates are based on the crop irrigation requirement and the application efficiency of the distribution system. Nitrogen loading must be checked to avoid excess nitrogen.

In *crop irrigation systems*, the objective is to supply the wastewater to the soil in quantities compatible with the *nutrient* requirements of the crops. However, initially the microbiological and biochemical characteristics of the sewage should be evaluated, taking into consideration the type of crop, soil, irrigation system and the form in which the product will be used or consumed. Only after the verification that the sewage meets the conditions specified by the health standards should the evaluation of the chemical components be considered (Mattos, 1998).

Figure 4.15 presents a flowsheet of a slow-rate system. The irrigation with wastewater can be done by flooding, furrows, sprinkler and dripping.

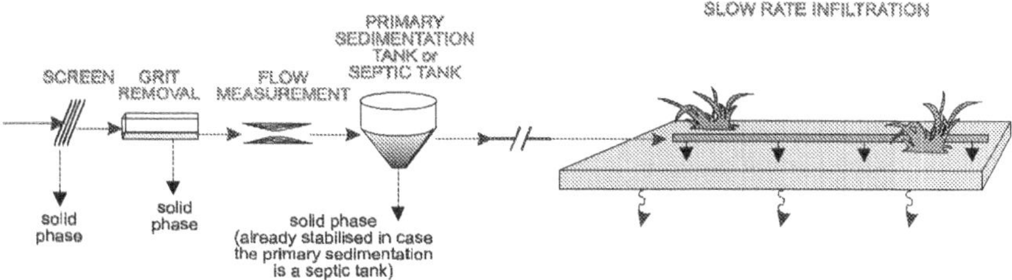


Figure 4.15.   Typical flowsheet of a slow-rate system

Loamy soils (medium texture) are indicated because they exhibit the best balance for wastewater renovation and drainage. The depth to the water table should be greater than 1.5 m to prevent groundwater contamination.

The application rates must be compatible with the evapotranspiration of the crop in the period, therefore depending on the type of crop and the climatic conditions. In arid zones, wastewater can be used for irrigation throughout the year. On the other hand, in wet areas, the application of wastewater in rainy periods can lead to anaerobic conditions and consequently odour and insect appearance problems.

Irrigation constitutes a treatment/disposal system that requires the largest surface area per unit of wastewater treated. On the other hand, it is the natural system with the highest efficiency. The plants are those mainly responsible for the removal of nutrients, such as phosphorus and nitrogen, while the microorganisms in the soil perform the removal of the organic substances. There is also a high removal of pathogenic organisms during the percolation through the soil (Mattos, 1998).

## b) Rapid infiltration

The objective of the rapid infiltration system is to use the soil as a filtering medium for the wastewater. This system is characterised by the percolation of the wastewater, which is purified by the filtering action of the porous medium. The percolated wastewater may be used for groundwater recharge or be collected by underdrains or wells. The rapid infiltration method requires the lowest area within all the land disposal processes.

Wastewater is applied in shallow infiltration basins, from which wastewater percolates through the soil. The application is intermittent, in order to allow a resting period for the soil, during which the soil dries and re-establishes aerobic conditions. Due to the higher application rates, evaporation losses are small and most of the liquid percolates through the soil, undergoing treatment.

The application can be done by direct discharge (furrows, channels, perforated pipes) or by high capacity sprinklers. Vegetation growth may or may not occur, does not interfere with the efficiency of the process and is not part of the treatment objective (Coraucci Filho et al, 1999).

Figure 4.16 presents a flowsheet of a rapid infiltration system.

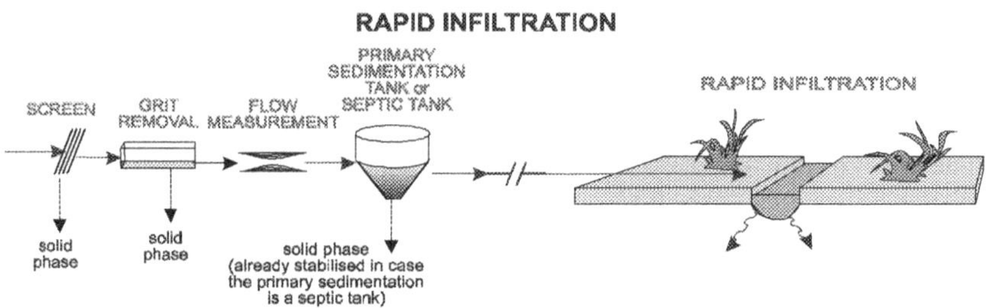


Figure 4.16.   Typical flowsheet of a rapid infiltration system (other types of pre-treatment may be applied)

### c) Subsurface infiltration

In subsurface infiltration systems, pre-settled or pre-treated sewage is applied below ground level. The infiltration sites are prepared in buried excavations, and filled in with a porous medium. The filling medium maintains the structure of the excavation, allows free sewage flow and provides sewage storage during peak flows. The sewage percolates through unsaturated soil, where additional treatment occurs. This treatment process is similar to rapid infiltration, the main difference residing in the application below ground level.

The subsurface infiltration systems have the following variants:

- *infiltration trenches or pits* (without final effluent: wastewater percolates to groundwater)
- *filtration trenches* (with final effluent: collection by underdrain system)

The subsurface infiltration systems are normally used following septic tanks (Figure 4.17) and, in some cases, after further treatment provided by systems such as anaerobic filters. The applicability is usually for small residential areas or rural dwellings.

### d) Overland flow

Overland flow systems consist of the application of untreated (at least screened) or pre-treated wastewater in the upper part of sloped terraces, planted with water resistant grasses. Wastewater flows gently downwards, having contact with the root-soil system, in which biochemical reactions take place. Some evapotranspiration occurs, and the final effluent is collected at the lower end by drainage channels. Application is intermittent.

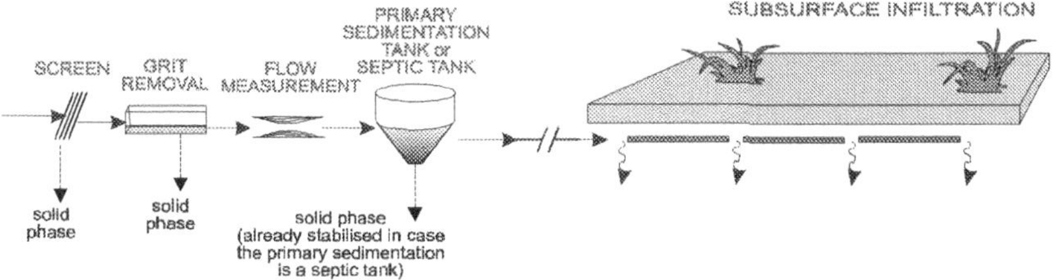



Figure 4.17.    Typical flowsheet of a subsurface infiltration system

The soils should have a low permeability (e.g. clay). The slope should be moderate (between 2 and 8% ).

The use of vegetation is essential to increase the absorption rate of the nutrients and the loss of water by transpiration. Besides this, the vegetation represents a barrier to the free surface flow of the liquid in the soil, increasing the retention of suspended solids and avoiding erosion. This gives better conditions for the development of the microorganisms that will carry out the various biochemical reactions. The vegetation should be perennial, water tolerant grasses. Local agricultural extension agents should be consulted.

Wastewater application can be done by sprinklers, gated pipes, slotted or perforated pipes or bubbling orifices (WPCF, 1990).

Figure 4.18 presents a flowsheet of an overland flow system.

## OVERLAND FLOW

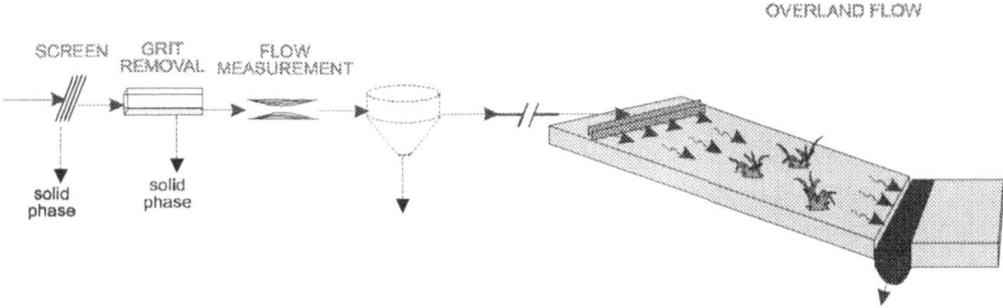



Figure 4.18.    Typical flowsheet of an overland flow system

## e) Constructed wetlands

Most of the following concepts were extracted from Marques (1999), OPS/OMS (1999) and mainly EPA (2000). **Natural** wetlands are areas inundated or saturated by surface or groundwater that support a vegetation adapted to these conditions. The natural wetlands include marshes, swamps and similar areas, that shelter diverse forms of aquatic life.

**Constructed** wetlands are purposely built wastewater treatment processes, which consist of ponds, basins or shallow canals (usually with a depth of less than 1.0 m) that shelter aquatic plants, and use biological, chemical and physical mechanisms to treat the sewage. The constructed wetlands usually have an impermeable layer of clay or synthetic membrane, and structures to control the flow direction, hydraulic detention time and water level. Depending on the system, they can contain an inert porous medium such as stones, gravel or sand.

Constructed wetlands are different from natural wetlands because of human interference, such as landfills, drainage, flow alterations and physical properties. The direct use of natural wetlands for sewage treatment has great environmental impacts and must not be encouraged.

There are basically two types of constructed wetlands:

- **Surface flow (free water surface) wetlands**. These resemble natural wetlands in appearance, because they have plants which can be floating and/or rooted (emergent or submerged) in a soil layer at the bottom, and water flows freely between the leaves and the stems of these plants. There can be open areas dominated by these plants or islands exerting habitat functions. Plant genera in use include: (a) emergent: *Typha*, *Phragmites*, *Scirpus*, (b) submerged: *Potamogeon*, *Elodea*, etc., (c) floating: *Eichornia* (water hyacinth), *Lemna* (duckweed). Native plants are preferred. These wetlands present a very complex aquatic ecology. They may or may not have a lined bottom, depending on the environmental requirements. Water depth is between 0.6 and 0.9 m for the vegetated zones (or less, in the case of certain emergent plants), and 1.2 to 1.5 m for free water zones. This type of wetlands is adequate to receive effluent from stabilisation ponds. In these conditions, they occupy an area between 1.5 to 3.0 $m^2$/inhab.
- **Subsurface flow wetlands (vegetated submersed bed systems)**. These do not resemble natural wetlands because there is no free water on the surface. There is a bed composed of small stones, gravel, sand or soil that gives support to the growth of aquatic plants. The water level stays below the surface of the bed, and sewage flows in contact with the roots and the rhizomes of the plants (where a bacterial biofilm is developed), not being visible or available for the aquatic biota. Plant genera that have been used are: *Typha*, *Scirpus*, *Carex* and *Phragmites*. The medium height is between 0.5 and 0.6 m and water depth is between 0.4 and 0.5 m. The gravel should have a size that allows the continuos flow of the sewage without clogging problems. A large part of the subsurface zone is anaerobic, with aerobic sites immediately adjacent to the rhizomes and roots. There is a lower potential for the generation of bad odours and the appearance of mosquitoes and rats. These wetlands are suited to receive effluents from septic tanks and anaerobic reactors, but not from stabilisation ponds, because of the presence of algae. For effluents from septic tanks, the land requirements are around 5.0 to 6.0 $m^2$/hab, and for effluents from anaerobic reactors, between 2.5 and 4.0 $m^2$/hab.

Regarding the direction of the water flow, the wetlands can be classified as:

- **Vertical flow**. Typically, a filter of sand or gravel planted with vegetation. At the bottom of the filter medium there is a series of underdrains that collect the treated sewage. The operation resembles the routine of a filter, with dosing and draining cycles, therefore, differing from the conventional conception of wetlands. With intermittent dosing, the flow is normally through unsaturated media.
- **Horizontal flow**. The most classical conception of constructed wetlands. May be with surface or subsurface flow.

Figure 4.19 illustrates the main variants of constructed wetlands.

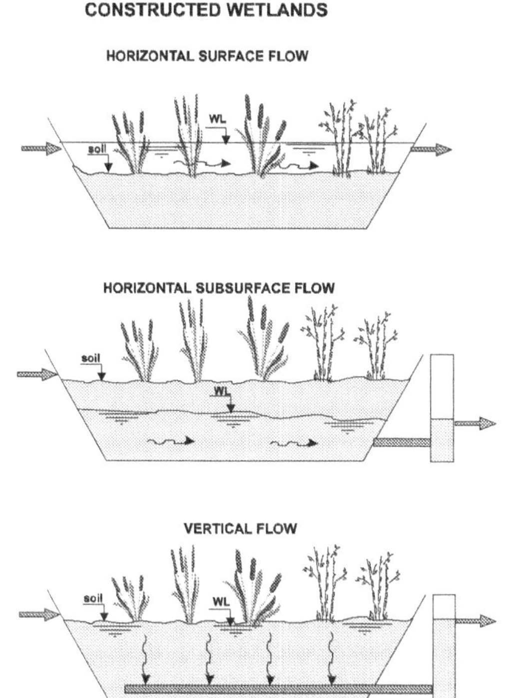


Figure 4.19.   Diagram showing the main variants of constructed wetlands

Constructed wetlands do not perform well in the treatment of raw sewage. Some form of primary or secondary treatment (e.g. stabilisation ponds or anaerobic reactors) must precede this process (Figure 4.20). In the case of having previous secondary treatment, low values of BOD, SS and nitrogen can be reached.

The layout of wetlands is usually in cells in series or in parallel.

A surface flow system receiving the effluent from a stabilisation pond can operate for 10 to 15 years without the need to remove the material composed of

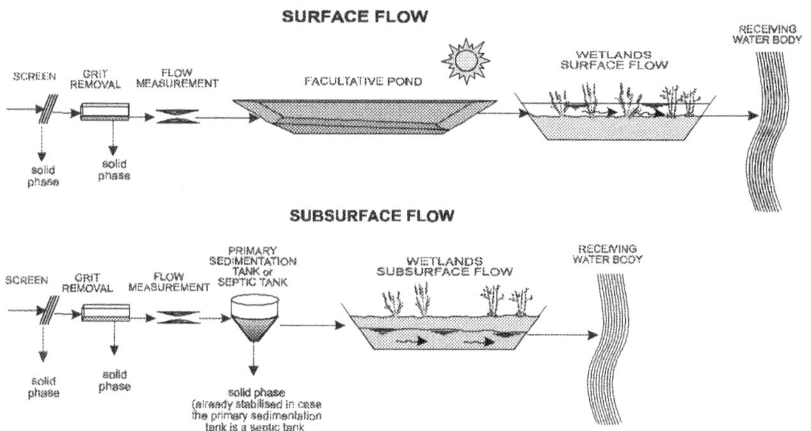



Figure 4.20.   Typical flowsheets of wetlands systems

plants and inert settled solids. Of these solids, the largest part tends to accumulate at the inlet end of the unit.

The operation and maintenance of constructed wetlands is very simple. Besides the activities related with the preceding treatment, the maintenance of the wetlands is usually associated with the control of undesired aquatic plants and mosquitoes (which are not normally a problem in well designed and operated subsurface flow systems). The removal of the plants is not normally necessary, but a certain pruning or replanting can be necessary to maintain the desired flow conditions and treatment.

## 4.5.4 Anaerobic reactors

There are many variants of anaerobic reactors. This section presents only the two most widely applied for domestic sewage treatment:

- anaerobic filter (frequently treating septic tank effluents)
- UASB (upflow anaerobic sludge blanket) reactor

### a) Septic tank – anaerobic filter system

The system of septic tanks followed by anaerobic filters (Figure 4.21) has been widely used in rural areas and in small sized communities. The septic tanks remove most of the suspended solids, which settle and undergo anaerobic digestion at the bottom of the tank.

The septic tank can be a single-chamber tank or a two-compartment tank (called an Imhoff tank). In the single chamber tank, there is no physical separation between the regions of the raw sewage solids sedimentation and bottom sludge digestion. The single chamber tanks can be single or in series.

In the Imhoff tank, settling occurs in the upper compartment (settling compartment). The settled solids pass through an opening at the bottom of the compartment and are directed to the bottom compartment (digestion compartment). The accumulated sludge then undergoes anaerobic digestion. The gases originating from the anaerobic digestion do not interfere with the settling process, as they cannot penetrate inside the sedimentation chamber.

Because septic tanks are sedimentation tanks (no biochemical reactions in the liquid phase), BOD removal is limited. The effluent, still with high organic matter concentration, goes to the anaerobic filter, where further removal takes place under anaerobic conditions. The filter is a biofilm reactor: the biomass grows attached to a support medium, usually stone. The following points are characteristic of anaerobic filters, differing from the trickling filters, which are also biofilm reactors (see Section 4.5.6):

- the liquid flow is upwards, i.e. the inlet is at the bottom and the outlet at the top of the anaerobic filter
- the anaerobic filter works submerged, i.e. the free spaces are filled with liquid
- the unit is closed
- the BOD load applied per unit volume is very high, which guarantees anaerobic conditions

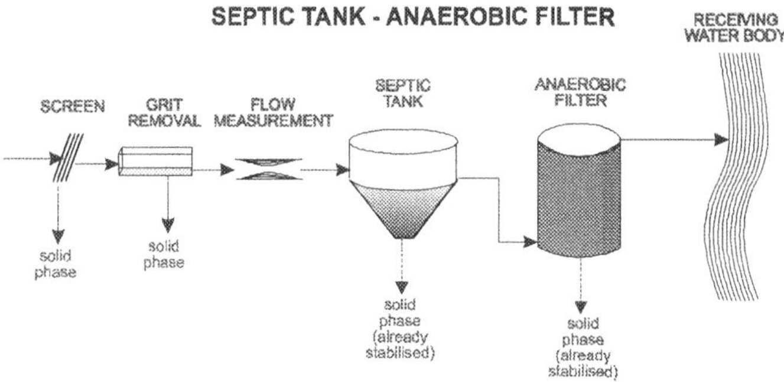


Figure 4.21.   Typical flowsheet of a system of a septic tank followed by an anaerobic filter (liquid phase)

The efficiency of a septic tank – anaerobic filter is usually less compared with fully aerobic systems, although in most situations sufficient. The system has been widely used for small populations, but there has been a trend in terms of anaerobic treatment favouring the use of anaerobic sludge blanket reactors (described below).

Sludge production in anaerobic systems is very low. The excess sludge is already digested and can go directly to dewatering (in this system, typically by drying beds).

Being an anaerobic system, there is always a risk of generation of bad odours. However, proper design and operational procedures can contribute to the reduction of these risks. It should also be remembered that the septic tank and the anaerobic reactors are closed units.

## b) Upflow anaerobic sludge blanket (UASB) reactors

The *upflow anaerobic sludge blanket* (UASB) reactors are currently the main trend in wastewater treatment in some warm-climate countries, either as single units, or followed by some form of post treatment.

In the UASB reactors, the biomass grows dispersed in the liquid, and not attached to a support medium, as in the case of anaerobic filters. When biomass grows it can form small granules, which are a result of the agglutination of various microorganisms. These small granules tend to serve as a support medium for other organisms. The granulation increases the efficiency of the system, but it is not essential for the working of the reactor, and is actually difficult to be obtained with domestic wastewater.

The concentration of the biomass in the reactor is very high, justifying the name of sludge blanket. Owing to this high concentration, the volume required for the UASB reactor is greatly reduced in comparison with all other treatment systems.

The liquid enters at the bottom, where it meets the sludge blanket, leading to the adsorption of the organic matter by the biomass. The flow is upward. As a result of the anaerobic activity, gases are formed (mainly methane and carbon dioxide) and the bubbles also present a rising tendency. The upper part of the anaerobic sludge blanket reactor presents a structure, whose functions are the separation and accumulation of the gas and the separation and return of the solids (biomass). In this way, the biomass is kept in the system (leading to high concentrations in the reactor), and only a minor fraction leaves with the effluent. This structure is called a three-phase separator, as it separates the liquid, solids, and gases. The form of the separator is frequently that of an inverted cone or pyramid.

The gas is collected in the upper part of the separator, in the gas compartment, from where it can be removed for reuse (energy from methane) or burning.

The solids settle in the upper part of the separator, in the settling compartment, and drain down the steeply inclined walls until they return to the reactor body. In this way, a large part of the biomass is retained by the system by simple gravitational return (differently from the activated sludge process, which requires pumping of the return sludge). Owing to the high solids retention, the hydraulic detention time can be low (in the order of 6 to 10 h). Because the gas bubbles do not penetrate the settling zone, the separation of the solids-liquid is not impaired. The effluent is relatively clarified when it leaves the settling compartment, and the concentration of the biomass in the reactor is maintained at a high level.

Figure 4.22 presents a schematic view of a UASB reactor. Various configurations are possible, including circular, square or rectangular tanks.

The sludge production is very low. The sludge wasted from the reactor is already digested and thickened, and may be simply dewatered in drying beds or other dewatering process. The dewaterability of the sludge is very good.

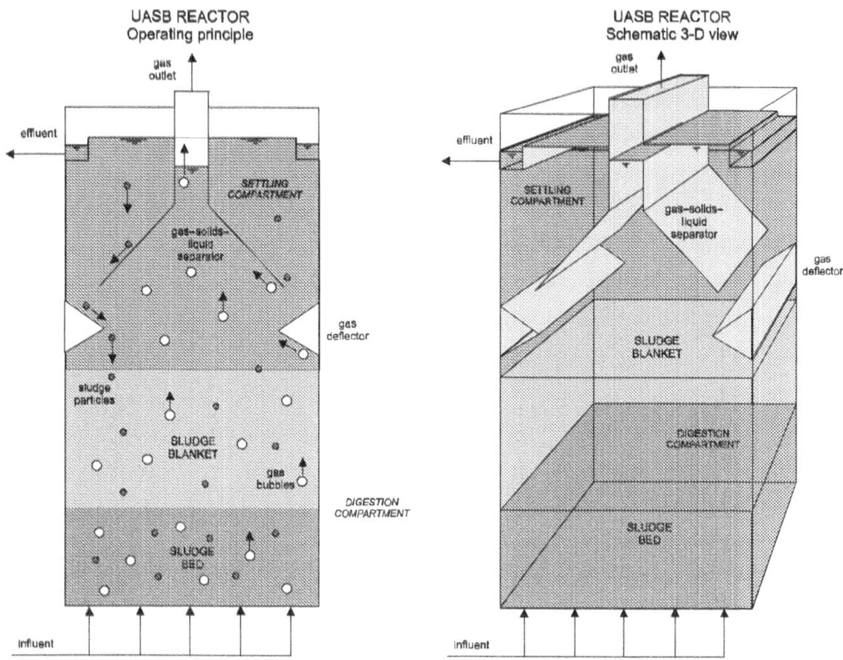



Figure 4.22.    Schematics of a UASB reactor (working principle and schematic view)

The plant flowsheet is simplified even more by the fact that, differently from anaerobic filters and other systems, there is no need for primary settling.

Figure 4.23 presents the flowsheet of a wastewater treatment system comprised by a UASB reactor.

The risk of generation or release of malodours can be greatly reduced by a careful design, not only in the kinetics calculations, but mainly in the hydraulic aspects. The complete sealing-off of the reactor, including a submerged exit of the effluent and the reduction of weirs, contributes noticeably to the reduction of these risks. The adequate operation of the reactor also contributes to this.

A characteristic aspect of this process is the limitation in the BOD removal efficiency, which is around 70% , therefore lower than in most of the other systems. This must not be considered a disadvantage in itself, but as a characteristic of the process. To reach the desired efficiency, some form of post-treatment must follow the UASB reactors. The post-treatment process can be any of the secondary processes (aerobic or anaerobic) covered in this chapter, or a physical–chemical process, such as dissolved air flotation. The difference is that the post treatment stage is much more compact, since around 70% of the organic load has been previously removed in the anaerobic stage. Besides this, in the case of post-treatment processes that incorporate aeration, the consumption of energy is less, by virtue of the lower influent organic load to the aerated tank. Overall sludge production will be also lower. The total size (volume) of all the treatment units in the

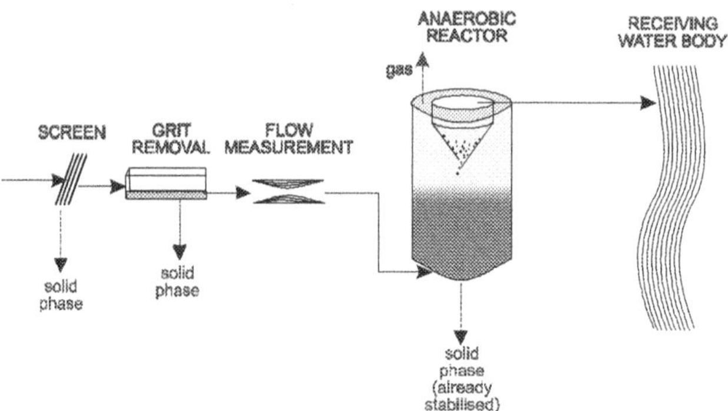



Figure 4.23.   Typical flowsheet of a UASB reactor system (liquid phase)

UASB – post-treatment system will be slightly smaller compared with the alternative of no UASB reactor. Therefore, an economy in the construction and operation costs is usually obtained, in comparison with conventional systems not preceded by an anaerobic stage.

Figure 4.24 illustrates some of the main possible combinations of UASB reactors with post-treatment systems. It can be observed that in the UASB – activated sludge and UASB – biofilm aerobic reactor systems, the aerobic biological excess sludge is simply returned to the UASB reactor, where it undergoes digestion and thickening with the anaerobic sludge, dispensing with the separate digestion and thickening units for the aerobic sludge. Thus a large simplification in the overall flowsheet is obtained, including the liquid (sewage) and solid (sludge) phases.

## 4.5.5  Activated sludge system

There are many variants of the activated sludge process, and the present section covers only the main ones. Under this perspective, activated sludge systems may be classified according to the following categories:

- *Division according to the sludge age* (see concept of sludge age in item a below):
    - Conventional activated sludge
    - Extended aeration

- *Division according to flow:*
    - Continuous flow
    - Intermittent flow (sequencing batch reactors and variants)

- *Division regarding the treatment objectives:*
    - Removal of carbon (BOD)
    - Removal of carbon and nutrients (N and/or phosphorus)

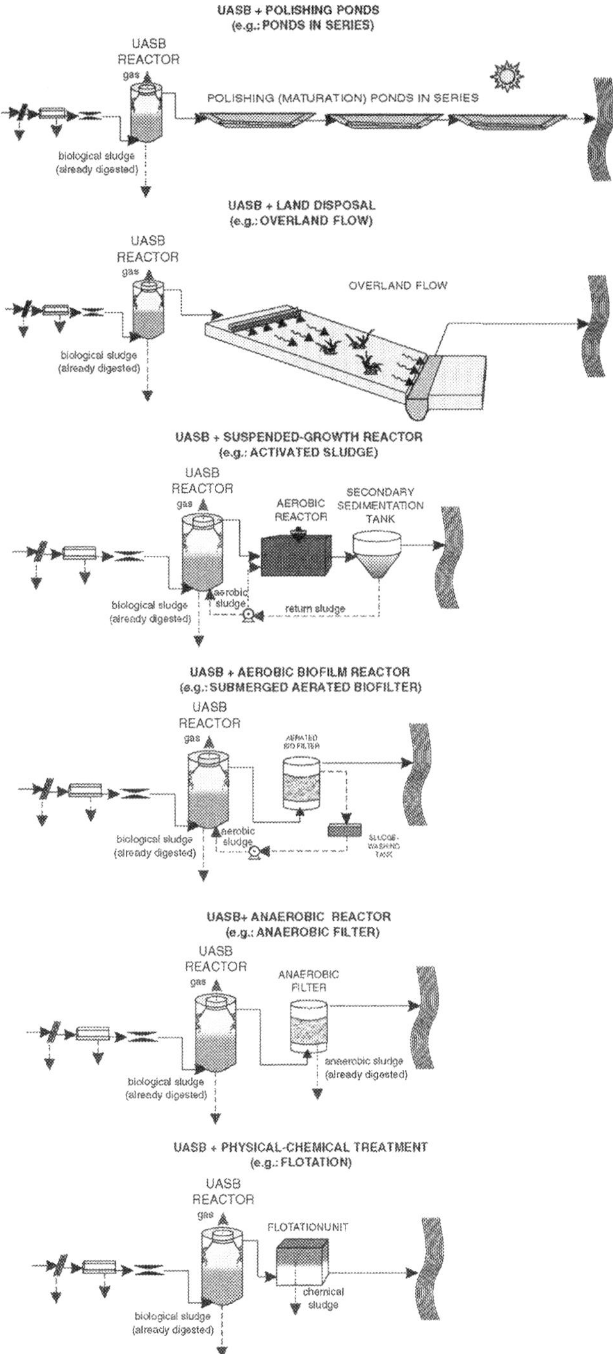


Figure 4.24. Examples of flowsheets of UASB reactors followed by post-treatment processes (liquid phase)

This section presents a brief description of the main variants of the activated sludge process, which are a combination of the above divisions:

- conventional activated sludge (continuous flow)
- extended aeration (continuous flow)
- sequencing batch reactors(intermittent operation)
- activated sludge with biological nitrogen removal
- activated sludge with biological nitrogen and phosphorus removal

All the systems above may be used as post-treatment of the effluent from anaerobic (UASB) reactors. In this case, primary sedimentation tanks (if existing) are substituted by the anaerobic reactor, and the excess sludge from the aerobic stage, if not yet stabilised, is pumped back to the anaerobic reactor, where it undergoes thickening and digestion. Biological nutrient removal is less efficient with the anaerobic pre-treatment, and adaptations or incorporation of physical-chemical treatment may be necessary.

## a) Conventional activated sludge

When analysing the aerated pond systems described in the previous items, it becomes evident that a reduction of the volume required could be reached by increasing the biomass concentration in suspension in the liquid. The more bacteria there are in suspension, the greater the food consumption is going to be, thus the greater the assimilation of the organic matter present in the raw sewage.

Within this concept, analysing the previously described aerated ponds – settling ponds system, it can be observed that there is a storage of bacteria still active in the settling unit. If part of these bacteria is returned to the aeration unit, the concentration of the bacteria in this unit will be greatly increased. This is the basic principle of the activated sludge system, in that the solids are recycled by pumping, from the bottom of the settling unit, to the aeration unit. The following items are therefore essential in the activated sludge system (liquid flow) (see Figure 4.25):

- *aeration tank (reactor)*
- *settling tank (secondary sedimentation tank, also called final clarifiers)*
- *pumps for the sludge recirculation*
- *removal of the biological excess sludge*

The biomass can be separated in the secondary sedimentation tank because of its property of flocculating. This is due to the fact that many bacteria have a gelatinous matrix that permits their agglutination. The floc has larger dimensions, which facilitates settling (see Figure 4.26).

The concentration of the suspended solids in the aeration tank of the activated sludge system is more than 10 times greater than in a complete-mix aerated pond. The detention time of the liquid is very low, of the order of 6 to 8 hours in the conventional activated sludge system, which implies that the aeration tank volume is very small. However, owing to the sludge recirculation, the solids (biomass)

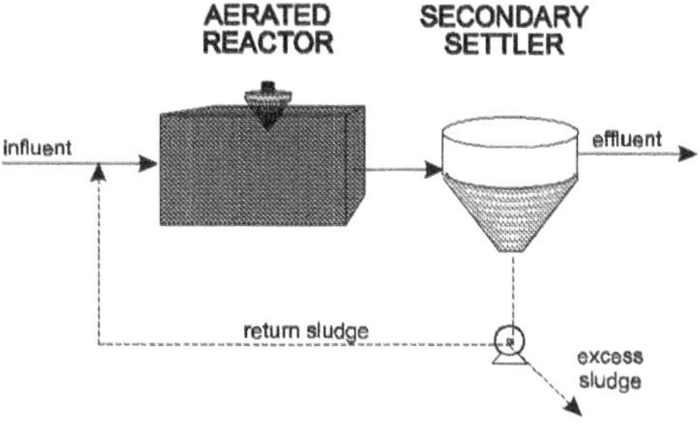


Figure 4.25.   Schematics of the units of the biological stage of the activated sludge system

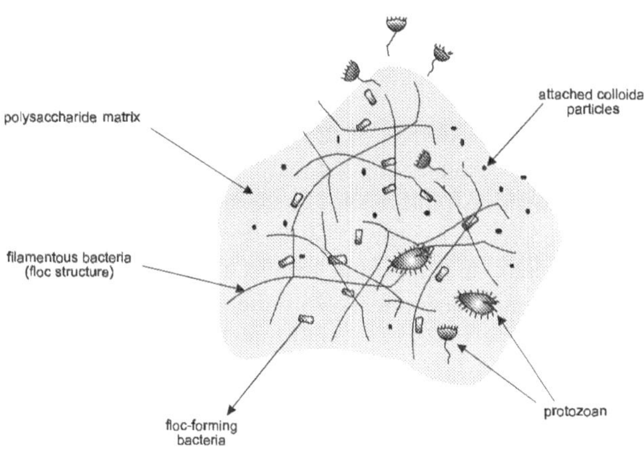


Figure 4.26.   Schematics of bacteria and other microorganisms forming an activated sludge floc

stay in the system for a time longer than that of the liquid. The retention time of the solids in the system is called **sludge age** or **solids retention time**, which is of the order of 4 to 10 days in the conventional activated sludge system. It is this longer retention of the solids in the system that guarantees the high efficiency of the activated sludge, as the biomass has sufficient time to metabolise practically all of the organic matter in the sewage. In the UASB reactor described in the previous section, the biomass is returned to the digestion compartment by gravity from the settling compartment situated on top of the digestion compartment and, therefore, the solids retention time is also greater than the hydraulic detention time.

In the activated sludge system, the tanks are typically made of concrete, different from stabilisation ponds. To save in terms of energy for the aeration, part of the organic matter (in suspension, settleable) of the sewage is removed before the aeration tank, in the primary sedimentation tank. Therefore, the conventional activated sludge systems have as an integral part also the *primary treatment* (Figure 4.27).

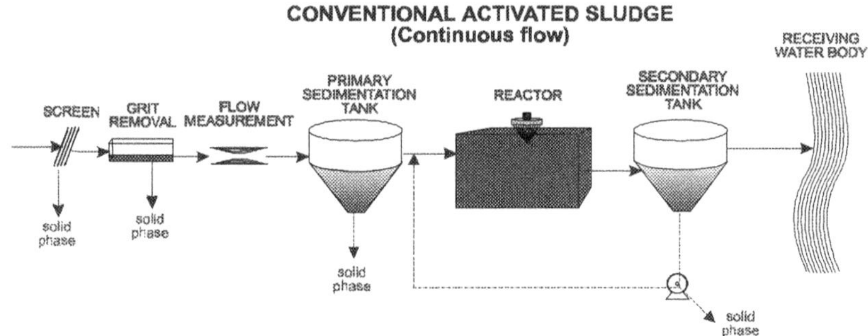


Figure 4.27.   Typical flowsheet of the conventional activated sludge system (liquid phase)

Owing to the continuous arrival of food in the form of BOD to the aeration tank, bacteria grow and reproduce continuously. If an indefinite population growth were allowed, the bacteria would reach excessive concentrations in the aeration tank, making the transfer of oxygen to all bacterial cells difficult. Besides this, the secondary sedimentation tank would become overloaded, the solids would not settle well and they would start to leave with the final effluent, thus deteriorating its quality. To maintain the system in equilibrium, it is necessary to draw approximately the same quantity of biomass that has increased by reproduction. This is, therefore, the *biological excess sludge* that can be wasted directly from the reactor or the recirculation line. The excess sludge must undergo additional treatment in the sludge treatment line.

The conventional activated sludge system has low land requirements and has very good removal efficiencies. However, the flowsheet of the system is more complex than in most other treatment systems, requiring more skill for its control and operation. Energy costs for aeration are higher than for aerated ponds.

**b) Extended aeration**

In the *conventional* activated sludge system, the average retention time of the sludge in the system is between 4 to 10 days. With this sludge age, the biomass removed in the excess sludge still requires a stabilisation stage in the sludge treatment. This is due to the high level of biodegradable organic matter in their cell composition.

However, if the biomass is retained in the system for a longer period, with a sludge age around 18 to 30 days (thus the name *extended aeration*), receiving the same BOD load as the conventional system, there is a lower food availability for the bacteria. Owing to the higher sludge age, the reactor has a greater volume (the liquid detention time is around 16 to 24 hours). Therefore, there is less organic matter per unit volume of the aeration tank, and per unit microbial mass. Consequently, in order to survive, the bacteria start to use in their metabolic processes the organic matter from their cellular material. This cellular organic matter is converted into carbon dioxide and water through respiration. This corresponds to a stabilisation (digestion) of the biomass, taking place in the aeration tank. While in the conventional system the sludge stabilisation is carried out separately (in sludge digesters in the sludge treatment line), in extended aeration systems the digestion is done concurrently with the BOD stabilisation in the reactor.

As there is no need to stabilise separately the excess biological sludge, the generation of another type of sludge in the system that would require subsequent treatment is also avoided. Consequently, extended aeration systems do not usually have primary sedimentation tanks. A great simplification in the flowsheet is obtained: there are no primary sedimentation tanks and no sludge digestion units (Figure 4.28).

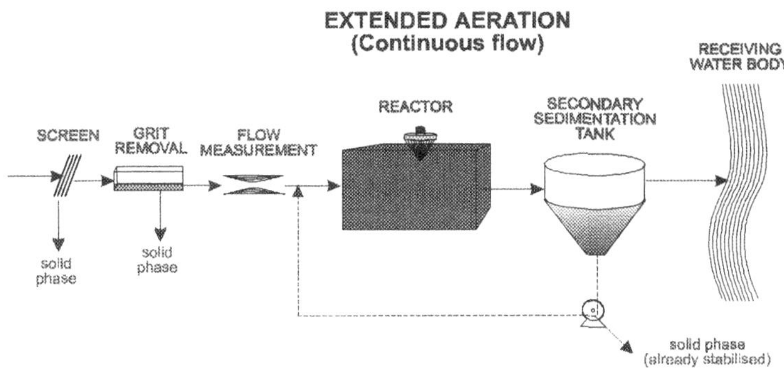


Figure 4.28.   Flowsheet of the extended aeration system (liquid phase)

The consequence of this simplification in the system is the energy expenditure for aeration, which is due, not only to the removal of the incoming BOD, but also for the aerobic digestion of the sludge in the reactor. On the other hand, the reduction in the availability of food and its practically complete assimilation by the biomass makes extended aeration one of the most efficient sewage treatment processes in terms of BOD removal.

### c) Intermittent operation (sequencing batch reactors)

The activated sludge systems described above are of continuous flow with relation to the sewage, that is, the sewage is always entering and leaving the reactor. However, there is a variant of the system which has *intermittent* flow and operation, also called a *sequencing batch reactor* (SBR).

The principle of the activated sludge process with intermittent operation consists in the incorporation of all the units, processes and operations normally associated to the conventional activated sludge (primary sedimentation, biological oxidation, secondary sedimentation, sludge pumping) within a single tank. Using only a single tank, these processes and operations become sequences in time and not separated units, such as in conventional processes with continuous flow. The process of activated sludge with intermittent flow can be used in the conventional or in the extended aeration sludge ages, although the latter is more common, due to its greater operational simplicity. In the extended aeration mode, the tank also incorporates the role of the sludge digestion (aerobic) unit.

The process consists of a complete-mix reactor where all the treatment stages occur. This is accomplished by the establishment of operating cycles with defined duration. The biological mass stays in the reactor, eliminating the need for separate sedimentation and sludge pumping. The retention of biomass occurs because it is not withdrawn with the supernatant (final effluent) after the sedimentation stage, remaining in the tank. The normal treatment cycle is composed of the following stages:

- *Fill* (entrance of the influent in the reactor)
- *React* (aeration/mixture of the liquid/biomass contained in the reactor)
- *Settle* (sedimentation and separation of the suspended solids from the treated sewage)
- *Draw* (removal of the supernatant, which is the treated effluent from the reactor)
- *Idle* (cycle adjustment and removal of the excess sludge)

The usual duration of each stage within the cycle can be altered as a function of the influent flow variations, the treatment needs and the sewage and biomass characteristics.

The wasting of excess sludge generally occurs during the last stage (idle), whose purpose is to allow the adjustment of the stages within the operating cycles of each reactor. However, as this stage is optional or may be short, sludge wasting can happen in other phases of the process. The sludge wasting quantity and frequency are established in function with the performance requirements, in the same way as in the conventional continuous flow processes.

The flowsheet of the process is greatly simplified due to the elimination of various units, compared with the continuous flow activated sludge systems. The only units in an SBR operating in the extended aeration mode are: *screens, grit chamber, reactors, sludge thickener (optional) and sludge drying* (Figure 4.29). With domestic sewage, which arrives at the treatment plant 24 hours per day, more than one tank is necessary, since only the tank in the fill stage is apt to receive the incoming sewage.

There are some variants of the sequencing batch reactor systems related to its operation (continuous feeding and discontinuous supernatant withdrawal) as well as in the sequence and duration of the stages within each cycle. These variations may lead to additional simplifications in the process or to biological nutrient removal.

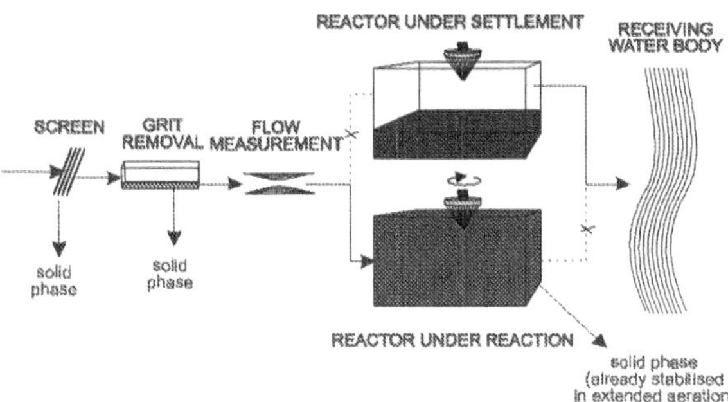


Figure 4.29.    Flowsheet of a sequencing batch reactor system in the extended aeration mode (liquid phase)

## d) Activated sludge with biological nitrogen removal

The activated sludge system is capable of producing, without process alterations, a satisfactory conversion of ammonia to nitrate (**nitrification**). In this case, *only ammonia and not nitrogen is removed*, as there is only a conversion of the nitrogen form. Nitrification occurs almost systematically in warm-climate regions unless there is some environmental problem in the aeration tank, such as lack of dissolved oxygen, low pH, little biomass or the presence of toxic or inhibiting substances.

*Biological nitrogen removal* is achieved in the absence of dissolved oxygen, but presence of nitrates (called *anoxic conditions*). In these conditions, a group of bacteria uses the nitrates in their respiration process, converting them to nitrogen gas, which escapes into the atmosphere. This process is called **denitrification**. To achieve denitrification in the activated sludge, process modifications are necessary, including the creation of anoxic zones and possible internal recycles.

In spite of nitrogen removal being considered as tertiary treatment, biological removal is presented in this item of secondary treatment, as it consists of essentially biological processes and can be achieved through adaptations in the flowsheet of the activated sludge process at a secondary level.

In activated sludge systems where nitrification occurs (mainly in warm-climate regions), it is interesting to induce denitrification to take place intentionally in the reactor. The reasons are usually associated to purely operational aspects, as well as to the final effluent quality:

- *Savings in oxygen* (energy economy in the aeration). Under anoxic conditions, facultative bacteria remove BOD by using the nitrate in their respiratory processes, therefore leading to an economy of oxygen, or in other words, in the energy used for aeration. This economy partially compensates the energy expenditure for nitrification, which occurs, necessarily, under aerobic conditions.

- *Savings in alkalinity* (preservation of the buffering capacity). During nitrification, $H^+$ ions are generated and alkalinity is consumed, which can lead to a decrease in the pH in the aeration tank. Conversely, denitrification consumes $H^+$ and generates alkalinity, partially compensating the pH reduction mechanisms that occur in nitrification.
- *Operation of the secondary sedimentation tank* (to avoid rising sludge). If denitrification occurs in the anoxic conditions in the secondary sedimentation tanks, there will be a production of small nitrogen gas bubbles. These bubbles tend to adhere to the settling flocs, dragging them to the surface and causing a loss of biomass and deterioration in the final effluent quality.
- *Nutrient control* (eutrophication). The reduction of the nitrogen levels is important when the effluent is discharged into sensitive water bodies that are subjected to eutrophication (see Chapter 3).

The main process variants for nitrification and denitrification combined in a single reactor are listed below.

- Pre-denitrification (nitrogen removal with carbon from the raw sewage)
- Post-denitrification (nitrogen removal with carbon from endogenous respiration)
- Bardenpho four-stage process
- Oxidation ditches
- Sequencing batch reactors

**e) Activated sludge with biological nitrogen and phosphorus removal**

Although phosphorus removal can be considered as a tertiary treatment, biological removal is presented in this section on secondary treatment because it consists of essentially biological mechanisms and can be achieved through adaptations of the activated sludge process flowsheet at a secondary level.

It is *essential to have anaerobic and aerobic zones* in the treatment line for the biological removal of phosphorus. The anaerobic zone is considered a *biological selector* for the phosphorus accumulating organisms. This zone allows an advantage in terms of competition for the phosphorus accumulating organisms, since they can assimilate the substrate of this zone before the other microorganisms. In this way, the anaerobic zone gives good conditions for the development or selection of a large population of phosphorus accumulating organisms in the system, which absorb substantial quantities of phosphorus from the liquid, much higher than the normal metabolic requirements. When the biological excess sludge is wasted from the system, phosphorus is removed, since it is present at high concentrations in the phosphorus accumulating organisms that are part of the withdrawn sludge.

Some of the main processes used for both nitrogen and phosphorus removal in the activated sludge system are listed below. Processes employed to remove just phosphorus are not listed due to the difficulties that these undergo with the presence of nitrates in the anaerobic zone. Nitrification occurs almost systematically in the activated sludge plants in warm-climate regions. If efficient denitrification is not provided in the reactor, substantial quantities of nitrates are returned to the anaerobic zone through the recycle lines, impairing the maintenance of strict anaerobic conditions in the anaerobic zone. For this reason, nitrogen removal is encouraged, even if, in terms of the water body requirements, there is only the need to remove phosphorus.

The literature presents a diverging nomenclature in relation to some processes, as a function of variations between commercial and scientific designations.

- $A^2O$ process (three-stage Phoredox)
- Five-stage Bardenpho process (Phoredox)
- UCT process
- Modified UCT process
- Sequencing batch reactors

If higher efficiencies are still desired for phosphorus removal, effluent polishing can adopted. Methods employed are:

- *addition of coagulants* (metallic ions): phosphorus precipitation
- *effluent filtration*: removal of the phosphorus present in the suspended solids in the effluent
- *combination of the addition of coagulants and filtration*

These physical–chemical polishing methods can also be employed for P removal from other biological wastewater treatment process, and not only from the activated sludge process.

## 4.5.6 Aerobic biofilm reactors

In this section, the aerobic units are biofilm reactors, in which the biomass grows attached to a support medium. There are many variants within this broad concept, and the following ones are presented in this section:

- Low rate trickling filter
- High rate trickling filters
- Submerged aerated biofilters
- Rotating biological contactors

All systems may be used as post-treatment of the effluent from anaerobic reactors. In this case, primary sedimentation tanks are substituted by the anaerobic reactor, and the excess sludge from the aerobic stage, if not yet stabilised, is pumped back to the anaerobic reactor, where it undergoes thickening and digestion.

### a) Low rate trickling filter

A trickling filter consists of a coarse material bed, such as stones, gravel, blast furnace slag, plastic material or other, over which the wastewater is applied, in the form of drops or jets. After application, the wastewater percolates in the direction of the drains at the bottom. This percolation allows the bacterial growth on the surface of the support medium, forming an attached biofilm. With the passage of the wastewater, there is a contact between the microorganisms and the organic matter.

The trickling filters are aerobic systems because the air circulates between the empty spaces between the stones, supplying the oxygen for the respiration of the microorganisms. The ventilation is usually natural.

Wastewater is usually applied over the medium through rotating distributors, moved by the hydraulic head of the wastewater. The liquid percolates rapidly through the support medium. However, the microbial film adsorbs the organic matter, which stays adhered for a time sufficient for its stabilisation (see Figure 4.30).

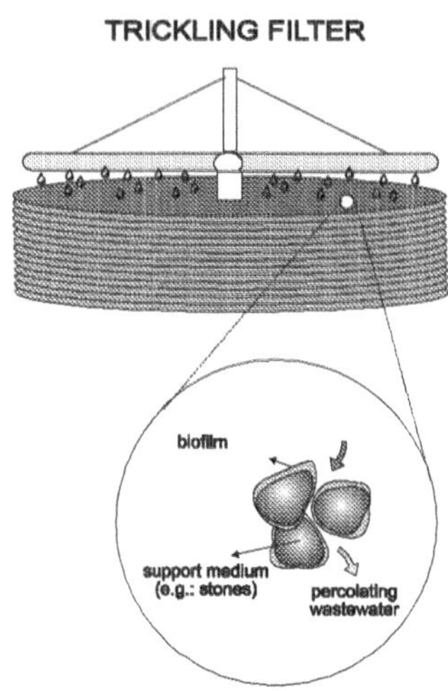


Figure 4.30.  Schematics of a trickling filter

The filters are normally circular and can be of various sizes in diameter (several metres). Contrary to what the name suggests, the primary function of the filter is not to filter. The diameters of the stones used are of the order of a few centimetres, which allows a large void space that is inefficient for the act of filtration by screening. The function of the medium is only to supply support for the formation of the microbial film. There are also synthetic media of various materials and forms, which present the advantage of being lighter than stone, besides having a higher surface area. However, the synthetic media are more expensive. The savings in construction costs must be analysed together with the greater expenditure in purchasing the synthetic media.

With the biomass growth on the surface of the stones, the empty spaces tend to decrease, increasing the liquid velocity through the pores. When growth reaches a certain level, the velocity causes a shearing stress that dislodges part of the attached material. This is a natural form of controlling the microbial population in the medium. The dislodged sludge must be removed by the secondary sedimentation tanks to reduce the level of suspended solids in the final effluent.

The applied BOD load per unit area and volume is lower in the *low rate* trickling filters. Therefore, food availability is low, which results in a partial self digestion of the sludge (self consumption of the cellular organic matter) and a higher BOD removal efficiency in the system. This is analogous to what happens in the extended aeration activated sludge system. This lower BOD load per surface unit of the tank is associated with higher area requirements when compared with high rate systems, which are described in the following item. The low rate trickling filters are still more efficient in the removal of ammonia by nitrification.

The low rate system is conceptually simple. Although the efficiency of the system is comparable with the conventional activated sludge system, the operation is simpler, although less flexible. The trickling filters have a lower capacity to adjust to influent variations, besides requiring a slightly higher total area. In terms of energy consumption, the filters present a very low consumption in relation to the activated sludge system. Figure 4.31 presents a typical flowsheet of low rate trickling filters.

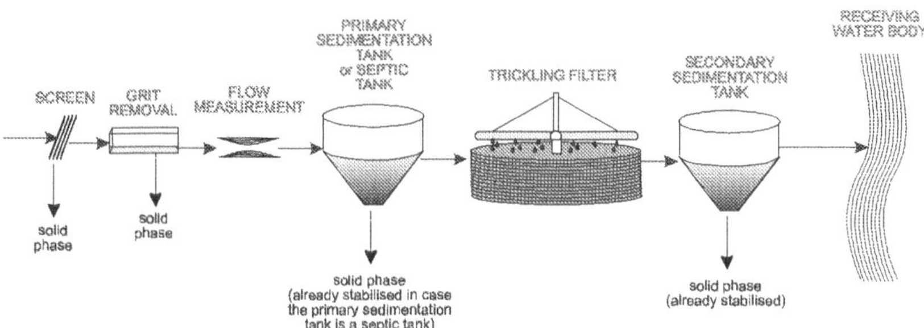


Figure 4.31.  Typical flowsheet of a low rate trickling filter (liquid phase)

## b) High rate trickling filters

High rate trickling filters are conceptually similar to the low rate filters. However, because the high rate units receive a higher BOD load per unit volume of the bed, there are the following main differences: (a) the area requirements are lower; (b) there is a slight reduction in the organic matter removal efficiency; (c) sludge is not digested in the filter.

Another difference is with respect to the existence of a recirculation of the final effluent. This is done with the main objectives of: (a) maintaining an approximately uniform flow during all the day (at night, the distributors could not rotate, due to the low flow, eventually drying the sludge); (b) balancing the influent load; (c) giving a new contact chance of the effluent organic matter with the biomass; and (d) bringing dissolved oxygen into the incoming liquid. The difference from the activated sludge system is that the recirculation of the high rate filters is of the liquid effluent and not of the sludge from the secondary sedimentation tanks (Figure 4.32).

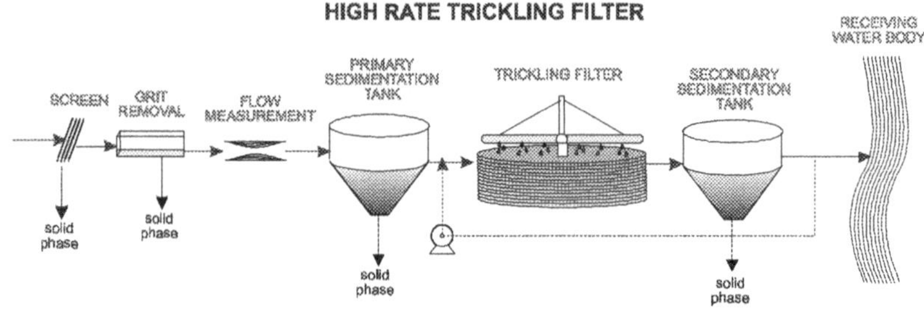


Figure 4.32.    Typical flowsheet of a high rate trickling filter (liquid phase)

Another way of improving the efficiency of trickling filters or to treat wastewaters with high concentrations of organic waste is by using two filters in series. This is called a two-stage trickling filter system. There are various possible configurations with different forms of effluent recirculation.

Some of the limitations of stone-bed trickling filters when operating with high organic loads refer to clogging of the void spaces, due to the excessive growth of the biofilm. In these conditions, flooding (ponding) and system failures may occur.

If land availability is of concern, a careful consideration of the filter media must be exercised. The most commonly used material is still stones and gravel. However, the empty volume is limited in a trickling filter with stones, thus restricting the air circulation in the filter and consequently the quantity of oxygen available for the microorganisms and the quantity of wastewater that can be treated. The specific surface area (exposure area per unit volume of the medium) is also low, reducing the available sites for biofilm attachment and growth.

To overcome these limitations, other materials can be used. These materials include corrugated plastic modules, plastic rings and others. These materials offer larger surface areas for the bacterial growth (approximately double that of typical stones), besides the significant increase in the empty spaces for air circulation. These materials are much lighter than stones (around 30 times), which allows the filters to be much higher without causing structural problems. While filters with stones are usually less than 3 metres in height, filters with synthetic media can be more than 6 metres high, substantially reducing the land required for the installation of the filters.

### c) Submerged aerated biofilters

A submerged aerated biofilter consists of a tank filled with a porous material, through which wastewater and air permanently flow. In almost all of the existing processes, the porous medium is maintained under total immersion. The biofilter is a three-phase reactor composed of (Gonçalves, 1996):

- *Solid phase:* consists of a support medium and microorganism colonies that develop in the form of a biofilm

- *Liquid phase:* consists of the liquid in permanent flow through the porous medium
- *Gas phase:* formed by artificial aeration and, in a reduced scale, by the gaseous by-products of the biological activity

The airflow in the submerged aerated biofilter is always upflow, while the liquid flow can be upflow or downflow.

Biofilters with granular media remove, in the same reactor, soluble organic compounds and suspended solids from the wastewater. Besides serving as support medium for the microorganisms, the granular material performs as an effective filter. In this type of biofilter, periodic washing is necessary to eliminate the accumulated biomass, reducing the hydraulic head losses through the medium. During washing, the feeding with the wastewater is interrupted, and various sequential hydraulic discharges are made with air and cleaning water (Gonçalves, 1996).

The flowsheet of a system composed of a submerged aerated filter is presented in Figure 4.33. The two sources of sludge generation are the primary sedimentation tanks and the washing of the biofilter. The sludge from the washing is collected in a storage tank and pumped to the primary sedimentation tank for clarification outside peak flow times. Therefore, the sludge sent to the sludge treatment stage is a mixed sludge, comprising primary sludge and biological sludge (Gonçalves, 1996). Submerged aerated biofilters are also being successfully applied for the post-treatment of UASB reactors. The aerobic sludge is returned to the UASB reactor, where it undergoes thickening and digestion, thereby simplifying substantially the overall flowsheet (see Figure 4.24) (Chernicharo et al, 2001).

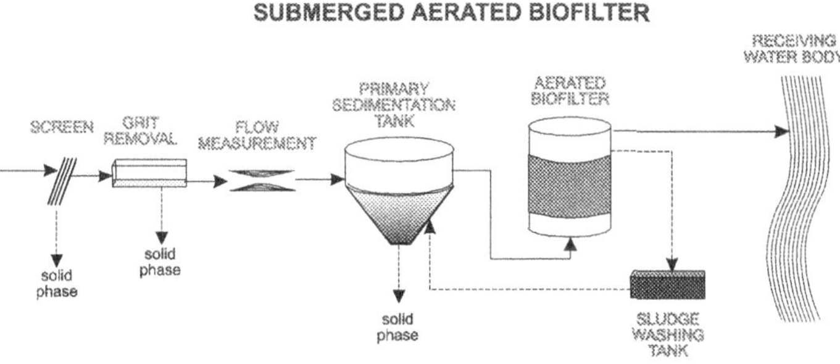


Figure 4.33.   Typical flowsheet of a conventional submerged aerated biofilter system (liquid phase)

Submerged aerated biofilters achieve good nitrification efficiencies and can be modified for the biological removal of nitrogen, through the incorporation of an anoxic zone in the reactor (zone below the air injection).

## d) Rotating biological contactors

The most widely version of rotating biological contactors are the biodiscs, a process that consists of a series of spaced discs, mounted on a horizontal axis. The discs rotate slowly and maintain at each instant around half the surface immersed in the sewage and the other half exposed to the air. Biomass grows attached to the discs, forming a biofilm (see Figure 4.34).

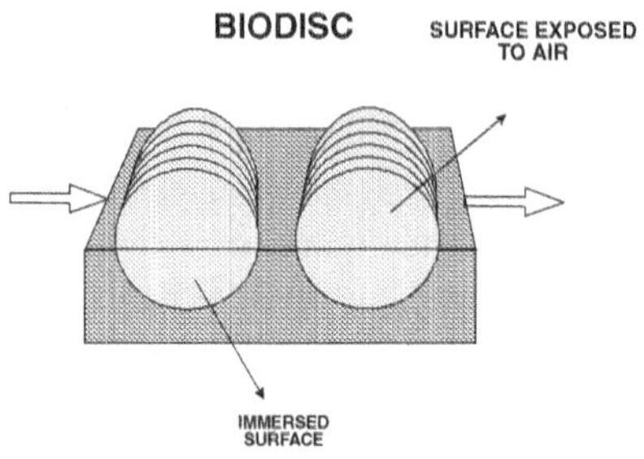


Figure 4.34.   Schematics of a tank with biodiscs

The discs usually are less than 3.6 metres in diameter and are generally constructed of low weight plastic. When the system is put into operation, the microorganisms of the sewage start to adhere to the rotating surfaces, where they grow until the entire disk surface is covered with a fine biological layer, a few millimetres thick. As the discs rotate, the part of the disc exposed to the air brings a thin layer of wastewater, allowing oxygen absorption through the drops and percolation on the surface of the discs. After the discs complete a rotation, this film mixes itself with the wastewater, bringing still some oxygen and mixing the partially and fully treated sewage. With the passage of the microorganisms attached to the disc surface through the wastewater, they absorb a new quantity of organic matter that is used as food.

When the biological layer reaches an excessive thickness, it detaches from the discs. Part of these detached microorganisms is maintained in suspension in the liquid due to the movement of the disk, which increases the efficiency of the system.

The main purposes of the discs are:

- to serve as the surface for microbial film growth;
- to promote the contact between the microbial film and the sewage;
- to maintain the biomass that detached from the discs in suspension in the liquid;
- to promote the aeration of the sewage that is adhered to the disc and the sewage immersed in the liquid.

The growth of the biofilm is similar in concept to the trickling filter, with the difference that the microorganisms pass through the sewage, instead of the sewage passing through the microorganisms, like in the filters. Like the trickling filter process, secondary sedimentation tanks are also necessary, with the objetive of removing the suspended solids.

Biodisc systems are mainly used for the treatment of sewage from small communities. Due to the limitations in the diameter of the discs, it would be necessary to have a large number of discs, often impractical, for the treatment of high flows. The system presents good BOD removal efficiency, although it sometimes shows signs of instability. DO in the effluent may be high. The operational level is moderate and the construction costs are usually high. The flowsheet of the system is presented in Figure 4.35.

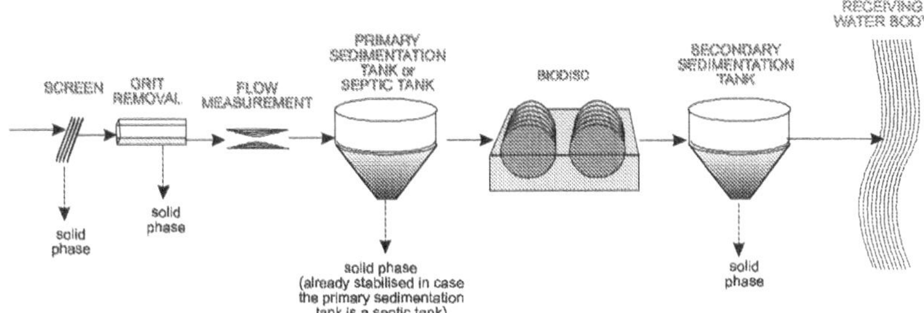


Figure 4.35.   Typical flowsheet of a biodisc system (liquid phase)

## 4.6  REMOVAL OF PATHOGENIC ORGANISMS

The main processes used for removal of pathogenic organisms are listed in Table 4.6. Only short comments are made, since the removal of pathogenic organisms, especially by artifical methods, is outside the scope of this book.

The processes listed above are capable of reaching a coliform removal of 99.99% or more. Regarding pathogenic organisms, bacteria removal efficiency is very high (equal to or higher than coliform removal), and the other pathogens (protozoa, virus, helminths) are usually high, but variable, depending on the removal mechanism and the resistance of each species.

## 4.7  ANALYSIS AND SELECTION OF THE WASTEWATER TREATMENT PROCESS

### 4.7.1  Criteria for the analysis

The decision regarding the wastewater treatment process to be adopted should be derived from a balance between technical and economical criteria, taking into account quantitative and qualitative aspects of each alternative. If the decision regarding economic aspects may seem relatively simple, the same may not be the case with the financial aspects. Besides, the technical points are in many cases intangible and in a large number of situations, the final decision can still have

Table 4.6. Main processes for the removal of pathogenic microorganisms in sewage treatment

| Type | Process | Comment |
|---|---|---|
| Natural | Maturation ponds | • Shallow ponds, where the penetration of solar ultraviolet radiation and unfavourable environmental conditions causes a high mortality of the pathogens.<br>• The maturation ponds do not need chemical products or energy, but require large areas.<br>• They are highly recommended systems (if there is area available), owing to their great simplicity and low costs. |
| | Land treatment (infiltration in soil) | • The unfavourable environmental conditions in the soil favour the mortality of the pathogens.<br>• In slow-rate systems, there is the possibility of plant contamination, depending on the type of application.<br>• Chemical products are not needed.<br>• Requires large areas. |
| Artificial | Chlorination | • Chlorine kills pathogenic microorganisms (although protozoan cysts and helminth eggs are not much affected).<br>• High dosages are necessary, which may increase operational costs. The larger the previous organic matter removal, the lower the chlorine dosage required.<br>• There is a concern regarding the generation of toxic by-products to human beings. However, the great benefit to public health in the removal of pathogens must be taken into consideration.<br>• The toxicity caused by the residual chlorine in the water bodies is also of concern. The residual chlorine must have very low levels, frequently requiring dechlorination.<br>• There is much experience with chlorination in the area of water treatment in various developing countries. |
| | Ozonisation | • Ozone is a very effective agent for the removal of pathogens.<br>• Ozonisation is usually expensive, although the costs are reducing, making this alternative a competitive option in certain specific circumstances.<br>• There is less experience with ozonisation in most developing countries. |
| | Ultraviolet radiation | • Ultraviolet radiation, generated by special lamps, affects the reproduction of the pathogenic agents.<br>• Toxic by-products are not generated.<br>• Ideally, the effluent must be well clarified for the radiation to penetrate well in the liquid mass.<br>• This process has recently shown substantial development, which has made it more competitive or more advantageous than chlorination in various applications. |

Table 4.6 (*Continued*)

| Type | Process | Comment |
|------|---------|---------|
|  | Membranes | • The passage of treated sewage through membranes of minute dimensions (e.g. ultrafiltration, nanofiltration) constitutes a physical barrier for the pathogenic microorganisms, which have larger dimensions than the pores.<br>• The process is highly interesting and does not introduce chemical products into the liquid.<br>• The costs are still high, but they have been reducing significantly in recent years. |

subjectivity. Criteria or weightings can be attributed to the various aspects connected essentially with the reality in focus, so that the selection really leads to the most adequate alternative for the system under analysis. There are no such generalised formulas for this, and common sense and experience when attributing the relative importance of each technical aspect are essential. While the economic side is fundamental, it needs to be remembered that the best alternative is not always the one that simply presents the lowest cost in economic–financial studies.

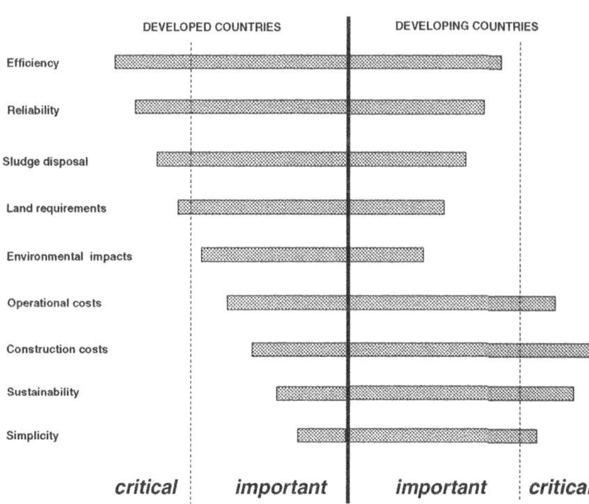


Figure 4.36. Critical and important aspects in the selection of wastewater treatment systems in developed and developing regions (von Sperling, 1996)

Figure 4.36 presents a comparison between important aspects in the selection of treatment systems, analysed in terms of developed and developing regions (von Sperling, 1996). The comparison is unavoidably general, owing to the specificity of each region or country and the high contrasts usually observed in developing countries. The items are organised in a decreasing order of importance for the developed regions. In these regions, the critical items are usually: efficiency, reliability, sludge disposal aspects and land requirements. In developing regions, these first items are organised in a similar manner of decreasing importance, but have a lower magnitude, in comparison with the developed regions. The main difference resides in what are considered the critical items for the developing regions: construction costs, sustainability, simplicity and operational costs. These items are of course important in developed regions, but cannot be usually considered critical.

Table 4.7 presents general factors to be taken into account when selecting and evaluating unit operations and processes in wastewater treatment, while Table 4.8 presents environmental aspects to be considered in the selection of processes for wastewater treatment and sludge management.

Each of these factors must be evaluated in terms of the local conditions and the technology employed. The reliability of the monitoring system must also be considered.

## 4.7.2 Comparison between the wastewater treatment systems

Presented below is a comparative analysis between the main wastewater treatment systems (liquid and solid phases) applied to domestic sewage. The analysis is summarised in various tables and figures:

- *Quantitative comparison* (Table 4.9): average effluent concentrations and typical removal efficiencies of the main pollutants of interest in domestic sewage
- *Quantitative comparison* (Table 4.10): typical characteristics of the main sewage treatment systems, expressed in per-capita values
- *Diagrammatic comparison* (Tables 4.11 to 4.13): capacity of the various sewage treatment systems in consistently reaching different quality levels in terms of BOD, COD, SS, ammonia, total nitrogen, total phosphorus, faecal coliforms and helminth eggs (based on von Sperling & Chernicharo, 2002)
- *Diagrammatic comparison* (Tables 4.14 to 4.18): per capita values of land requirement, power for aeration, production of sludge to be disposed of, construction costs and operation and maintenance costs, for various sewage treatment processes.
- *Qualitative comparison* (Table 4.19): a qualitative comparative analysis that covers various relevant aspects in the evaluation of the sewage treatment systems. The aspects of efficiency, economy, process and environmental problems are analysed.
- *Description* (Table 4.20): a list of the basic equipment usually necessary in the main sewage treatment systems.
- *Advantages and disadvantages* (Table 4.21): main advantages and disadvantages of the various sewage treatments systems. This analysis is principally oriented for the comparison of the processes within the same system, although it still permits, within certain limitations, the comparison between distinct systems.

Table 4.7. Important factors to be considered when evaluating and selecting unit operations and processes

| Condition | Factor |
| --- | --- |
| Process applicability | The applicability of a process is evaluated based on past experience, published data, data from operating works and from pilot plants. If new or unusual conditions are found, pilot scale studies are necessary. |
| Applicable flow | The process must be adequate for the expected flow range. |
| Acceptable flow variation | The majority of the operations and processes must be designed to operate over a wide flow range. The highest efficiency is usually obtained with a constant flow, although some variation can be tolerated. Equalisation of the flow could be necessary if the variation is very large. |
| Influent characteristics | The characteristics of the influent wastewater affect the process types to be used (e.g. chemical or biological) and the requirements for their adequate operation. |
| Inhibiting or refractory compounds | What are the constituents in the wastewater that could be inhibitory or toxic, and under what conditions? What constituents are not affected during the treatment? |
| Climatic aspects | Temperature affects reaction rates of most chemical and biological processes. Temperature can also affect the physical operation of the units. High temperatures can accelerate odour generation. |
| Process kinetics and reactor hydraulics | The design of the reactor is based on reaction kinetics. Kinetic data are normally obtained from experience, literature or pilot studies. Reactor configuration also plays an important role in the removal of some constituents. |
| Performance | Performance is normally measured in terms of the quality of the effluent, which should be consistent with the discharge requirements and/or the discharge standards. |
| Treatment residuals | The type, quality and quantity of the solids, liquids and gaseous by-products need to be known or estimated. If necessary, undertake a pilot study. |
| Sludge processing | Are there limitations that could make the sludge processing and disposal expensive or unfeasible? What is the influence in the liquid phase of the loads recycled from the sludge treatment units? The selection of the sludge-processing system must be done in parallel with the selection of the treatment processes of the liquid phase. |
| Environmental constraints | Environmental factors, such as prevailing winds and proximity to residential areas could restrict the use of certain processes, especially when odours are released. Noise and traffic could affect the selection of the works location. |
| Chemical product requirements | What resources and quantities must be guaranteed for the satisfactory operation of the unit for a long period of time? |
| Energy requirements | The energy requirements, together with the probable future energy costs, need to be estimated if it is desired to design cost-effective treatment systems. |
| Requirements of other resources | What additional resources are necessary to guarantee a satisfactory implementation and operation of the system? |

(*Continued*)

Table 4.7 (*Continued*)

| Condition | Factor |
|---|---|
| Personnel requirements | How many people and what levels of skills are necessary to operate the system? Are the skills easily found? What level of training will be necessary? |
| Operating and maintenance requirements | What are the special operational requirements that need to be provided? Which and how many spare parts will be required, and what is their availability and cost? |
| Ancillary processes | What support processes are necessary? How do they affect the effluent quality, especially when they become inoperative? |
| Reliability | What is the reliability of the operation and process in consideration? Is the unit likely to present frequent problems? Can the process resist periodical shock loads? If yes, is the effluent quality affected? |
| Complexity | What is the complexity of the process in routine and emergency operation? What is the level of training that an operator needs to operate the process? |
| Compatibility | Can the unit operation or process be used satisfactorily with the existing units? Can plant expansion be easily accomplished? |
| Area availability | Is there space availability to accommodate, not only the currently required units, but possible future expansions? Is there a buffer zone available to provide landscaping to minimise the aesthetical and environmental impacts in the neighbourhood? |

Source: adapted from Metcalf & Eddy (1991)

Table 4.8. Some environmental impacts to be considered in wastewater treatment and sludge management

| Item | Comment |
|---|---|
| Odours | Must be considered in the wastewater treatment and in the processing and disposal of the sludge. Important factor, mainly in urbanised areas. |
| Vector attraction | Vector (e.g. insects) attraction is connected with odour and can be one of the biggest problems in the sludge processing and disposal. |
| Noise | Important factor, principally in urbanised areas. |
| Sludge transportation | Transportation form and route need to be considered. |
| Sanitary risks | Although difficult to be evaluated objectively, the risk is related to the number of people exposed to the sewage, receiving body and sludge, their qualities and the infection routes. |
| Air contamination | Air can be contaminated by particulated material from aerosols and sprinkling. |
| Soil and subsoil contamination | Highly variable in function of the type of wastewater treatment and sewage and sludge disposals, and the processes employed. |
| Surface or ground water contamination | One of the main aspects of the disposal of wastewater and sludge. Risk depends on the technology employed. |
| Devaluation of nearby areas | The cost of land and property may be affected by the implementation of a wastewater treatment plant or a disposal site. |
| Inconvenience to the nearby population | Besides affecting many people, some solutions can generate opposition groups against the implementation of a certain system. |

*Source:* adapted from Fernandes et al (2001)

Table 4.9.  Average effluent concentrations and typical removal efficiencies of the main pollutants of interest in domestic sewage

| System | Average quality of the effluent | | | | | | | | Average removal efficiency | | | | | | |
|---|---|---|---|---|---|---|---|---|---|---|---|---|---|---|---|
| | $BOD_5$ (mg/L) | COD (mg/L) | SS (mg/L) | Ammonia (mg/L) | Total N (mg/L) | Total P (mg/L) | FC (FC/100mL) | Helminth eggs (eggs/L) | $BOD_5$ (%) | COD (%) | SS (%) | Ammonia (%) | Total N (%) | Total P (%) | FC (log units) |
| Primary treatment (septic tanks) | 200–250 | 400–450 | 100–150 | >20 | >30 | >4 | $10^7$–$10^8$ | >1 | 30–35 | 25–35 | 55–65 | <30 | <30 | <35 | <1 |
| Conventional primary treatment | 200–250 | 400–450 | 100–150 | >20 | >30 | >4 | $10^7$–$10^8$ | >1 | 30–35 | 25–35 | 55–65 | <30 | <30 | <35 | <1 |
| Advanced primary treatment (chemically enhanced) | 60–150 | 150–250 | 30–90 | >20 | >30 | <2 | $10^6$–$10^7$ | >1 | 45–80 | 55–75 | 60–90 | <30 | <30 | 75–90 | ≈1 |
| Facultative pond | 50–80 | 120–200 | 60–90 | >15 | >20 | >4 | $10^6$–$10^7$ | <1 | 75–85 | 65–80 | 70–80 | <50 | <60 | <35 | 1–2 |
| Anaerobic pond + facultative pond | 50–80 | 120–200 | 60–90 | >15 | >20 | >4 | $10^6$–$10^7$ | <1 | 75–85 | 65–80 | 70–80 | <50 | <60 | <35 | 1–2 |
| Facultative aerated lagoon | 50–80 | 120–200 | 60–90 | >20 | >30 | >4 | $10^6$–$10^7$ | >1 | 75–85 | 65–80 | 70–80 | <30 | <30 | <35 | 1–2 |
| Complete-mix aerated lagoon + sedimentation pond | 50–80 | 120–200 | 40–60 | >20 | >30 | >4 | $10^6$–$10^7$ | >1 | 75–85 | 65–80 | 80–87 | <30 | <30 | <35 | 1–2 |
| Anaerobic pond + facult. pond + maturation pond | 40–70 | 100–180 | 50–80 | 10–15 | 15–20 | <4 | $10^2$–$10^4$ | <1 | 80–85 | 70–83 | 73–83 | 50–65 | 50-65 | >50 | 3–5 |
| Anaerobic pond + facultative pond + high rate pond | 40–70 | 100–180 | 50–80 | 5–10 | 10–15 | 3-4 | $10^4$–$10^5$ | >1 | 80–85 | 70–83 | 73–83 | 65–85 | 75-90 | 50–60 | 3–4 |
| Anaerobic pond – facultative pond + algae removal | 30–50 | 100–150 | <30 | >15 | >20 | >4 | $10^4$–$10^5$ | >1 | 85–90 | 75–83 | >90 | <50 | <60 | <35 | 3–4 |
| Slow rate treatment | <20 | <80 | <20 | <5 | <10 | <1 | $10^2$–$10^4$ | <1 | 90–99 | 85–95 | >93 | >80 | >75 | >85 | 3–5 |
| Rapid infiltration | <20 | <80 | <20 | <10 | <15 | <4 | $10^3$–$10^4$ | <1 | 85–98 | 80–93 | >93 | >65 | >65 | >50 | 4–5 |
| Overland flow | 30–70 | 100–150 | 20–60 | 10–20 | >15 | >4 | $10^4$–$10^6$ | <1 | 80–90 | 75–85 | 80–93 | 35–65 | <65 | <35 | 2–3 |
| Constructed wetlands | 30–70 | 100–150 | 20–40 | >15 | >20 | >4 | $10^4$–$10^5$ | <1 | 80–90 | 75–85 | 87–93 | <50 | <60 | <35 | 3–4 |
| Septic tank + anaerobic filter | 40–80 | 100–200 | 30–60 | >15 | >20 | >4 | $10^6$–$10^7$ | >1 | 80–85 | 70–80 | 80–90 | <45 | <60 | <35 | 1–2 |
| Septic tank + infiltration | <20 | <80 | <20 | <10 | <15 | <4 | $10^3$–$10^4$ | <1 | 90–98 | 85–95 | >93 | >65 | >65 | >50 | 4–5 |

(*Continued*)

Table 4.9. *(Continued)*

| System | Average quality of the effluent | | | | | | | | Average removal efficiency | | | | | | |
|---|---|---|---|---|---|---|---|---|---|---|---|---|---|---|---|
| | $BOD_5$ (mg/L) | COD (mg/L) | SS (mg/L) | Ammonia (mg/L) | Total N (mg/L) | Total P (mg/L) | FC (FC/100mL) | Helminth eggs (eggs/L) | $BOD_5$ (%) | COD (%) | SS (%) | Ammonia (%) | Total N (%) | Total P (%) | FC (log units) |
| UASB reactor | 70–100 | 180–270 | 60–100 | >15 | >20 | >4 | $10^6$–$10^7$ | >1 | 60–75 | 55–70 | 65–80 | <50 | <60 | <35 | 1–2 |
| UASB + activated sludge | 20–50 | 60–150 | 20–40 | 5–15 | >20 | >4 | $10^6$–$10^7$ | >1 | 83–93 | 75–88 | 87–93 | 50–85 | <60 | <35 | 1–2 |
| UASB + submerged aerated biofilter | 20–50 | 60–150 | 20–40 | 5–15 | >20 | >4 | $10^6$–$10^7$ | >1 | 83–93 | 75–88 | 87–93 | 50–85 | <60 | <35 | 1–2 |
| UASB + anaerobic filter | 40–80 | 100–200 | 30–60 | >15 | >20 | >4 | $10^6$–$10^7$ | >1 | 75–87 | 70–80 | 80–90 | <50 | <60 | <35 | 1–2 |
| UASB + high rate trickling filter | 20–60 | 70–180 | 20–40 | >15 | >20 | >4 | $10^6$–$10^7$ | >1 | 80–93 | 73–88 | 87–93 | <50 | <60 | <35 | 1–2 |
| UASB + dissolved-air flotation | 20–50 | 60–100 | 10–30 | >20 | >30 | 1–2 | $10^6$–$10^7$ | >1 | 83–93 | 83–90 | 90–97 | <30 | <30 | 75–88 | 1–2 |
| UASB + maturation ponds | 40–70 | 100–180 | 50–80 | 10–15 | 15-20 | <4 | $10^2$–$10^4$ | <1 | 77–87 | 70–83 | 73–83 | 50–65 | 50–65 | >50 | 3–5 |
| UASB + facultative aerated pond | 50–80 | 120–200 | 60–90 | >20 | >30 | >4 | $10^6$–$10^7$ | >1 | 75–85 | 65–80 | 70–80 | <30 | <30 | <35 | 1–2 |
| UASB + compl. mix. aerated lagoon + sedim. pond | 50–80 | 120–200 | 40–60 | >20 | >30 | >4 | $10^6$–$10^7$ | >1 | 75–85 | 65–80 | 80–87 | <30 | <30 | <35 | 1–2 |
| UASB + overland flow | 30–70 | 90–180 | 20–60 | 10–20 | >15 | >4 | $10^4$–$10^6$ | <1 | 77–90 | 70–85 | 80–93 | 35–65 | <65 | <35 | 2–3 |
| Conventional activated sludge | 15–40 | 45–120 | 20–40 | <5 | >20 | >4 | $10^6$–$10^7$ | >1 | 85–93 | 80–90 | 87–93 | >80 | <60 | <35 | 1–2 |
| Activated sludge – extended aeration | 10–35 | 30–100 | 20–40 | <5 | >20 | >4 | $10^6$–$10^7$ | >1 | 90–97 | 83–93 | 87–93 | >80 | <60 | <35 | 1–2 |
| Sequencing batch reactor (extended aeration) | 10–35 | 30–100 | 20–40 | <5 | >20 | >4 | $10^6$–$10^7$ | >1 | 90–97 | 83–93 | 87–93 | >80 | <60 | <35 | 1–2 |
| Convent. activated sludge with biological N removal | 15–40 | 45–120 | 20–40 | <5 | <10 | >4 | $10^6$–$10^7$ | >1 | 85–93 | 80–90 | 87–93 | >80 | >75 | <35 | 1–2 |
| Convent. activated sludge with biolog. N/P removal | 15–40 | 45–120 | 20–40 | <5 | <10 | 1–2 | $10^6$–$10^7$ | >1 | 85–93 | 80–90 | 87–93 | >80 | >75 | 75–88 | 1–2 |
| Conventional activated sludge + tertiary filtration | 10–20 | 30–60 | 10–20 | <5 | >20 | 3–4 | $10^2$–$10^4$ | <1 | 93–98 | 90–95 | 93–97 | >80 | <60 | 50–60 | 3–5 |

| Low rate trickling filter | 15–40 | 30–120 | 20–40 | 5–10 | >20 | >4 | $10^6$–$10^7$ | >1 | 85–93 | 80–90 | 87–93 | 65–85 | <60 | <35 | 1–2 |
| High rate trickling filter | 30–60 | 80–180 | 20–40 | >15 | >20 | >4 | $10^6$–$10^7$ | >1 | 80–90 | 70–87 | 87–93 | <50 | <60 | <35 | 1–2 |
| Submerged aerated biofilter with nitrification | 15–35 | 30–100 | 20–40 | <5 | >20 | >4 | $10^6$–$10^7$ | >1 | 88–95 | 83–90 | 87–93 | >80 | <60 | <35 | 1–2 |
| Submerged aerated biofilter with biolog. N removal | 15–35 | 30–100 | 20–40 | <5 | <10 | >4 | $10^6$–$10^7$ | >1 | 88–95 | 83–90 | 87–93 | >80 | >75 | <35 | 1–2 |
| Rotating biological contactor | 15–35 | 30–100 | 20–40 | 5–10 | >20 | >4 | $10^6$–$10^7$ | >1 | 88–95 | 83–90 | 87–93 | 65–85 | <60 | <35 | 1–2 |

- Chemical precipitation of phosphorus with any of the technologies above: $P < 1$ mg/L
- Disinfection: e.g. chlorination, ozonisation, UV radiation; Barrier: e.g. membranes (provided the disinfection/barrier process is compatible with the quality of the effluent from the preceding treatment): $CF < 10^3$ FC/100ml; helminth eggs: variable
- Advanced primary treatment: the removal efficiencies vary depending on the coagulant dosage

Table 4.10. Typical characteristics of the main sewage treatment systems, expressed as per capita values

| System | Land requirements ($m^2$/inhab) | Power for aeration | | Sludge volume | | Costs | |
| --- | --- | --- | --- | --- | --- | --- | --- |
| | | Installed power (W/inhab.) | Consumed power (kWh/inhab.year) | Liquid sludge to be treated (L/ inhab.year) | Dewatered sludge to be disposed of (L/ inhab.year) | Construction (US$/inhab.) | Operation and maintenance (US$/inhab.year) |
| Primary treatment (septic tanks) | 0.03–0.05 | 0 | 0 | 110–360 | 15–35 | 12–20 | 0.5–1.0 |
| Conventional primary treatment | 0.02–0.04 | 0 | 0 | 330–730 | 15–40 | 12–20 | 0.5–1.0 |
| Advanced primary treatment (chemically enhanced) (a) | 0.04–0.06 | 0 | 0 | 730–2500 | 40–110 | 15–25 | 3.0–6.0 |
| Facultative pond | 2.0–4.0 | 0 | 0 | 35–90 | 15–30 | 15–30 | 0.8–1.5 |
| Anaerobic pond + facultative pond | 1.2–3.0 | 0 | 0 | 55–160 | 20–60 | 12–30 | 0.8–1.5 |
| Facultative aerated lagoon | 0.25–0.5 | 1.2–2.0 | 11–18 | 30–220 | 7–30 | 20–35 | 2.0–3.5 |
| Complete-mix aerated lagoon + sedimentation pond | 0.2–0.4 | 1.8–2.5 | 16–22 | 55–360 | 10–35 | 20–35 | 2.0–3.5 |
| Anaerobic pond + facultative pond + maturation pond | 3.0–5.0 | 0 | 0 | 55–160 | 20–60 | 20–40 | 1.0–2.0 |
| Anaerobic pond + facultative pond + high rate pond | 2.0–3.5 | <0.3 | <2 | 55–160 | 20–60 | 20–35 | 1.5–2.5 |
| Anaerobic pond – facultative pond + algae removal | 1.7–3.2 | 0 | 0 | 60–190 | 25–70 | 20–35 | 1.5–2.5 |
| Slow rate treatment | 10–50 | 0 | 0 | – | – | 8–25 | 0.4–1.2 |
| Rapid infiltration | 1.0–6.0 | 0 | 0 | – | – | 12–30 | 0.5–1.5 |
| Overland flow | 2.0–3.5 | 0 | 0 | – | – | 15–30 | 0.8–1.5 |
| Constructed wetlands | 3.0–5.0 | 0 | 0 | – | – | 20–30 | 1.0–1.5 |
| Septic tank + anaerobic filter | 0.2–0.35 | 0 | 0 | 180–1000 | 25–50 | 30–50 | 2.5–4.0 |
| Septic tank + infiltration | 1.0–1.5 | 0 | 0 | 110–360 | 15–35 | 25–40 | 1.2–2.0 |
| UASB reactor | 0.03–0.10 | 0 | 0 | 70–220 | 10–35 | 12–20 | 1.0–1.5 |
| UASB + activated sludge | 0.08–0.2 | 1.8–3.5 | 14–20 | 180–400 | 15–60 | 30–45 | 2.5–5.0 |
| UASB + submerged aerated biofilter | 0.05–0.15 | 1.8–3.5 | 14–20 | 180–400 | 15–55 | 25–40 | 2.5–5.0 |
| UASB + anaerobic filter | 0.05–0.15 | 0 | 0 | 150–300 | 10–50 | 20–30 | 1.5–2.2 |
| UASB + high rate trickling filter | 0.1–0.2 | 0 | 0 | 180–400 | 15–55 | 25–35 | 2.0–3.0 |

| | | | | | | | |
|---|---|---|---|---|---|---|---|
| UASB + dissolved-air flotation | 0.05–0.15 | 1.0–1.5 | 8–12 | 300–470 | 25–75 | 25–35 | 2.5–3.5 |
| UASB + maturation ponds | 1.5–2.5 | 0 | 0 | 150–250 | 10–35 | 15–30 | 1.8–3.0 |
| UASB + facultative aerated pond | 0.15–0.3 | 0.3–0.6 | 2–5 | 150–300 | 15–50 | 15–35 | 2.0–3.5 |
| UASB + compl.mix aerated lagoon + sediment. pond | 0.1–0.3 | 0.5–0.9 | 4–8 | 150–300 | 15–50 | 15–35 | 2.0–3.5 |
| UASB + overland flow | 1.5–3.0 | 0 | 0 | 70–220 | 10–35 | 20–35 | 2.0–3.0 |
| Conventional activated sludge | 0.12–0.25 | 2.5–4.5 | 18–26 | 1100–3000 | 35–90 | 40–65 | 4.0–8.0 |
| Activated sludge – extended aeration | 0.12–0.25 | 3.5–5.5 | 20–35 | 1200–2000 | 40–105 | 35–50 | 4.0–8.0 |
| Sequencing batch reactor (extended aeration) (b) | 0.12–0.25 | 4.5–6.0 | 20–35 | 1200–2000 | 40–105 | 35–50 | 4.0–8.0 |
| Conventional activated sludge with biological N removal | 0.12–0.25 | 2.2–4.2 | 15–22 | 1100–3000 | 35–90 | 45–70 | 4.0–9.0 |
| Convention. activated sludge with biological N/P removal | 0.12–0.25 | 2.2–4.2 | 15–22 | 1100–3000 | 35–90 | 50–75 | 6.0–10.0 |
| Conventional activated sludge + tertiary filtration | 0.15–0.30 | 2.5–4.5 | 18–26 | 1200–3100 | 40–100 | 50–75 | 6.0–10.0 |
| Low rate trickling filter | 0.15–0.3 | 0 | 0 | 360–1100 | 35–80 | 50–60 | 4.0–6.0 |
| High rate trickling filter | 0.12–0.25 | 0 | 0 | 500–1900 | 35–80 | 50–60 | 4.0–6.0 |
| Submerged aerated biofilter with nitrification | 0.1–0.15 | 2.5–4.5 | 18–26 | 1100–3000 | 35–90 | 30–50 | 3.0–6.0 |
| Submerged aerated biofilter with biological N removal | 0.1–0.15 | 2.2–4.2 | 15–22 | 11000–3000 | 35–90 | 30–50 | 3.0–6.0 |
| Rotating biological contactor | 0.1–0.2 | 0 | 0 | 330–1500 | 20–75 | 50–60 | 4.0–6.0 |

- Costs based on Brazilian experience (basis: year 2002 – US$1.00 = R$2.50)
- Per capita costs are applicable inside the typical population ranges within which each treatment system is usually applied (usually, for a certain system, the lower the population, the greater the per capita costs)
- Additional disinfection: construction costs – increase US$2.0 to 4.0/inhab.; operational and maintenance costs: increase US$0.2 to 0.6/inhab.year
- In compact aerated systems (e.g.: activated sludge, submerged aerated biofilters) or after treatment with a UASB reactor, aeration control allows a certain economy (not all the installed power is consumed)

(a) Advanced primary treatment: the operational costs depend on the dosage of the chemical product
(b) Sequencing batch reactors (activated sludge) have a greater installed power compared with the consumed power, because all reactors have aerators, but not all aerators are turned on simultaneously

Table 4.11. Capacity of various wastewater treatment technologies in consistently achieving the indicated levels of effluent quality in terms of BOD, COD and SS

| System | BOD | | | | | COD | | | SS | | |
|---|---|---|---|---|---|---|---|---|---|---|---|
| | 100 mg/L | 80 mg/L | 60 mg/L | 40 mg/L | 20 mg/L | 200 mg/L | 150 mg/L | 100 mg/L | 90 mg/L | 60 mg/L | 30 mg/L |
| Facultative pond | | | | | | | | | | | |
| Anaerobic pond + facultative pond | | | | | | | | | | | |
| Facultative aerated lagoon | | | | | | | | | | | |
| Complete-mix aerated lagoon + sedimentation pond | | | | | | | | | | | |
| Anaerobic pond + facultat. pond + maturation pond | | | | | | | | | | | |
| Anaerobic pond + facultative pond + high rate pond | | | | | | | | | | | |
| Anaerobic pond – facultative pond + algae removal | | | | | | | | | | | |
| Slow rate treatment | | | | | | | | | | | |
| Rapid infiltration | | | | | | | | | | | |
| Overland flow | | | | | | | | | | | |
| Constructed wetlands | | | | | | | | | | | |
| Septic tank + anaerobic filter | | | | | | | | | | | |
| Septic tank + infiltration | | | | | | | | | | | |
| UASB reactor | | | | | | | | | | | |
| UASB + activated sludge | | | | | | | | | | | |
| UASB + submerged aerated biofilter | | | | | | | | | | | |
| UASB + anaerobic filter | | | | | | | | | | | |
| UASB + high rate trickling filter | | | | | | | | | | | |
| UASB + maturation ponds | | | | | | | | | | | |
| UASB + overland flow | | | | | | | | | | | |
| Conventional activated sludge | | | | | | | | | | | |
| Extended aeration | | | | | | | | | | | |
| Sequencing batch reactor | | | | | | | | | | | |
| Activated sludge with biological N removal | | | | | | | | | | | |
| Activated sludge with biological N/P removal | | | | | | | | | | | |
| Activated sludge + tertiary filtration | | | | | | | | | | | |
| Low rate trickling filter | | | | | | | | | | | |
| High rate trickling filter | | | | | | | | | | | |
| Submerged aerated biofilter | | | | | | | | | | | |
| Submerged aerated biofilter with biological N removal | | | | | | | | | | | |
| Rotating biological contactor | | | | | | | | | | | |

Table 4.12.   Capacity of various wastewater treatment technologies in consistently achieving the indicated levels of effluent quality in terms of Ammonia, total N and total P

| System | Ammonia – N | | | Total N | | | Total P | | | |
|---|---|---|---|---|---|---|---|---|---|---|
| | 15 mg/L | 10 mg/L | 5 mg/L | 20 mg/L | 15 mg/L | 10 mg/L | 4.0 mg/L | 3.0 mg/L | 2.0 mg/L | 1.0 mg/L |
| Facultative pond | | | | | | | | | | |
| Anaerobic pond + facultativepond | | | | | | | | | | |
| Facultative aerated lagoon | | | | | | | | | | |
| Complete-mix aerated lagoon + sedimentation pond | | | | | | | | | | |
| Anaerobic pond + facultative pond + maturation pond | ▨ | | | ▨ | ▨ | | | | | |
| Anaerobic pond + facultative pond + high rate pond | ▨ | ▨ | | ▨ | ▨ | ▨ | ▨ | | | |
| Anaerobic pond – facultative pond + algae removal | | | | | | | | | | |
| Slow rate treatment | ▨ | ▨ | ▨ | ▨ | ▨ | ▨ | ▨ | ▨ | ▨ | ▨ |
| Rapid infiltration | ▨ | ▨ | ▨ | ▨ | ▨ | | ▨ | ▨ | | |
| Overland flow | ▨ | ▨ | | ▨ | ▨ | | | | | |
| Constructed wetlands | | | | | | | | | | |
| Septictank + anaerobic filter | | | | | | | | | | |
| Septic tank + infiltration | ▨ | ▨ | ▨ | ▨ | ▨ | ▨ | | | | |
| UASB reactor | | | | | | | | | | |
| UASB + activated sludge | ▨ | ▨ | ▨ | | | | | | | |
| UASB + submerged aerated biofilter | ▨ | ▨ | ▨ | | | | | | | |
| UASB + anaerobic filter | | | | | | | | | | |
| UASB + high rate trickling filter | | | | | | | | | | |
| UASB + maturation ponds | ▨ | ▨ | | ▨ | ▨ | | | | | |
| UASB + overland flow | ▨ | ▨ | ▨ | ▨ | ▨ | | | | | |
| Conventional activated sludge | ▨ | ▨ | ▨ | | | | | | | |
| Extended aeration | ▨ | ▨ | ▨ | | | | | | | |
| Sequencing batch reactor | ▨ | ▨ | ▨ | | | | | | | |
| Activated sludge with biological N removal | ▨ | ▨ | ▨ | ▨ | ▨ | ▨ | | | | |
| Activated sludge with biological N/P removal | ▨ | ▨ | ▨ | ▨ | ▨ | ▨ | ▨ | ▨ | ▨ | ▨ |
| Activated sludge + tertiary filtration | ▨ | ▨ | ▨ | | | | ▨ | ▨ | | |
| Low rate trickling filter | ▨ | ▨ | | | | | | | | |
| High rate trickling filter | | | | | | | | | | |
| Submerged aerated biofilter | ▨ | ▨ | ▨ | | | | | | | |
| Submerged aerated biofilter with biological N removal | ▨ | ▨ | ▨ | ▨ | ▨ | ▨ | | | | |
| Rotating biological contactor | ▨ | ▨ | | | | | | | | |
| Any of the technologies above + chemical P precipitation | | | | | | | ▨ | ▨ | ▨ | ▨ |

Table 4.13. Capacity of various wastewater treatment technologies in consistently achieving the indicated levels of effluent quality in terms of Faecal (thermotolerant) Coliforms and Helminth Eggs

| System | Faecal (thermotolerant) Coliforms (FC/100mL) | | | | Helminth eggs |
| --- | --- | --- | --- | --- | --- |
| | $1 \times 10^6$ | $1 \times 10^5$ | $1 \times 10^4$ | $1 \times 10^3$ | $\leq 1$ egg/L |
| Facultative pond | ▓ | | | | ▨ |
| Anaerobic pond + facultative pond | ▓ | ▓ | | | ▨ |
| Facultative aerated lagoon | ▓ | | | | |
| Complete-mix aerated lagoon + sedimentation pond | ▓ | | | | |
| Anaerobic pond + facultative pond + maturation pond | ▓ | ▓ | ▓ | ▓ | ▨ |
| Anaerobic pond + facultative pond + high rate pond | ▓ | ▓ | | | |
| Anaerobic pond – facultative pond + algae removal | ▓ | ▓ | | | |
| Slow rate treatment | ▓ | ▓ | ▓ | ▓ | ▨ |
| Rapid infiltration | ▓ | ▓ | ▓ | | ▨ |
| Overland flow | ▓ | | | | ▨ |
| Constructed wetlands | ▓ | ▓ | | | ▨ |
| Septic tank + anaerobic filter | ▓ | | | | |
| Septic tank + infiltration | ▓ | ▓ | ▓ | ▓ | ▨ |
| UASB reactor | | | | | |
| UASB + activated sludge | ▓ | | | | |
| UASB + submerged aerated biofilter | ▓ | | | | |
| UASB + anaerobic filter | ▓ | | | | |
| UASB + high rate trickling filter | ▓ | | | | |
| UASB + maturation ponds | ▓ | ▓ | ▓ | ▓ | ▨ |
| UASB + overland flow | ▓ | ▓ | | | ▨ |
| Conventional activated sludge | | | | | |
| Extended aeration | ▓ | | | | |
| Sequencing batch reactor | | | | | |
| Activated sludge with biological N removal | | | | | |
| Activated sludge with biological N/P removal | | | | | |
| Activated sludge + tertiary filtration | ▓ | ▓ | ▓ | ▓ | ▨ |
| Low rate trickling filter | | | | | |
| High rate trickling filter | | | | | |
| Submerged aerated biofilter | | | | | |
| Submerged aerated biofilter with biological N removal | | | | | |
| Rotating biological contactor | | | | | |
| Any of the technologies above + disinfection/barrier | ▓ | ▓ | ▓ | ▓ | Variable |

Table 4.14.   Graphical representation of the per capita land requirements for various sewage treatment systems.

| System | Land requirement ($m^2$/inhab.) | Land requirements ($m^2$/inhab.) |
| --- | --- | --- |
| Primary treatment (septic tanks) | 0.03–0.05 | |
| Conventional primary treatment | 0.02–0.04 | |
| Advanced primary treatment (chemically enhanced) | 0.04–0.06 | |
| Facultative pond | 2.0–4.0 | |
| Anaerobic pond + facultative pond | 1.2–3.0 | |
| Facultative aerated lagoon | 0.25–0.5 | |
| Complete-mix aerated lagoon + sedimentation pond | 0.2–0.4 | |
| Anaerobic pond + facultative pond + maturation pond | 3.0–5.0 | |
| Anaerobic pond + facultative pond + high rate pond | 2.0–3.5 | |
| Anaerobic pond + facultative pond + algae removal | 1.7–3.2 | |
| Slow rate treatment | 10–50 | |
| Rapid infiltration | 1.0–6.0 | |
| Overland flow | 2.0–3.5 | |
| Constructed wetlands | 3.0–5.0 | |
| Septic tank + anaerobic filter | 0.2–0.35 | |
| Septic tank + infiltration | 1.0–1.5 | |
| UASB reactor | 0.03–0.10 | |
| UASB + activated sludge | 0.08–0.2 | |
| UASB + submerged aerated biofilter | 0.05–0.15 | |
| UASB + anaerobic filter | 0.05–0.15 | |
| UASB + high rate trickling filter | 0.1–0.2 | |
| UASB + dissolved-air flotation | 0.05–0.15 | |
| UASB + maturation ponds | 1.5–2.5 | |
| UASB + facultative aerated pond | 0.15–0.3 | |
| UASB + compl.mix aerated lagoon + sediment. pond | 0.1–0.3 | |
| UASB + overland flow | 0.12–0.25 | |
| Conventional activated sludge | 0.12–0.25 | |
| Activated sludge – extended aeration | 0.12–0.25 | |
| Sequencing batch reactor (extended aeration) | 0.12–0.25 | |
| Conventional activated sludge with biological N removal | 0.12–0.25 | |
| Convention. activated sludge with biological N/P removal | 0.12–0.25 | |
| Conventional activated sludge + tertiary filtration | 0.15–0.30 | |
| Low rate trickling filter | 0.15–0.3 | |
| High rate trickling filter | 0.12–0.25 | |
| Submerged aerated biofilter with nitrification | 0.1–0.15 | |
| Submerged aerated biofilter with biological N removal | 0.1–0.15 | |
| Rotating biological contactor | 0.1–0.2 | |

Table 4.15.  Graphical representation of the per capita installed power for aeration in various sewage treatment systems.

| System | Installed power for aeration(W/inhab.) | Installed power for aeration (W/inhab.) | | | | | | | |
|---|---|---|---|---|---|---|---|---|---|
| | | 0.5 | 1.0 | 1.5 | 2.0 | 2.5 | 3.0 | 3.5 | 4.0 |
| Primary treatment (septic tanks) | 0 | | | | | | | | |
| Conventional primary treatment | 0 | | | | | | | | |
| Advanced primary treatment (chemically enhanced) | 0 | | | | | | | | |
| Facultative pond | 0 | | | | | | | | |
| Anaerobic pond + facultative pond | 0 | | | | | | | | |
| Facultative aerated lagoon | 1.2–2.0 | | | | | | | | |
| Complete-mix aerated lagoon + sedimentation pond | 1.8–2.5 | | | | | | | | |
| Anaerobic pond + facultative pond + maturation pond | 0 | | | | | | | | |
| Anaerobic pond + facultative pond + high rate pond | <0.3 | | | | | | | | |
| Anaerobic pond + facultative pond + algae removal | 0 | | | | | | | | |
| Slow rate treatment | 0 | | | | | | | | |
| Rapid infiltration | 0 | | | | | | | | |
| Overland flow | 0 | | | | | | | | |
| Constructed wetlands | 0 | | | | | | | | |
| Septic tank + anaerobic filter | 0 | | | | | | | | |
| Septic tank + infiltration | 0 | | | | | | | | |
| UASB reactor | 0 | | | | | | | | |
| UASB + activated sludge | 1.8–3.5 | | | | | | | | |
| UASB + submerged aerated biofilter | 1.8–3.5 | | | | | | | | |
| UASB + anaerobic filter | 0 | | | | | | | | |
| UASB + high rate trickling filter | 0 | | | | | | | | |
| UASB + dissolved-air flotation | 1.0–1.5 | | | | | | | | |
| UASB + maturation ponds | 0 | | | | | | | | |
| UASB + facultative aerated pond | 0.3–0.6 | | | | | | | | |
| UASB + compl.mix aerated lagoon + sediment. pond | 0.5–0.9 | | | | | | | | |
| UASB + overland flow | 0 | | | | | | | | |
| Conventional activated sludge | 2.5–4.5 | | | | | | | | |
| Activated sludge – extended aeration | 3.5–5.5 | | | | | | | | |
| Sequencing batch reactor (extended aeration) | 4.5–6.0 | | | | | | | | |
| Conventional activated sludge with biological N removal | 2.2–4.2 | | | | | | | | |
| Convention. activated sludge with biological N/P removal | 2.2–4.2 | | | | | | | | |
| Conventional activated sludge + tertiary filtration | 2.5–4.5 | | | | | | | | |
| Low rate trickling filter | 0 | | | | | | | | |
| High rate trickling filter | 0 | | | | | | | | |
| Submerged aerated biofilter with nitrification | 2.5–4.5 | | | | | | | | |
| Submerged aerated biofilter with biological N removal | 2.2–4.2 | | | | | | | | |

Table 4.16. Graphical representation of the per capita dewatered sludge to be disposed of in various sewage treatment systems.

| System | Volume of sludge to be disposed of (L/inhab.year) | Volume of dewatered sludge to be disposed of (L/inhab.year) (≈kg of sludge/inhab.year, on a wet basis) |
|---|---|---|
| Primary treatment (septic tanks) | 15–35 | |
| Conventional primary treatment | 15–40 | |
| Advanced primary treatment (chemically enhanced) | 40–110 | |
| Facultative pond | 15–30 | |
| Anaerobic pond + facultative pond | 20–60 | |
| Facultative aerated lagoon | 7–30 | |
| Complete-mix aerated lagoon + sedimentation pond | 10–35 | |
| Anaerobic pond + facultative pond + maturation pond | 20–60 | |
| Anaerobic pond + facultative pond + highrate pond | 20–60 | |
| Anaerobic pond + facultative pond + algae removal | 25–70 | |
| Slow rate treatment | – | |
| Rapid infiltration | – | |
| Overland flow | – | |
| Constructed wetlands | – | |
| Septic tank + anaerobic filter | 25–50 | |
| Septic tank + infiltration | 15–35 | |
| UASB reactor | 10–35 | |
| UASB + activated sludge | 15–60 | |
| UASB + submerged aerated biofilter | 15–55 | |
| UASB + anaerobic filter | 10–50 | |
| UASB + high rate trickling filter | 15–55 | |
| UASB + dissolved-air flotationt | 25–75 | |
| UASB + maturation ponds | 10–35 | |
| UASB + facultative aerated pond | 15–50 | |
| UASB + compl.mix aerated lagoon + sediment.pond | 15–50 | |
| UASB + overland flow | 10–35 | |
| Conventional activated sludge | 35–90 | |
| Activated sludge – extended aeration | 40–105 | |
| Sequencing batch reactor (extended aeration) | 40–105 | |
| Conventional activated sludge with biological N removal | 35–90 | |
| Convention. activated sludge with biological N/P removal | 35–90 | |
| Conventional activated sludge + tertiary filtration | 40–100 | |
| Low rate trickling filter | 35–80 | |
| High rate trickling filter | 35–80 | |
| Submerged aerated biofilter with nitrification | 35–90 | |
| Submerged aerated biofilter with biological N removal | 35–90 | |
| Rotating biological contactor | 20–75 | |

Table 4.17.  Graphical representation of the per capita construction costs in various sewage treatment systems.

| System | Construction costs (US$/inhab.) | Construction costs (US$/inhab.) |
|---|---|---|
| Primary treatment (septic tanks) | 12–20 | |
| Conventional primary treatment | 12–20 | |
| Advanced primary treatment (chemically enhanced) | 15–25 | |
| Facultative pond | 15–30 | |
| Anaerobic pond + facultative pond | 20–30 | |
| Facultative aerated lagoon | 20–35 | |
| Complete-mix aerated lagoon + sedimentation pond | 20–35 | |
| Anaerobic pond + facultative pond + maturation pond | 20–40 | |
| Anaerobic pond + facultative pond + high rate pond | 20–35 | |
| Anaerobic pond + facultative pond + algae removal | 20–35 | |
| Slow rate treatment | 8–25 | |
| Rapid infiltration | 12–30 | |
| Overland flow | 15–30 | |
| Constructed wetlands | 20–30 | |
| Septic tank + anaerobic filter | 30–50 | |
| Septic tank + infiltration | 25–40 | |
| UASB reactor | 12–20 | |
| UASB + activated sludge | 30–45 | |
| UASB + submerged aerated biofilter | 25–40 | |
| UASB + anaerobic filter | 20–30 | |
| UASB + high rate trickling filter | 25–35 | |
| UASB + dissolved-air flotation | 25–35 | |
| UASB + maturation ponds | 15–30 | |
| UASB + facultative aerated pond | 15–35 | |
| UASB + compl.mix aerated lagoon + sediment. pond | 15–35 | |
| UASB + overland flow | 20–35 | |
| Conventional activated sludge | 40–65 | |
| Activated sludge – extended aeration | 35–50 | |
| Sequencing batch reactor (extended aeration) | 35–50 | |
| Conventional activated sludge with biological N removal | 45–70 | |
| Convention. activated sludge with biological N/P removal | 50–75 | |
| Conventional activated sludge + tertiary filtration | 50–75 | |
| Low rate trickling filter | 50–60 | |
| High rate trickling filter | 50–60 | |
| Submerged aerated biofilter with nitrification | 30–50 | |
| Submerged aerated biofilter with biological N removal | 30–50 | |

The graphical column is headed "Construction costs (US$/inhab.)" with a scale marked 10, 20, 30, 40, 50, 60, 70, 80.

Table 4.18.   Graphical representation of the per capita annual operation and maintenance costs in various sewage treatment systems.

| System | Operation and maintenance costs (US$/inhab.year) | Operation and maintenance costs (US$/inhab.year) |
|---|---|---|
| Primary treatment (septic tanks) | 0.5–1.0 | |
| Conventional primary treatment | 0.5–1.0 | |
| Advanced primary treatment (chemically enhanced) | 3.0–6.0 | |
| Facultative pond | 0.8–1.5 | |
| Anaerobic pond + facultative pond | 0.8–1.5 | |
| Facultative aerated lagoon | 2.0–3.5 | |
| Complete-mix aerated lagoon + sedimentation pond | 2.0–3.5 | |
| Anaerobic pond + facultative pond + maturation pond | 1.0–2.0 | |
| Anaerobic pond + facultative pond + high rate pond | 1.5–2.5 | |
| Anaerobic pond + facultative pond + algae removal | 1.5–2.5 | |
| Slow rate treatment | 0.4–1.2 | |
| Rapid infiltration | 0.5–1.5 | |
| Overland flow | 0.8–1.5 | |
| Constructed wetlands | 1.0–1.5 | |
| Septic tank + anaerobic filter | 2.5–4.0 | |
| Septic tank + infiltration | 1.2–2.0 | |
| UASB reactor | 1.0–1.5 | |
| UASB + activated sludge | 2.5–5.0 | |
| UASB + submerged aerated biofilter | 2.5–5.0 | |
| UASB + anaerobic filter | 1.5–2.2 | |
| UASB + high rate trickling filter | 2.0–3.0 | |
| UASB + dissolved-air flotation | 2.5–3.5 | |
| UASB + maturation ponds | 1.8–3.0 | |
| UASB + facultative aerated pond | 2.0–3.5 | |
| UASB + compl.mix aerated lagoon + sediment. pond | 2.0–3.5 | |
| UASB + overland flow | 2.0–3.0 | |
| Conventional activated sludge | 4.0–8.0 | |
| Activated sludge – extended aeration | 4.0–8.0 | |
| Sequencing batch reactor (extended aeration) | 4.0–8.0 | |
| Conventional activated sludge with biological N removal | 4.0–9.0 | |
| Convention. activated sludge with biological N/P removal | 6.0–10.0 | |
| Conventional activated sludge + tertiary filtration | 6.0–10.0 | |
| Low rate trickling filter | 4.0–6.0 | |
| High rate trickling filter | 4.0–6.0 | |
| Submerged aerated biofilter with nitrification | 3.0–6.0 | |
| Submerged aerated biofilter with biological N removal | 3.0–6.0 | |
| Rotating biological contactor | 4.0–6.0 | |

Table 4.19.  Relative evaluation of the main domestic sewage treatment systems (liquid phase)

| | Removal efficiency | | | Economy | | | | | Resistance capacity to influent variations and shock loads | | | Reliability | Simpli-city in O&M. | Independence of other charact.for good perform. | | Lower possibility of environmental problems | | | |
|---|---|---|---|---|---|---|---|---|---|---|---|---|---|---|---|---|---|---|---|
| | | | | Requirements | | Costs | | Generation | | | | | | | | | | | |
| | BOD | Nutrients | Coliforms | Land | Energy | Constr. | O & M | Sludge | Flow | Quality | Toxic | | | Climate | Soil | Bad odours | Noise | Aerosols | Insects and worms |
| Preliminary treatment | 0 | 0 | 0 | +++++ | +++++ | +++++ | ++++ | +++++ | +++++ | +++++ | +++++ | +++++ | +++ | +++++ | +++++ | + | ++++ | +++++ | +++ |
| Primary treatment | + | + | + | +++++ | ++++ | ++++ | +++ | +++ | ++++ | +++++ | +++++ | ++++ | +++ | ++++ | +++++ | ++ | ++++ | +++++ | +++ |
| Advanced primary treatment | ++ | +/++++ | ++ | +++++ | ++++ | +++ | ++ | + | ++++ | +++++ | ++++ | ++++ | +++ | +++++ | +++++ | +++ | ++++ | +++++ | +++ |
| Facultative pond | +++ | ++ | ++/++++ | + | +++++ | +++ | +++++ | +++++ | ++++ | ++++ | +++ | ++++ | +++++ | ++ | +++ | +++ | +++++ | +++++ | ++ |
| Anaerobic pond – facultative pond | +++ | ++ | ++/++++ | ++ | +++++ | ++++ | +++++ | +++++ | ++++ | ++++ | +++ | ++++ | +++++ | ++ | +++ | + | +++++ | +++++ | ++ |
| Facultative aerated lagoon | +++ | ++ | ++/++++ | ++ | +++ | +++ | ++++ | +++++ | ++++ | ++++ | +++ | ++++ | ++++ | +++ | +++ | ++++ | + | + | +++ |
| Compl. mix aerated – sedim. pond | +++ | ++ | ++/++++ | +++ | +++ | +++ | +++ | +++ | +++ | ++++ | +++ | +++ | +++ | +++ | ++++ | +++ | + | + | ++ |
| Pond – maturation pond | +++ | +++ | +++++ | + | +++++ | +++ | +++++ | +++++ | ++++ | ++++ | +++ | ++++ | +++++ | ++ | +++ | +++ | +++++ | +++++ | ++ |
| Pond – high rate pond | +++ | ++++ | ++++ | ++ | ++++ | +++ | ++++ | +++++ | ++++ | ++++ | +++ | ++++ | +++ | +++ | +++ | +++ | ++ | ++ | ++ |
| Pond – algae removal | ++++ | ++ | ++/++++ | ++ | +++++ | +++ | ++++ | +++ | ++++ | ++++ | +++ | ++++ | +++ | +++ | +++ | +++ | +++++ | +++++ | ++ |
| Slow rate treatment | +++++ | ++++ | ++++ | + | +++++ | +++ | +++++ | +++++ | ++++ | ++++ | ++++ | ++++ | ++++ | ++ | + | ++ | +++++ | +/+++++ | ++ |
| Rapid infiltration | +++++ | ++++ | ++++ | + | +++++ | ++++ | +++++ | +++++ | ++++ | ++++ | ++++ | ++++ | ++++ | ++ | + | ++ | +++++ | +++++ | ++ |
| Overland flow | ++++ | +++ | ++/+++ | + | +++++ | +++ | ++++ | +++++ | ++++ | ++++ | +++ | ++++ | +++++ | +++ | ++ | ++ | +++++ | +/+++++ | ++ |
| Constructed wetlands | ++++ | ++ | +++ | + | +++++ | +++ | ++++ | +++++ | ++++ | ++++ | +++ | ++++ | +++++ | ++ | ++ | ++ | +++++ | +++++ | ++ |
| Septic tank – anaerobic filter | +++ | + | ++ | +++++ | +++++ | +++ | +++ | ++++ | +++ | +++ | ++ | +++ | ++++ | ++ | +++++ | ++ | ++++ | +++++ | ++++ |
| UASB reactor | +++ | + | ++ | +++++ | +++++ | ++++ | ++++ | ++++ | ++ | ++ | ++ | +++ | ++++ | ++ | +++++ | ++ | ++++ | +++++ | ++++ |
| UASB reactor – post-treatment | (a) | (a) | (a) | (a) | (a) | (a) | (a) | (a) | (b) | (b) | (b) | (a) | (a) | (a) | (a) | (b) | (a) | (a) | (a) |
| Conventional activated sludge | ++++ | ++/++++ | ++ | ++++ | ++ | + | ++ | + | +++ | +++ | ++ | ++++ | + | +++ | +++++ | ++++ | + | +/+++++ | ++++ |
| Activated sludge (extended aeration) | +++++ | ++/++++ | ++ | ++++ | + | ++ | + | ++ | ++++ | ++++ | +++ | ++++ | ++ | ++++ | +++++ | +++++ | + | +/+++++ | ++++ |
| Sequencing batch reactor | ++++ | ++/++++ | ++ | ++++ | +/++ | + | + | +/++ | ++++ | ++++ | +++ | ++++ | +++ | ++++ | +++++ | +++ | + | +/+++++ | ++++ |
| Trickling filter (low rate) | ++++ | ++/++++ | ++ | +++ | ++++ | + | +++ | ++ | +++ | ++ | ++ | ++++ | +++ | ++ | +++++ | ++++ | ++++ | ++++ | ++ |
| Trickling filter (high rate) | ++++ | ++/+++ | ++ | ++++ | +++ | ++ | +++ | + | ++++ | +++ | +++ | ++++ | +++ | ++ | +++++ | ++++ | ++++ | +++++ | +++ |
| Submerged aerated biofilter | +++++ | ++/+++ | ++ | +++++ | ++ | | +++ | + | +++ | +++ | ++ | ++++ | ++ | ++++ | +++++ | +++++ | ++ | ++++++ | ++++ |
| Rotating biological contactor | ++++ | ++/+++ | ++ | ++++ | +++ | + | +++ | + | +++ | +++ | ++ | +++ | +++ | ++ | +++++ | ++++ | ++++ | ++++++ | +++ |

*Notes:* the grading is only relative in each column and is not generalised for all the items. The grading can vary widely with the local conditions

+++++ : most favourable   + : least favourable   ++++, +++, ++: intermediate grades, in decreasing order   0 : zero effect

+ / +++++: variable with the type of process, equipment, variant or design

UASB reactor + post-treatment: (a) post-treatment characteristics prevail; (b) UASB reactor characteristics prevail

Table 4.20. Minimum equipment necessary for the main wastewater treatment processes

| Treatment process | Basic equipment required |
|---|---|
| *Preliminary treatment* | Screens; grit chamber; flowmeter |
| *Primary treatment* | Sludge scraper (larger systems); mixers in the digesters; gas equipment |
| *Facultative pond* | – |
| *Anaerobic pond – facult. pond* | Effluent recycle pump (optional) |
| *Facultative aerated lagoon* | Aerators |
| *Compl. mix. aerated – sedim. pond* | Aerators |
| *High rate pond* | Rotors for movement of liquid |
| *Maturation pond* | – |
| *Slow rate treatment* | Sprinklers (optional) |
| *Rapid infiltration* | – |
| *Subsurface infiltration* | – |
| *Overland flow* | Sprinklers (optional) |
| *Constructed wetlands* | – |
| *Septic tank – anaerobic filter* | – |
| *UASB reactor* | – |
| *UASB reactor + post-treatment* | The equipment depends on the post-treatment process used. However, there is no need for equipment relating to primary sedimentation tanks, thickeners and sludge digesters. |
| *Conventional activated sludge* | Aerators; sludge recycle pumps; sludge scrapers in sedimentation tanks; sludge scrapers in thickeners; mixers in digesters; gas equipment; pumps for the return of supernatants and drained liquids from sludge treatment. |
| *Activated sludge (extended aeration)* | Aerators; sludge recycle pumps; sludge scrapers in sedimentation tanks; thickening equipment; pumps for the return of supernatants and drained liquids from sludge treatment. |
| *Sequencing batch reactors* | Aerators; thickening equipment; pumps for the return of supernatants and drained liquids from sludge treatment. |
| *Trickling filter (low rate)* | Rotating distributor; sludge scrapers in sedimentation tanks; sludge scrapers in thickeners; pumps for the return of supernatants and drained liquids from sludge treatment. |
| *Trickling filter (high rate)* | Rotating distributor; effluent recycle pumps; sludge scrapers in sedimentation tanks; sludge scrapers in thickeners; mixers in digesters; gas equipment; pumps for the return of supernatants and drained liquids from sludge treatment. |
| *Submerged aerated biofilters* | Aeration system; filter washing system; sludge scrapers in sedimentation tanks; thickening equipment; mixers in digesters; gas equipment; pumps for the return of supernatants and drained liquids from sludge treatment. |
| *Rotating biological contactors* | Motor for the rotation of the discs; thickening equipment; pumps for the return of supernatants and drained liquids from sludge treatment. |

Table 4.21. Comparative analysis of the main wastewater treatment systems. Balance of the advantages and disadvantages

| STABILISATION PONDS SYSTEMS | | |
|---|---|---|
| System | Advantages | Disadvantages |
| *Facultative pond* | • Satisfactory BOD removal efficiency<br>• Reasonable pathogen removal efficiency<br>• Simple construction, operation and maintenance<br>• Reduced construction and operating costs<br>• Absence of mechanical equipment<br>• Practically no energy requirements<br>• Satisfactory resistance to load variations<br>• Sludge removal only necessary after periods greater than 20 years | • High land requirements<br>• Difficulty in satisfying restrictive discharge standards<br>• Operational simplicity can bring a disregard to maintenance (e.g. vegetation growth)<br>• Possible need for removing algae from effluent to comply with stringent discharge standards<br>• Variable performance with climatic conditions (temperature and sunlight)<br>• Possible insect growth |
| *Anaerobic pond – facultative pond system* | • The same as facultative ponds<br>• Lower land requirements than single facultative ponds | • The same as facultative ponds<br>• Possibility of bad odours in the anaerobic pond<br>• Occasional need for effluent recycling to control bad odours<br>• Need for a safe distance from surrounding neighbourhoods<br>• Need for periodic (few years interval) removal of sludge from anaerobic pond |
| *Facultative aerated lagoon* | • Relatively simple construction, operation and maintenance<br>• Lower land requirements than the facultative and anaerobic-facultative pond systems<br>• Greater independence from climatic conditions than the facultative and anaerobic-facultative pond systems<br>• Satisfactory resistance to load variations<br>• Reduced possibilities of bad odours | • Introduction of equipment<br>• Slight increase in the sophistication level<br>• Land requirements still high<br>• Relatively high energy requirements<br>• Low coliform removal efficiency<br>• Need for periodic (some years interval) removal of sludge from aerated pond |
| *Completely-mixing aerated lagoon – sedimentation pond system* | • Same as facultative aerated lagoons<br>• Lowest land requirements for all the ponds systems | • Same as facultative aerated lagoons (exception: land requirements)<br>• Rapid filling of the sedimentation pond with sludge (2 to 5 years)<br>• Need for continuous or periodic (few years interval) removal of sludge from sedimentation pond |

Table 4.21 (*Continued*)

| System | Advantages | Disadvantages |
|---|---|---|
| *Ponds – maturation pond system* | • Same as the preceding ponds<br>• High pathogen removal efficiency<br>• Reasonable nutrient removal efficiency | • Same as the preceding ponds<br>• Very high land requirements |
| *Ponds – high rate pond* | • Same as the preceding ponds<br>• Good pathogen removal efficiency<br>• High nutrient removal efficiency | • Same as the preceding ponds |
| LAND DISPOSAL SYSTEMS | | |
| *Slow rate treatment* | • High removal efficiency of BOD and coliforms<br>• Satisfactory removal efficiency of N and P<br>• Combined treatment and final disposal methods<br>• Practically no energy requirements<br>• Simple construction, operation and maintenance<br>• Reduced construction and operation costs<br>• Good resistance to load variations<br>• No sludge to be treated<br>• Provides soil fertilisation and conditioning<br>• Financial return from irrigation in agricultural areas<br>• Recharge of groundwater | • Very high land requirements<br>• Possibility of bad odours<br>• Possibility of vector attraction<br>• Relatively dependent on the climate and the nutrient requirements of the plants<br>• Dependent on the soil characteristics<br>• Contamination risk to the plants to be consumed if applied indiscriminately<br>• Possibility of the contamination of the farm workers (e.g. in application by sprinklers)<br>• Possibility of chemical effects in the soil, plants or groundwater (in the case of industrial wastewater)<br>• Difficult inspection and control of the irrigated vegetables<br>• The application must be suspended or reduced in rainy periods |
| *Rapid infiltration* | • The same as slow rate treatment (although the removal efficiency of pollutants is lower)<br>• Much lower land requirements than slow rate treatment<br>• Reduced dependence on the slope of the ground<br>• Application during all the year | • Same as slow rate treatment (but with lower land requirements and the possibility of application during all the year)<br>• Potential contamination of groundwater with nitrates |
| *Subsurface infiltration* | • Same as rapid infiltration<br>• Possible economy in the implementation of interceptors<br>• Absence of bad odours<br>• The above ground can be used as green area or parks<br>• Independent of climatic conditions<br>• Absence of problems related to the contamination of plants and workers | • Same as rapid infiltration<br>• Requires spare units to allow switching between units (operation and rest)<br>• The larger systems require very permeable soil to reduce land requirements |

(*Continued*)

Table 4.21 (*Continued*)

| System | Advantages | Disadvantages |
| --- | --- | --- |
| *Overland flow* | • Same as rapid infiltration (but with the generation of a final effluent and with a greater dependence on the ground slope)<br>• Lowest dependence on the soil characteristics among the land disposal systems | • Same as rapid infiltration<br>• Greater dependence on the ground slope<br>• Generation of a final effluent |
| *Constructed wetlands* | • High removal efficiency of BOD and coliforms<br>• Practically no energy requirements<br>• Simple construction, operation and maintenance<br>• Reduced construction and operational costs<br>• Good resistance to load variations<br>• No sludge to be treated<br>• Possibility of using the produced plant biomass | • High land requirements<br>• Wastewater requires previous treatment (primary or simplified secondary)<br>• Need for a substrate, such as gravel or sand<br>• Susceptible to clogging<br>• Need of macrophytes handling<br>• Possibility of mosquitoes in surface flow systems |
| | ANAEROBIC REACTORS | |
| *UASB reactor* | • Reasonable BOD removal efficiency<br>• Low land requirements<br>• Low construction and operational costs<br>• Tolerance to influents highly concentrated in organic matter<br>• Practically no energy consumption<br>• Possibility of energy use of the biogas<br>• Support medium not required<br>• Simple construction, operation and maintenance<br>• Very low sludge production<br>• Sludge stabilisation in the reactor itself<br>• Sludge with good dewaterability<br>• Sludge requires only dewatering and final disposal<br>• Rapid start up after periods of no use (biomass preservation for various months) | • Difficulty in complying with restrictive discharge standards<br>• Low coliform removal efficiency<br>• Practically no N and P removal<br>• Possibility of the generation of an effluent with an unpleasant aspect<br>• Possibility of the generation of bad odours, although controllable<br>• Initial start up is generally slow (but can be accelerated with the use of seeding)<br>• Relatively sensitive to load variations and toxic compounds<br>• Usually needs post-treatment |
| *Septic tank – anaerobic filter* | • Same as UASB reactors (exception: support medium required)<br>• Good adaptation to different wastewater types and concentrations<br>• Good resistance to load variations | • Difficulty in complying with restrictive discharge standards<br>• Low coliform removal efficiency<br>• Practically no N and P removal<br>• Possibility of the generation of an effluent with an unpleasant aspect<br>• Possibility of the generation of bad odours, although controllable |

Table 4.21 (*Continued*)

| System | Advantages | Disadvantages |
|---|---|---|
| | | • Risks of clogging<br>• Restricted to the treatment of influents without high solids concentrations |
| *UASB reactor – post-treatment system* | • Maintenance of the inherent advantages of the UASB reactor<br>• Maintenance of the inherent advantages of the post-treatment system<br>• Reduction in the volume in the biological reactors in the post-treatment system (and frequently in the overall volume of the whole system)<br>• Reduction in the energy consumption for aerated post-treatment systems<br>• Reduction in the sludge production in the post-treatment system | • Maintenance of the inherent disadvantages of the UASB reactor (with the exception of the effluent quality, that assumes the characteristics of the post-treatment system)<br>• Maintenance of the inherent disadvantages of the post-treatment system<br>• Greater difficulty in the biological removal of nutrients in the post-treatment system |
| | ACTIVATED SLUDGE SYSTEMS | |
| *Conventional activated sludge* | • High BOD removal efficiency<br>• Nitrification usually obtained<br>• Biological removal of N and P is possible<br>• Low land requirements<br>• Reliable process, as long as it is supervised<br>• Reduced possibilities of bad odours, insects and worms<br>• Operational flexibility | • Low coliform removal efficiency<br>• High construction and operational costs<br>• High energy consumption<br>• Sophisticated operation required<br>• High mechanisation level<br>• Relatively sensitive to toxic discharges<br>• Requires complete treatment and final disposal of the sludge<br>• Possible environmental problems with noise and aerosols |
| *Extended aeration* | • Same as conventional activated sludge<br>• Variant with the highest BOD removal efficiency<br>• Consistent nitrification<br>• Conceptually simpler than conventional activated sludge (simpler operation)<br>• Lower sludge production than conventional activated sludge<br>• Sludge digestion in the reactor itself<br>• High resistance to load variations and toxic loads<br>• Satisfactory independence from climatic conditions | • Low coliform removal efficiency<br>• High construction and operational costs<br>• System with the highest energy consumption<br>• High mechanisation level (although less than conventional activated sludge)<br>• Thickening / dewatering and final disposal of the sludge required |
| *Sequencing batch reactors* | • High BOD removal efficiency<br>• Satisfactory removal of N and possibly P<br>• Low land requirement | • Low coliform removal efficiency<br>• High construction and operational costs |

(*Continued*)

Table 4.21 (*Continued*)

| System | Advantages | Disadvantages |
|---|---|---|
| | • Conceptually simpler than the other activated sludge systems<br>• Less equipment than the other activated sludge systems<br>• Operational flexibility (through cycle variation)<br>• Secondary sedimentation tanks and sludge recycle pumps are not necessary (operation as extended aeration: primary clarifiers and sludge digesters also not necessary) | • Greater installed power than the other activated sludge systems<br>• Treatment and disposal of the sludge is required (variable with the conventional or extended aeration mode, although the latter is more frequent)<br>• Usually economically more competitive for small to medium-size populations |
| *Activated sludge with biological nutrient removal* | • Same as conventional activated sludge<br>• High nutrient removal efficiency | • Same as conventional activated sludge<br>• Requirement of internal recycles<br>• Increase in the operational complexity |

| AEROBIC BIOFILM REACTORS | | |
|---|---|---|
| System | Advantages | Disadvantages |
| *Low rate trickling filter* | • High BOD removal efficiency<br>• Frequent nitrification<br>• Relatively low land requirements<br>• Conceptually simpler than activated sludge<br>• Relatively low mechanisation level<br>• Simple mechanical equipment<br>• Sludge digestion in the filter itself | • Low coliform removal efficiency<br>• Lower operational flexibility than activated sludge<br>• High construction costs<br>• Land requirements higher than high rate trickling filters<br>• Relative dependence from the air temperature<br>• Relatively sensitive to toxic discharges<br>• Thickening / dewatering and final disposal of the sludge required<br>• Possible problems with flies<br>• High head loss |
| *High rate trickling filter* | • Good BOD removal efficiency (although slightly less than the low rate filters)<br>• Low land requirements<br>• Conceptually simpler than activated sludge<br>• Greater operational flexibility than low rate filters<br>• Better resistance to load variations than low rate filters<br>• Reduced possibilities of bad odours | • Low coliform removal efficiency<br>• Operation slightly more sophisticated than low rate filters<br>• High construction costs<br>• Relative dependence from the air temperature<br>• Complete sludge treatment and final disposal required<br>• High head loss |
| *Submerged aerated biofilters* | • High BOD removal efficiency<br>• Optional nitrification (frequent, when desired)<br>• Very low land requirements<br>• Reduced possibilities of bad odours<br>• Reduced head loss | • Low coliform removal efficiency<br>• Relatively high construction and operational costs<br>• High energy consumption<br>• Requirement of a slightly more careful operation compared to trickling filters (aeration and washing of the filters)<br>• Complete sludge treatment and final disposal required |

Table 4.21 (*Continued*)

| System | Advantages | Disadvantages |
|---|---|---|
| *Rotating biological contactors* | • High BOD removal efficiency<br>• Frequent nitrification<br>• Very low land requirements<br>• Conceptually simpler than activated sludge<br>• Simple mechanical equipment<br>• Reduced possibilities of bad odours<br>• Reduced head loss | • Low coliform removal efficiency<br>• High construction and operational costs<br>• Mainly indicated for small populations (avoid excessive number of discs)<br>• Usually the discs need to be covered (protection against rain, wind and vandalism)<br>• Relative dependence from the air temperature<br>• Complete sludge treatment (possibly without digestion if the discs are installed on top of septic tanks) and final disposal required |

# 5

# Overview of sludge treatment and disposal

## 5.1 INTRODUCTION

The main solid by-products produced in wastewater treatment are:

- *screened material*
- *grit*
- *scum*
- *primary sludge*
- *secondary sludge*
- *chemical sludge (if a physical–chemical stage is included)*

The treatment of the solid by-products generated in the various units is an essential stage in wastewater treatment. Even though the sludge, in most of its handling stages, is constituted by more than 95% water, it is only by convention that it is called a **solid phase**, with the aim at distinguishing it from the wastewater, or the liquid flow being treated (**liquid phase**). Owing to the greater volume and mass generated, compared with the other solid by-products, the present book covers the problems of sludge in greater depth.

The following aspects need to be taken into consideration and quantified when planning the sludge management:

- **production** of the sludge in the *liquid phase*

- **wastage** of the sludge from the *liquid phase* (removal to the sludge processing line)
- **wastage** of the sludge from the *solid phase* (removal from the wastewater treatment plant to the sludge disposal or reuse site)

The sludge *production* is a function of the wastewater treatment system used for the liquid phase. In principle, all the biological treatment processes generate sludge. The processes that receive raw wastewater in primary settling tanks generate the **primary sludge**, which is composed of the settleable solids of the raw wastewater. In the biological treatment stage, there is the so-called **biological sludge** or **secondary sludge**. This sludge is the biomass that grows at the expense of the food supplied by the incoming sewage. If the biomass is not removed, it tends to accumulate in the system and eventually leaves with the final effluent, deteriorating its quality in terms of suspended solids and organic matter. Depending on the treatment system, the primary sludge can be sent for treatment together with the secondary sludge. In this case, the resultant sludge of the mixture is called **mixed sludge**. In treatment systems that incorporate a physical–chemical stage for improving the performance of primary or secondary settling tanks, a **chemical sludge** is produced.

Since sludge is produced, its *wastage* from the liquid phase is necessary. However, not all the wastewater treatment systems need the continuous removal of this biomass. Some treatment systems can store the sludge for all the operating horizon of the works (e.g. facultative ponds), others require only an occasional withdrawal (e.g. anaerobic reactors) and others need the continuous or very frequent removal (e.g. activated sludge). The *biological sludge* withdrawn is also called **excess sludge**, **surplus sludge**, **waste sludge** or **secondary sludge**.

Finally, the sludge is treated and processed in the solid phase stage, from where it is removed or wasted, going to *final disposal* or *reuse* routes.

Table 5.1 presents a summary description of the main types of solid by-products and their origin in wastewater treatment. All of the treatment processes start with preliminary treatment, where there is, necessarily, the generation of *screened material* and *grit*. The *scum* is variable from process to process and can or cannot occur systematically. The *primary sludge* is only generated in plants that have a primary treatment stage (primary sedimentation tank). The *secondary sludge* is generated in all biological treatment processes. The sludge type varies, and the table makes a distinction between aerobic sludge (non-stabilised), aerobic sludge (stabilised) and anaerobic sludge (stabilised). The biological treatment processes are described in Chapter 4. The *chemical sludge* is only produced in plants that explicitly incorporate a physical–chemical stage in the treatment of the liquid phase.

Wastewater treatment processes based on land application also generate biomass, which is, in this case, mainly composed by the plant biomass related to the irrigated culture. This plant biomass can be used or disposed of after cutting or harvesting and possible processing. The analysis of the management of this biomass is outside the scope of the present text.

Table 5.1. Origin and description of the main solid by-products generated in wastewater treatment

| Solid by-product | Origin | Description |
|---|---|---|
| Coarse solids | • Screen | The solids removed in the screens include all of the organic and inorganic solids with dimensions greater than the free space between the bars. The organic material varies in function of the characteristics of the sewerage system and the season of the year. The removal can be manual or mechanical. |
| Grit | • Grit chamber | The grit usually consists of heavier inorganic solids that settle with relatively high velocities. The grit is removed in units called grit chambers that are settling tanks with a low hydraulic detention time, which is only sufficient for the grit to settle. However, depending on the operating conditions, organic matter, mainly fats and grease, can also be removed. |
| Scum | • Grit chamber<br>• Primary settling tank<br>• Secondary settling tank<br>• Stabilisation pond<br>• Anaerobic tank | The scum removed from primary settling tanks consists of floating material that has been scraped from the surface; this includes grease, vegetable and mineral oils, animal fats, soaps, food wastes, vegetable and fruits peelings, hair, paper, cotton, cigarette tips and similar materials. The specific gravity of scum is less than 1.0 (generally around 0.95). The grit chambers do not usually have scum removal equipment. In secondary treatment, biological reactors also produce scum, which includes scum-forming microorganisms that develop under specific environmental conditions. This scum is usually removed in the secondary settling tanks by scraping the surface. Stabilisation ponds and anaerobic reactors can also present scum. |
| Primary sludge | • Septic tank<br>• Primary settling tank | The solids removed by settling from primary sedimentation tanks constitute the primary sludge. Primary sludge can have a strong odour, principally if retained in the primary settling tank for a long time in high temperature conditions. The primary sludge removed from septic tanks stays a time long enough for its anaerobic digestion, under controlled conditions (closed tanks). |

Table 5.1  (*Continued*)

| Solid by-product | Origin | Description |
|---|---|---|
| Aerobic biological sludge (non-stabilised) | • Conventional activated sludge<br>• Aerobic biofilm reactors – high rate (high rate trickling filter, submerged aerated biofilter, rotating biological contactor) | The excess biological sludge (secondary sludge) consists of a biomass of aerobic microorganisms generated at the expense of the removal of the organic matter (substrate) from the wastewater. This biomass is in constant growth, resulting from the continuous input of organic matter into the biological reactors. To maintain the system in equilibrium, approximately the same mass of biological solids generated must be removed from the system. If the residence time of the solids in the system is low and there is a satisfactory level of substrate available, the biological solids will contain greater levels of organic matter in their cellular composition. These solids are not stabilised (digested), requiring a subsequent separate digestion stage. If no digestion is included, release of bad odours by the sludge during its treatment and final disposal is likely to occur, because of the anaerobic decomposition of the organic matter under uncontrolled conditions. |
| Aerobic biological sludge (stabilised) | • Activated sludge – extended aeration<br>• Aerobic biofilm reactors – low rate (low rate trickling filter, rotating biological contactor, submerged aerated biofilter) | This biological sludge is also predominantly composed by aerobic microorganisms that grow and multiply themselves at the expense of the organic matter in the raw wastewater. However, in low loading rate systems, the availability of substrate is lower and the biomass is retained longer in the system, thus prevailing endogenous respiration conditions. Under these conditions, the biomass uses its own reserves of organic matter in the composition of the cellular protoplasm, which leads to a sludge with a lower level of organic matter (digested sludge) and higher level of inorganic solids. This sludge does not require a subsequent separate digestion stage. |
| Anaerobic biological sludge (stabilised) | • Stabilisation ponds (facultative ponds, anaerobic-facultative ponds, facultative aerated lagoons, complete-mix aerated lagoons – sedimentation lagoons) | Anaerobic reactors and the sludge at the bottom of stabilisation ponds are in predominantly anaerobic conditions. The anaerobic biomass also grows and multiplies itself at the expense of the organic matter. In these treatment processes, the biomass is usually retained for a long time, in which anaerobic digestion of their own cellular material occurs. In stabilisation ponds, the sludge |

(*Continued*)

Table 5.1 (*Continued*)

| Solid by-product | Origin | Description |
|---|---|---|
| | • Anaerobic reactors (UASB reactors, anaerobic filters) | is also composed by settled solids from the raw sewage, together with dead algae. This sludge does not require a subsequent digestion stage. |
| Chemical sludge | • Primary settling tanks with chemical precipitation<br>• Activated sludge with chemical phosphorus precipitation | This sludge is usually a result of chemical precipitation with metallic salts or lime. The concern with odours is less than with the primary sludge, although it still can occur (only in the case of the use of lime as a coagulant). The decomposition rate of the chemical sludge in the tanks is less than the primary sludge. |

*Source:* adapted from Metcalf & Eddy (1991) and von Sperling and Gonçalves (2001)

The term '**sludge**' has been used to designate the solid by-products from wastewater treatment. In the biological treatment processes, part of the organic matter is absorbed and converted into microbial biomass, generically called biological or secondary sludge. This is principally composed of biological solids, and for this reason it is also called a **biosolid**. To adopt this term, it is still necessary that its chemical and biological characteristics are compatible with a productive use, such as for example in agriculture. The term 'biosolid' is a way of emphasising its beneficial aspects, giving more value to productive uses, in comparison with the mere final non-productive disposal by means of landfills or incineration.

## 5.2 RELATIONSHIPS IN SLUDGE: SOLIDS LEVELS, CONCENTRATION AND FLOW

To express the characteristics of the sludge, as well as to calculate the sludge production in terms of mass and volume, the understanding of certain fundamental relations that are covered below are essential.

### a) Relation between solids levels and water content

The relation between the level of dry solids and the water content in the sludge is given by:

$$\boxed{\text{Water content (\%)} = 100 - \text{Dry solids level (\%)}} \qquad (5.1)$$

A sludge with a level of dry solids of 2% has a water content of 98%. Therefore, in every 100 kg of sludge, 98 kg are water and 2 kg are solids.

The water content influences the mechanical properties of the sludge and these influence the handling processes and the final disposal of the sludge. The relation between the water content and the mechanical properties in most forms of sludges is (van Haandel and Lettinga, 1994):

| Water content | Dry-solids content | Mechanical properties of the sludge |
| --- | --- | --- |
| 100% to 75% | 0% to 25% | fluid sludge |
| 75% to 65% | 25% to 35% | semi-solid cake |
| 65% to 40% | 35% to 60% | hard solid |
| 40% to 15% | 60% to 85% | sludge in granules |
| 15% to 0% | 85% to 100% | sludge disintegrating into a fine powder |

*In the present context, dry solids (**d.s.**) are equivalent to total solids (**TS**), which, in the case of sludges, are very similar to total suspended solids (**TSS** or simply **SS**). These variables may be used interchangeably in this book, when representing solids concentration in the sludge.*

The water content has a large influence on the volume to be handled, as detailed in item d below.

The water in the sludge can be divided into four distinct classes, with different degrees in the easiness of separation (van Haandel and Lettinga, 1994):

- *Free water*. Can be removed by gravity (thickening, flotation)
- *Adsorbed water*. Can be removed by mechanical forces or by the use of a flocculating agent
- *Capillary water*. Maintains itself adsorbed in the solid phase by capillary forces, and is distinguished from the adsorbed water by the need of a greater separation force
- *Cellular water*. Is part of the solid phase and can only be removed by the change of the water aggregation state, that is, through freezing or evaporation

## b) Sludge density

The density of the sludge during most of its processing is very close to water. Usual values are between 1.02 and 1.03 (1020 to 1030 kg/m$^3$) for the liquid sludge during its treatment, and between 1.05 and 1.08 (1050 to 1080 kg/m$^3$) for the dewatered sludge going to final disposal.

## c) Expression of the concentration of dry solids

The concentration of solids in the sludge is expressed in the form of dry solids, that is, excluding the water content of the sludge. The concentration can be in mg/L or in % (the latter being more frequent for sludge processing) and both are related by:

$$\text{Concentration } (\%) = \frac{\text{Concentration (mg/L)} \times 100}{1 \times 10^6 \text{(mg/kg)} \times \text{Density (kg/L)}} \qquad (5.2)$$

Since in most of the sludge processing stages the specific gravity is very close to 1.0 (except for the dewatered sludge), Equation 5.2 can be simplified to the following:

$$\text{Concentration } (\%) \approx \frac{\text{Concentration (mg/L)}}{10,000} \qquad (5.3)$$

For instance, a sludge with a concentration of 20,000 mg/L could have this same concentration expressed as 20,000/10,000 = 2.0% of dry solids. Thus, each 100 kg (or 100 litres) of sludge has 2 kg of dry solids (and 98 kg of water). To clarify the example even further, it can be said 1000 kg (or 1000 litres, or 1 m$^3$) of sludge have 20 kg of dry solids (or 20,000 g of dry solids). Hence, there are 20,000 g of dry solids in 1 m$^3$ of sludge or 20,000 gTS/m$^3$ or 20,000 mgTS/L (mg/L = g/m$^3$).

## d) Relation between flow, concentration and load

The design of the sludge treatment and final disposal stages is based on the sludge flow (volume per unit time) or in many cases, the dry solids load (mass per unit time). The sludge flow is related to the SS load and concentration by:

$$\text{Flow} = \text{Load} / \text{Concentration} \qquad (5.4)$$

$$\text{Sludge flow (m}^3/\text{d)} = \frac{\text{SS load (kgSS/d)}}{\dfrac{\text{Dry solids } (\%)}{100} \times \text{Sludge density (kg sludge/m}^3\text{sludge)}}$$

$$(5.5)$$

Considering that the density of the sludge in practically all of its processing stages is very close to 1000 kg/m$^3$, Equation 5.5 can be simplified to:

$$\text{Sludge flow (m}^3/\text{d)} = \frac{\text{SS load (kgSS/d)}}{\text{Dry solids } (\%) \times 10} \qquad (5.6)$$

A sludge with a solids load of 120 kgSS/d and a solids concentration of 2.0 % (20,000 mg/L) will have a flow of 120/(2.0 $\times$ 10) = 120/20 = 6.0 m$^3$/d.

To estimate the SS load from the sludge flow and SS concentration, the rearrangement of the previous equations can be used:

Load = Flow × Concentration

$$\text{Load (kgSS/d)} = \frac{\text{Flow (m}^3/\text{d)} \times \text{Concentration (g/m}^3)}{1000 \text{ (g/kg)}} \qquad (5.7)$$

The conversion of the units is based on the fact that mg/L is the same as $g/m^3$ (as seen above). A sludge with 20,000 mg/L is the same as a with 20,000 $g/m^3$. If the flow is 6 $m^3/d$, the solids load will be $6 \times 20,000/1000 = 120$ kg of dry solids per day (or 120 kgSS/d or 120 kgTS/d)

For an approximate calculation, it can still be said that the sludge volume (flow) varies inversely with the dry solids concentration (for a sludge with a specific gravity equal to 1.0):

$$\frac{\text{Flow sludge}_1}{\text{Flow sludge}_2} = \frac{\text{Conc. SS}_2 \text{ (\%)}}{\text{Conc. SS}_1 \text{ (\%)}} \qquad (5.8)$$

Therefore, a sludge with a SS concentration of 2.0% and a flow of 6 $m^3/d$ will have the following flow, if the SS concentration is raised to 5.0%: $(6 \text{ m}^3/\text{d} \times 2.0)/5.0 = 2.4 \text{ m}^3/\text{d}$. The other 3.6 $m^3/d$ $(= 6.0 - 2.4)$ are the removed liquid from this stage, which needs to be returned to the head of the works.

## 5.3 QUANTITY OF SLUDGE GENERATED IN THE WASTEWATER TREATMENT PROCESSES

Table 5.2 presents typical sludge removal intervals from the treatment units of the liquid phase, from where the sludge goes to the treatment stage. The intervals are expressed as: *continuous, hours, days, weeks, months, years*, and *decades*. For example, the classification of 'months' indicates that the sludge must be removed in the order of a few months from the treatment unit in the liquid phase to go on to the processing stage in the solid phase. The storage period has a large influence on the sludge characteristics and, as a result, on its treatment. Sludges removed in intervals of weeks, months, years or decades are usually thicker and already digested.

The quantity of sludge generated in sewage treatment, and that should be directed to the sludge treatment stage, can be expressed in terms of mass and volume. Various chapters in the present book detail the methodology of calculating the masses and volumes of the sludge produced in each treatment system. In the present section, a simplified approach of expressing the sludge production in *per capita* terms is adopted.

Table 5.2 also presents typical values for the production of liquid sludge (to be treated) and the dewatered sludge (to be disposed of or reused). As mentioned in

Table 5.2. Frequency of removal, treatment stages and characteristics of the sludge generated and to be disposed of, according to various sewage treatment processes

| System | Sludge removal interval from the liquid phase | | Usual stages of sludge processing | | | | Liquid sludge (to be treated) | | | Dewatered sludge (to be disposed of) | | |
|---|---|---|---|---|---|---|---|---|---|---|---|---|
| | Primary sludge | Biological sludge | Thickening | Digestion | Dewatering | Final disposal | Dry solids level (%) (a) | Sludge mass (gSS/inhab.d) (b) | Sludge volume (L/ inhab.d) (c) | Dry solids level (%) (a) (d) | Sludge mass (gSS/inhab.d) | Sludge volume (L/inhab.d) (c) |
| Primary treatment (septic tank) | Months | – | – | – | PS | PS | 3–6 | 20–30 | 0.3–1.0 | 30–40 | 20–30 | 0.05–0.10 |
| Conventional primary treatment | Hours | – | PS | PS | PS | PS | 2–4 | 35–45 | 0.9–2.0 | 25–45 | 25–28 | 0.05–0.11 |
| Advanced primary treatment | Hours | – | PS/CS | PS/CS | PS/CS | PS/CS | 1–3 | 60–70 | 2.0–7.0 | 20–35 | 40–60 | 0.11–0.30 |
| Facultative pond | – | Decades | – | – | BS (e) | BS | 5–15 | 20–25 | 0.1–0.25 | 30–40 | 20–25 | 0.05–0.08 |
| Anaerobic pond – facultative pond | – | Years | – | – | BS (e) | BS | – | 26–55 | 0.15–0.45 | 30–40 | 26–55 | 0.06–0.17 |
| Facultative aerated lagoon | – | Years | – | – | BS (e) | BS | 4–10 | 8–24 | 0.08–0.60 | 30–40 | 8–24 | 0.02–0.08 |
| Complete-mix aerated lagoon + sedim. pond | – | Years | – | – | BS (e) | BS | 3–8 | 12–30 | 0.15–1.0 | 30–40 | 12–30 | 0.03–0.10 |
| Anaerobic pond + facult. pond + maturation pond | – | Years | – | – | BS (e) | BS | – | 26–55 | 0.15–0.45 | 30–40 | 26–55 | 0.06–0.17 |
| Anaerobic pond + facultative pond + high rate pond | – | Years | – | – | BS (e) | BS | – | 26–55 | 0.15–0.45 | 30–40 | 26–55 | 0.06–0.17 |
| Anaerobic pond + facult. pond + algae removal | – | Years | – | – | BS (e) | BS | – | 30–60 | 0.17–0.52 | 30–40 | 30–60 | 0.07–0.20 |
| Slow rate treatment | – | – | – | – | – | – | – | – | – | – | – | – |
| Rapid infiltration | – | – | – | – | – | – | – | – | – | – | – | – |
| Overland flow | – | – | – | – | – | – | – | – | – | – | – | – |
| Wetlands | – | – | – | – | – | – | – | – | – | – | – | – |
| Septic tank + anaerobic filter | Months | Months | – | – | PS/BS | PS/BS | 1.4–5.4 | 27–39 | 0.5–2.8 | 30–40 | 27–39 | 0.07–0.13 |
| Septic tank + infiltration | Months | – | – | – | PS | PS | 3–6 | 20–30 | 0.3–1.0 | 30–40 | 20–30 | 0.05–0.10 |
| UASB reactor | – | Weeks | – | – | BS | BS | 3–6 | 12–18 | 0.2–0.6 | 20–45 | 12–18 | 0.03–0.09 |
| UASB + activated sludge | – | Weeks | – | – | BS | BS | 3–4 | 20–32 | 0.5–1.1 | 20–45 | 20–32 | 0.04–0.16 |
| UASB + submerged aerated biofilter | – | Weeks | – | – | BS | BS | 3–4 | 20–32 | 0.5–1.1 | 20–45 | 18–30 | 0.04–0.15 |
| UASB + anaerobic filter | – | Weeks | – | – | BS | BS | 3–4 | 15–25 | 0.4–0.8 | 20–45 | 15–25 | 0.03–0.13 |
| UASB + high rate trickling filter | – | Weeks | – | – | BS | BS | 3–4 | 20–32 | 0.5–1.1 | 20–45 | 18–30 | 0.04–0.15 |
| UASB + dissolved air flotation | – | Weeks | – | – | BS/CS | BS/CS | 3–4 | 33–40 | 0.8–1.3 | 20–45 | 33–40 | 0.07–0.20 |

| UASB + polishing pond | – | Weeks | – | – | BS | BS | 3–4 | 15–20 | 0.4–0.7 | 20–45 | 15–20 | 0.03–0.10 |
|---|---|---|---|---|---|---|---|---|---|---|---|---|
| UASB + facultative aerated lagoon | – | Weeks | – | – | BS | BS | – | 20–25 | 0.4–0.8 | 20–45 | 20–25 | 0.04–0.13 |
| UASB + compl.-mix. aerated lagoon + sedim. pond | – | Weeks | – | – | BS | BS | – | 20–25 | 0.4–0.8 | 20–45 | 20–25 | 0.04–0.13 |
| UASB + overland flow | – | Weeks | – | – | BS | BS | 3–6 | 12–18 | 0.2–0.6 | 20–45 | 12–18 | 0.03–0.09 |
| Conventional activated sludge | Hours | ~Continuous | MS | MS | MS | MS | 1–2 | 60–80 | 3.1–8.2 | 20–40 | 38–50 | 0.10–0.25 |
| Activated sludge – extended aeration | – | ~Continuous | BS | – | BS | BS | 0.8–1.2 | 40–45 | 3.3–5.6 | 15–35 | 40–45 | 0.11–0.29 |
| Sequencing batch reactor (extended aeration) | – | Hours | BS | – | BS | BS | 0.8–1.2 | 40–45 | 3.3–5.6 | 15–35 | 40–45 | 0.11–0.29 |
| Convent. activ. sludge with biological N removal | Hours | ~Continuous | MS | MS | MS | MS | 1–2 | 60–80 | 3.1–8.2 | 20–40 | 38–50 | 0.10–0.25 |
| Convent. activ. sludge with biological N/P removal | Hours | ~Continuous | MS | MS | MS | MS | 1–2 | 60–80 | 3.1–8.2 | 20–40 | 38–50 | 0.10–0.25 |
| Conventional activated sludge + tertiary filtration | Hours | ~Continuous | MS | MS | MS | MS | 1–2 | 65–85 | 3.2–8.5 | 20–40 | 43–55 | 0.11–0.28 |
| Low rate trickling filter | Hours | Hours | – | – | MS | MS | 1.5–4.0 | 38–47 | 1.0–3.1 | 20–40 | 38–47 | 0.09–0.22 |
| High rate trickling filter | Hours | Hours | MS | MS | MS | MS | 1.5–4.0 | 55–75 | 1.4–5.2 | 20–40 | 38–47 | 0.09–0.22 |
| Submerged aerated biofilter with nitrification | Hours | Hours | MS | MS | MS | MS | 1–2 | 60–80 | 3.1–8.2 | 20–40 | 38–50 | 0.10–0.25 |
| Submerged aerated biofilter with biol. N removal | Hours | Hours | MS | MS | MS | MS | 1–2 | 60–80 | 3.1–8.2 | 20–40 | 38–50 | 0.10–0.25 |
| Septic tank + rotating biological contactor | Months | Hours | – | – | PS/BS | PS/BS | 1–4 | 25–40 | 0.9–4.0 | 20–40 | 25–40 | 0.06–0.20 |

*Source:* Qasim (1985), EPA (1979, 1987), Metcalf & Eddy (1991), Jordão and Pessoa (1995), Franci (1996), Aisse et al (1999), Chernicharo (1997), Franci (1999), Alem Sobrinho and Jordão (2001), Alem Sobrinho (2001), von Sperling and Gonçalves (2001)

*Notes:* PS = primary sludge; BS = biological sludge; CS = chemical sludge; MS = mixed sludge

(a) Dry solids content (%) = 100 – water content (%)

(b) In units with a long sludge detention time (e.g. ponds, septic tanks, UASB reactors, anaerobic filters), the presented values include the digestion and the thickening that occurs in the unit (which reduce the mass and the volume of the sludge).

(c) Litres of sludge / inhab.d = [(gSS/inhab.d) / (dry solids (%))] x (100/1000) (assuming a density of 1000 kg/m$^3$)

(d) The broad variation of the ranges of dry solids reflects different technologies (natural and mechanised), operating in distinct climatic conditions

(e) The mass of solids in the dewatered sludge becomes lower than the sludge to be treated when the sludge undergoes digestion in the treatment process. In the UASB reactor process followed by aerobic post treatment (e.g. activated sludge, trickling filters and submerged aerated biofilters), the aerobic sludge is returned to the UASB reactor, where it undergoes digestion and thickening, together with the anaerobic sludge. In these cases, the presented values correspond to the combined sludge mixture extracted from the UASB reactor.

the previous item, the sludge mass, expressed as solids, represents the fraction of solids of the sludge generated. The rest of the sludge consists of pure water. The calculation of the sludge volume produced per capita per day is done based on the daily per capita load and the dry solids concentration.

## 5.4 SLUDGE TREATMENT STAGES

The main stages in sludge management, with their respective objectives are:

> - **Thickening**: removal of water (volume reduction)
> - **Stabilisation**: removal of organic matter – volatile solids (mass reduction)
> - **Conditioning**: preparation for dewatering (principally mechanical)
> - **Dewatering**: removal of water (volume reduction)
> - **Disinfection**: removal of pathogenic organisms
> - **Final disposal**: final destination of the by-products

The incorporation of each of these stages in the sludge-processing flowsheet depends on the characteristics of the sludge produced or, in other words, on the treatment system used for the liquid phase, as well as on the subsequent sludge-treatment stage and on the final disposal.

*Thickening* is a physical process of concentrating the sludge, with the aim of reducing its water content and, as a result, its volume, facilitating the subsequent sludge treatment stages.

*Stabilisation* aims at attenuating the inconveniences associated with the generation of bad odours during processing and disposing of the sludge. This is accomplished through the removal of the biodegradable organic matter of the sludge, what also brings about a reduction in the solids mass in the sludge.

*Conditioning* is a sludge preparation process, based on the addition of chemical products (coagulants, polyelectrolytes) to increase its dewatering capability and to improve the capture of solids in the sludge dewatering systems.

The next stage is the *dewatering* of the sludge, which can be done through natural or mechanical methods. The objective of this phase is to remove water and reduce the volume even further, producing a sludge with a mechanical behaviour close to solids. The dewatering of the sludge has an important impact in its transport and final disposal costs, besides influencing its subsequent handling, since the mechanical behaviour varies with the water content level.

The *disinfection* of the sludge is necessary if its destination is for agricultural recycling, since the anaerobic or aerobic digestion processes usually employed do not reduce the pathogens content to acceptable levels. Disinfection is not necessary if the sludge is to be incinerated or disposed of in landfills.

Table 5.2 presents the sludge management stages usually adopted for the most frequently used sewage treatment systems. There are process variants within each stage, with the main ones being presented in Table 5.3.

Table 5.3. Sludge management stages and main processes used

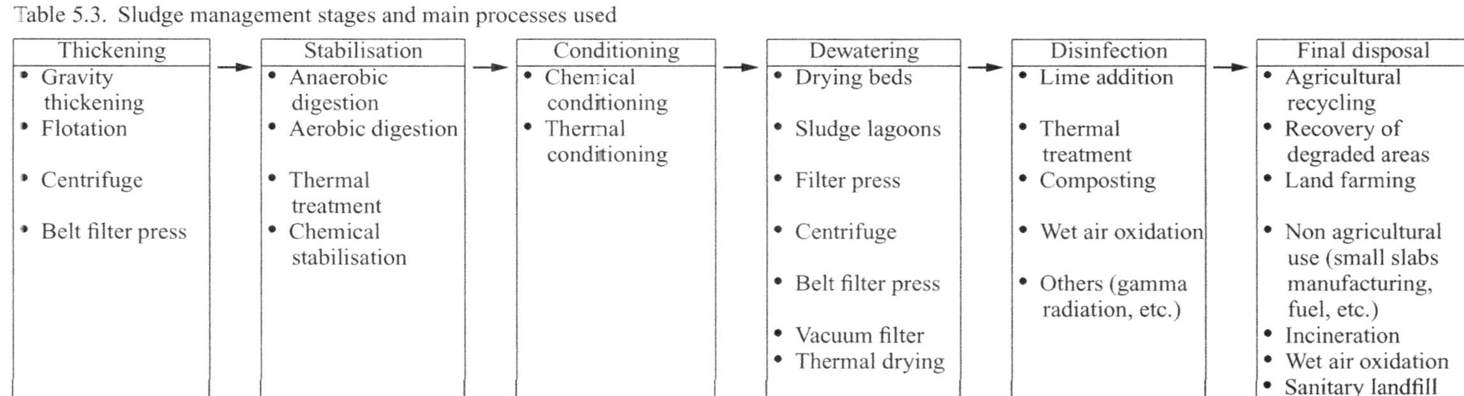

| Thickening | Stabilisation | Conditioning | Dewatering | Disinfection | Final disposal |
|---|---|---|---|---|---|
| • Gravity thickening<br>• Flotation<br>• Centrifuge<br>• Belt filter press | • Anaerobic digestion<br>• Aerobic digestion<br>• Thermal treatment<br>• Chemical stabilisation | • Chemical conditioning<br>• Thermal conditioning | • Drying beds<br>• Sludge lagoons<br>• Filter press<br>• Centrifuge<br>• Belt filter press<br>• Vacuum filter<br>• Thermal drying | • Lime addition<br>• Thermal treatment<br>• Composting<br>• Wet air oxidation<br>• Others (gamma radiation, etc.) | • Agricultural recycling<br>• Recovery of degraded areas<br>• Land farming<br>• Non agricultural use (small slabs manufacturing, fuel, etc.)<br>• Incineration<br>• Wet air oxidation<br>• Sanitary landfill |

The flowsheets of sludge treatment systems allow various combinations of unit operations and process, comprising different sequences, as a function of the sludge characteristics, sewage treatment processes and final disposal methods. Figure 5.1 shows examples of sludge treatment flowsheets frequently used.

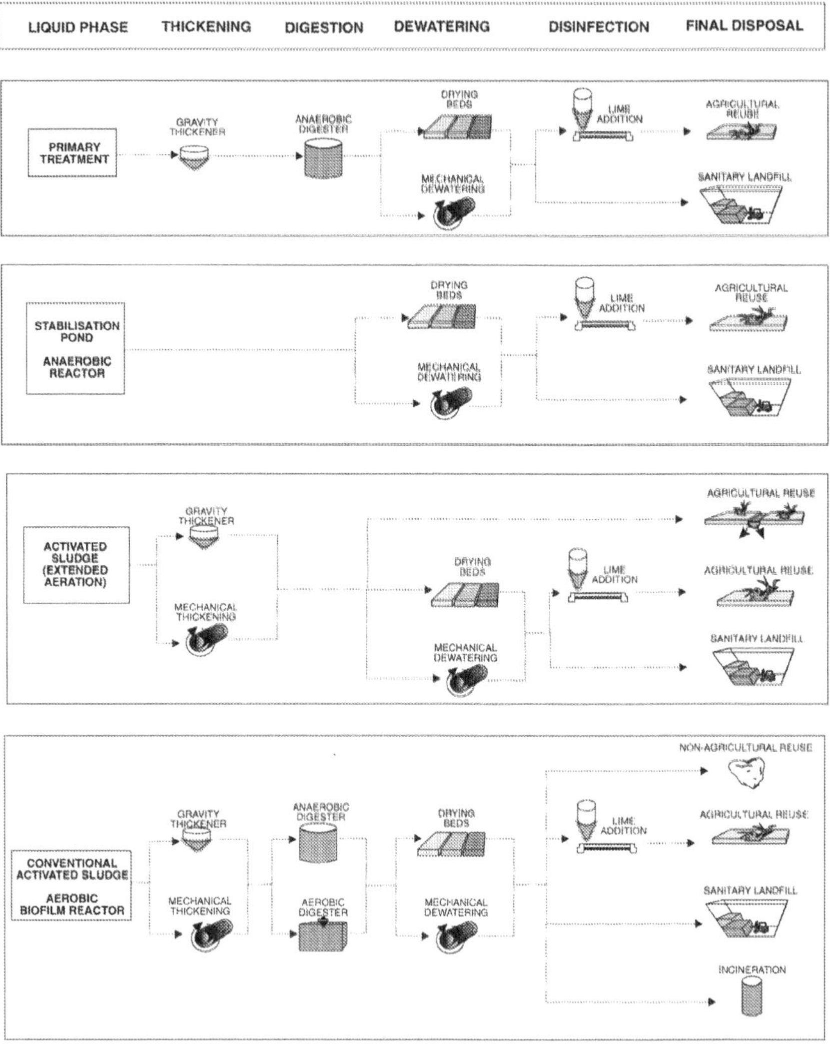


Figure 5.1. Usual sludge treatment and disposal flowsheets (schematic representation of processes frequently employed; certain stages can be optional and in each stage there are various process variants; see Tables 5.2 and 5.3)

Table 5.4. Typical uses of the main sludge thickening methods

| Thickening method | Sludge type | Comment |
| --- | --- | --- |
| Gravity | Primary | Frequently used, with excellent results |
| | Activated sludge | Less frequent, owing to the small increase in the solid levels |
| | Mixed sludge (primary sludge and activated sludge) | Frequently used |
| | Mixed sludge (primary sludge and sludge from the aerobic biofilm reactor) | Frequently used |
| Dissolved air flotation | Mixed sludge (primary sludge and activated sludge) | Less frequent use, since results are similar to gravity thickeners |
| | Activated sludge | Frequently used, with much better results than gravity thickening |
| Centrifuge | Activated sludge | Increasing use |
| Belt press | Activated sludge | Increasing use |

*Source:* adapted from Metcalf and Eddy (1991) and Jordão and Pessôa (1995)

## 5.5 SLUDGE THICKENING

The main processes used for sludge thickening are:

- gravity thickeners
- dissolved air flotation
- centrifuges
- belt presses

Other mechanical processes used for dewatering can also be adapted to sludge thickening. Table 5.4 presents the typical uses of these processes.

Table 5.5 presents typical thickened sludge concentrations, according to the thickening processes employed.

*Gravity thickeners* have a similar structure to settling tanks. The format is usually circular with central feeding, a bottom sludge exit and a supernatant side exit. The thickened sludge goes on to the next stage (normally digestion), while the supernatant is returned to the head of the works. Figure 5.2 presents a schematics of a gravity thickener.

In the process of *dissolved air flotation*, air is introduced in a solution maintained at high pressure. In these conditions, the air is dissolved. When there is a depressurisation, the dissolved air is released in the form of small bubbles. These bubbles, with an upward movement, tend to carry the sludge particles to the surface, from where they are removed. Thickening by flotation has a good applicability for activated sludge, which does not thicken well in gravity thickeners. Dissolved air flotation also has good applicability in WWTPs with biological phosphorus

Table 5.5. Dry solids levels in thickened sludges, according to process

| Sewage treatment process | Process | Dry solids level (%) |
|---|---|---|
| Primary treatment (conventional) | Gravity | 4–8 |
| Conventional activated sludge | | |
| • Primary sludge | Gravity | 4–8 |
| • Secondary sludge | Gravity | 2–3 |
| | Flotation | 2–5 |
| | Centrifuge | 3–7 |
| • Mixed sludge | Gravity | 3–7 |
| | Centrifuge | 4–8 |
| Activated sludge – extended aeration | Gravity | 2–3 |
| | Flotation | 3–6 |
| | Centrifuge | 3–6 |
| High rate trickling filter | | |
| • Primary sludge | Gravity | 4–8 |
| • Secondary sludge | Gravity | 1–3 |
| • Mixed sludge | Gravity | 3–7 |
| Submerged aerated biofilter | | |
| • Primary sludge | Gravity | 4–8 |
| • Secondary sludge | Gravity | 2–3 |
| | Flotation | 2–5 |
| | Centrifuge | 3–7 |
| • Mixed sludge | Gravity | 3–7 |
| | Centrifuge | 4–8 |

**GRAVITY THICKENER**

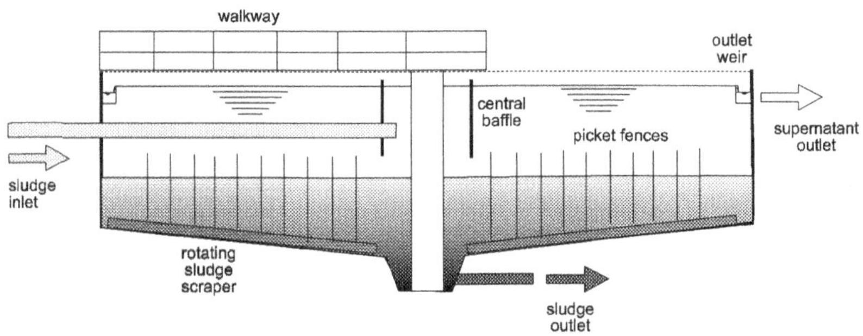


Figure 5.2.   Schematics of a gravity thickener

removal, in which the sludge needs to remain in aerobic conditions in order not to release the phosphorus into the liquid mass.

The other mechanised thickening processes are described in Section 5.7.

## 5.6 SLUDGE STABILISATION

Raw sewage sludge is rich in microorganisms, decomposes easily and quickly releases offensive odours. The stabilisation processes were developed with the

Table 5.6. Comparison between raw and anaerobically digested sludge

| Raw sludge | Digested sludge |
| --- | --- |
| Unstable organic matter | Stabilised organic matter |
| High biodegradable fraction in the organic matter | Low proportion of the biodegradable fraction |
| High potential in the generation of odours | Low potential in the generation of odours |
| High concentration of pathogens | Concentration of pathogens lower than in raw sludge |

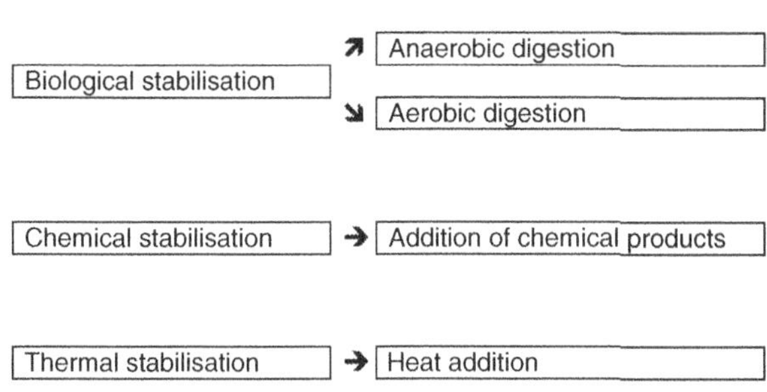


Figure 5.3.   Main sludge stabilisation processes

objective of stabilising (digesting) the biodegradable fraction of the organic matter present in the sludge, thus decreasing the risk of putrefaction, as well as reducing the concentration of pathogens. Table 5.6 shows the main differences between raw and anaerobically digested sludge.

The stabilisation processes can be divided into (see Figure 5.3):

- **biological stabilisation**: use of specific bacteria to promote the stabilisation of the biodegradable fraction of the organic matter
- **chemical stabilisation**: the stabilisation is achieved by the chemical oxidation of the organic matter
- **thermal stabilisation**: obtained from the action of heat on the volatile fraction in hermetically closed recipients.

Anaerobic digestion is the most frequently used sludge stabilisation process. Aerobic digestion is less diffused, but has a good applicability in the stabilisation of the excess activated sludge originating from WWTPs with biological nutrient removal. Composting processes are common in urban solid waste treatment works, but only in a limited number of small-scale WWTPs. Alkaline treatment and thermal drying are other processes used in sludge stabilisation.

The main methods of final sludge disposal associated with different stabilisation processes can be found in Table 5.7. The stabilisation of the sludge facilitates its final disposal and opens alternatives for its reuse as an agricultural soil conditioner.

From the various sewage treatment systems listed in Table 5.2, it is seen that the degree of stabilisation of the sludge produced varies according to the treatment process used.

Table 5.7. Sludge stabilisation technologies and final disposal methods

| Treatment process | Use or final disposal method |
|---|---|
| Anaerobic/aerobic digestion | Produces a biosolid that is suitable to be used with restrictions in agriculture, such as a soil conditioner and organic fertiliser. Usually followed by dewatering. Needs disinfection post-treatment for unrestricted use in agriculture. |
| Chemical treatment (lime stabilisation) | Agricultural use or in the daily cover of a sanitary landfill. |
| Composting | Agricultural humus-like product, appropriate for use in nurseries, horticulture and landscaping. Usually adopted after sludge dewatering. |
| Thermal drying | Product with a high level of solids, significant nitrogen concentration and free from pathogens. Indicated for unrestricted agricultural use. |

The process of anaerobic digestion has been known by sanitary engineers since the end of the $19^{th}$ century and is characterised by the stabilisation of the organic matter in an environment free from molecular oxygen. Owing to its robustness and high efficiency, anaerobic digestion is present from simple domestic septic tanks acting as an individual residential solution, up to completely automated plants, serving large metropolitan regions.

In a conventional activated sludge or trickling filter plant, the mixture between primary sludge and excess biological sludge is stabilised biologically under anaerobic conditions and converted into methane ($CH_4$) and carbon dioxide ($CO_2$). The process is done in closed biological reactors known as anaerobic digesters. The digester is fed in a continuous or batch form and the sludge is maintained inside it for a certain detention time.

The anaerobic digesters are constructed of concrete or steel. The raw sludge is mixed – and heated in temperate climate countries – with the gas produced, and the gas is stored in floating gasholders for processing or burning. The configuration of the digesters varies according to land availability, the need for maintaining a completely-mixed regime and the removal of grit and scum. Figure 5.4 illustrates cylindrical and oval anaerobic digesters.

## 5.7 SLUDGE DEWATERING

Dewatering is done with digested sludges and has an important impact on the sludge transportation and final disposal costs. The main reasons for sludge dewatering are:

- reduction of the transportation costs to the final disposal site;
- improvement of the handling conditions of the sludge, since dewatered sludge is easier to be handled and transported;
- increase of the calorific value of the sludge, through the reduction of the water aiming at preparing it for incineration;
- reduction in the volume for disposal in a landfill or for agricultural use;

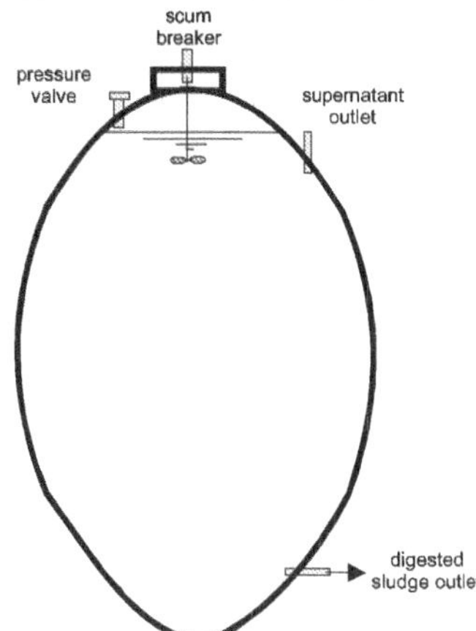


Figure 5.4.   Typical shapes of anaerobic digesters (adapted from WEF, 1996)

- reduction in the production of leachate when the sludge is disposed of in landfills.

Sludge dewatering can be done by *natural* or *mechanised* processes. Natural processes use evaporation and percolation as the main water removal mechanisms, thus requiring more time for dewatering. Although simpler and cheaper to operate, they need larger areas and volumes for installation. In contrast, the mechanised processes are based on mechanisms such as filtration, compaction, or centrifugation to accelerate dewatering, resulting in compact and sophisticated units, from an operational and maintenance point of view.

Many variables influence the selection of the dewatering process, but the sludge type and land availability are the most important ones. Natural processes such as drying beds are considered the best alternative for small scale WWTPs located in regions where there are no area restrictions. In the same way, medium and large scale WWTPs located in metropolitan areas tend to use mechanical dewatering.

The main sludge dewatering processes are listed below:

| *Natural* | *Mechanised* |
|---|---|
| • Drying beds | • Centrifuges |
| • Sludge lagoons | • Vacuum filters |
| | • Belt presses |
| | • Filter presses |

Typical dry solids levels obtained from dewatering processes applied to sludge originating from various wastewater treatment processes are presented in Table 5.8.

Table 5.8. Typical dry solids levels in dewatered sludges from various wastewater treatment processes

| Sewage treatment system | Dewatering process | Dry solids level in the dewatered sludge (%) |
| --- | --- | --- |
| Primary treatment (conventional) | Drying bed | 35–45 |
| | Filter press | 30–40 |
| | Centrifuge | 25–35 |
| | Belt press | 25–40 |
| Primary treatment (septic tank) | Drying bed | 30–40 |
| Facultative pond | Drying bed | 30–40 |
| Anaerobic pond – facultative pond | Drying bed | 30–40 |
| Facultative aerated lagoon | Drying bed | 30–40 |
| Completely-mixing aerated lagoon – sedimentation pond | Drying bed | 30–40 |
| Septic tank + anaerobic filter | Drying bed | 30–40 |
| Conventional activated sludge (mixed sludge) | Drying bed | 30–40 |
| | Filter press | 25–35 |
| | Centrifuge | 20–30 |
| | Belt press | 20–25 |
| Activated sludge – extended aeration | Drying bed | 25–35 |
| | Filter press | 20–30 |
| | Centrifuge | 15–20 |
| | Belt press | 15–20 |
| High rate trickling filter (mixed sludge) | Drying bed | 30–40 |
| | Filter press | 25–35 |
| | Centrifuge | 20–30 |
| | Belt press | 20–25 |
| Submerged aerated biofilter (mixed sludge) | Drying bed | 30–40 |
| | Filter press | 25–35 |
| | Centrifuge | 20–30 |
| | Belt press | 20–25 |
| UASB reactor | Drying bed | 30–45 |
| | Filter press | 25–40 |
| | Centrifuge | 20–30 |
| | Belt press | 20–30 |
| UASB reactor + activated sludge (combined sludge) | Drying bed | 30–45 |
| | Filter press | 25–40 |
| | Centrifuge | 20–30 |
| | Belt press | 20–30 |
| UASB reactor + aerobic biofilm reactor (combined sludge) | Drying bed | 30–45 |
| | Filter press | 25–40 |
| | Centrifuge | 20–30 |
| | Belt press | 20–30 |

- Mixed sludge = primary sludge + secondary sludge
- Combined sludge = anaerobic sludge + aerobic sludge resulting from post treatment and returned to the anaerobic reactor
- The wide ranges of dry solids reflect distinct climatic conditions and operational modes

Table 5.9. Main characteristics of sludge dewatering processes

| | Natural processes | | Mechanised processes | | | |
| Characteristics | Drying beds* | Sludge lagoons | Centrifuges | Vacuum filters | Belt presses | Filter presses |
|---|---|---|---|---|---|---|
| Land requirements | + + + | + + + | + | + + | + | + |
| Energy requirements | − | − | + + | + + + | + + | + + + |
| Implementation cost | + | + | + + + | + + | + + | + + |
| Operational complexity | + | + | + + | + + | + + | + + + |
| Maintenance requirements | + | + | + + | + + | + + + | + + + |
| Complexity of installation | + | + | + + | + + | + + | + + |
| Influence of climate | + + + | + + + | + | + | + | + |
| Sensitivity to the sludge quality | + | + | + + + | + + | + + | + + |
| Chemical products | + | − | + + + | + + + | + + + | + + + |
| Sludge removal complexity | + + | + + + | + | + | + | + |
| Level of dry solids in the cake | + + + | + + | + + | + | + + | + + + |
| Odours and vectors | + + | + + + | + | + | + | + |
| Noise and vibration | − | − | + + + | + + | + + | + + |
| Groundwater contamination | + + | + + + | + | + | + | + |

+ Little, reduced   + + + large, high, very
cake = dewatered sludge
(*) drying bed: a dewatering cycle of 30 days assumed

The main characteristics, advantages and disadvantages of the various dewatering methods are listed in Table 5.9.

To increase the dewatering capability and the solids capture (solids incorporated in the sludge), the sludge can be submitted to a conditioning stage before the dewatering stage itself. The conditioning can be accomplished using chemical products or physical processes; the most common of the latter is the heating of the sludge. The chemical products are applied to the sludge upstream of the dewatering unit, favouring the aggregation of the solids particles and the formation of flocs. The conditioning can be also employed upstream of the mechanised thickening units. The main coagulants used are metallic salts and polyelectrolytes (polymers). The most common metallic coagulants are:

- aluminium sulphate
- ferric chloride
- ferrous sulphate
- ferric sulphate
- quicklime/ hydrated lime

The polymers are organic compounds, usually synthetic, of high molecular weight that can be used as coagulants or flocculating aids. Depending on the

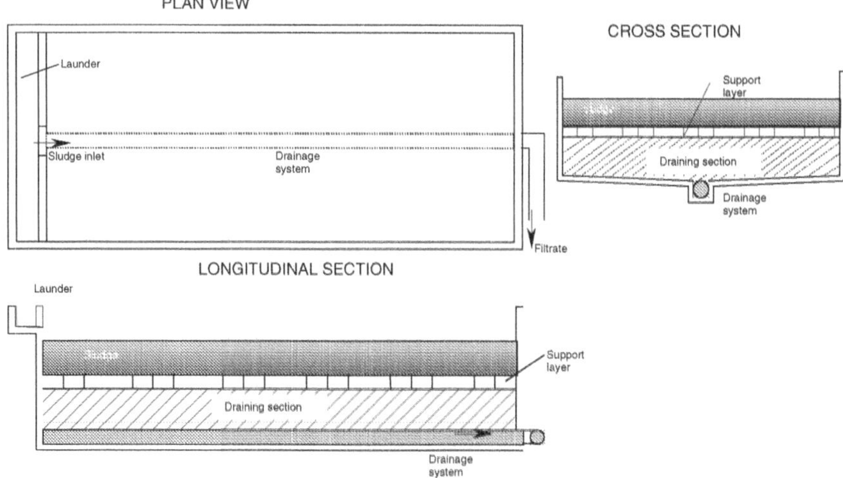


Figure 5.5.   Schematics of a sludge drying bed (Gonçalves, 1999)

prevailing surface charge, the polymers are classified into cationic, anionic and non ionic.

A brief description of the main dewatering processes is presented below.

## a)  Sludge drying beds

Drying beds are one of the oldest techniques and very much used for solids-liquid separation in sludge. The construction costs are generally low in comparison with mechanical dewatering options, especially for small-sized communities. The process generally has a rectangular tank with masonry or concrete walls and a concrete bottom. On the inside of the tank are the following devices to drain the water present in the sludge (Figure 5.5):

- support layer (bricks and coarse sand), on top of which the sludge is placed
- draining medium (fine to coarse sand followed by fine to coarse gravel)
- drainage system (open or perforated pipes)

Part of the liquid evaporates and part percolates through the sand and support layer. The dewatered sludge stays in the layer above the sand.

Drying beds are suggested for small and medium sized communities with the STWs treating a population equivalent of up to around 20,000 inhabitants. The main advantages and disadvantages of using drying beds are presented in Table 5.10.

## b)  Sludge drying lagoon

Sludge drying lagoons are used for thickening, complementary digestion, dewatering and even for the final disposal of sewage sludge. Drying lagoons are generally

Table 5.10. Advantages and disadvantages of sludge drying beds

| Advantages | Disadvantages |
|---|---|
| • Low construction costs<br>• Operational simplicity<br>• Low level of attention required<br>• Operator with a low qualification level required<br>• Low or non existent electrical energy consumption<br>• Low or non existent consumption of chemical products<br>• Low sensitivity to variations of the sludge characteristics<br>• Cake with high solids level | • Large area required<br>• Previous stabilisation of the sludge required<br>• Significant climate influence on the operational performance of the process<br>• Slow removal of the sludge cake<br>• Requires a high quantity of labour to remove the dry cake<br>• High risk of odour release and proliferation of flies<br>• Contamination risk of the groundwater, in case the bottom and the drainage system of the beds are not well executed |

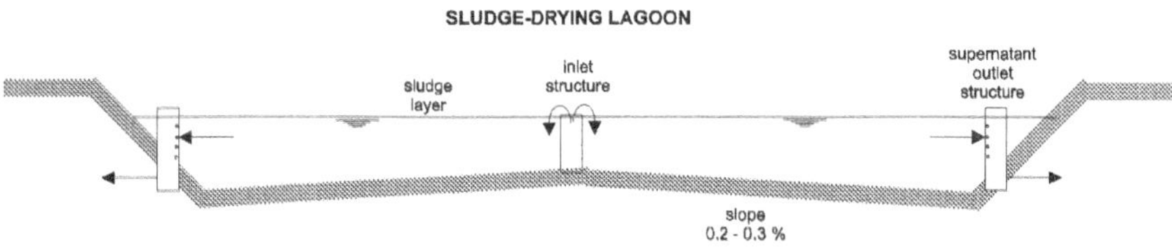


Figure 5.6. Schematics of a sludge drying lagoon (adapted from EPA, 1987)

excavated in the soil, located in natural depressions in the land, or put inside banks, where the discarded sludge from the WWTPs is accumulated for prolonged time periods (from 3 to 5 years). During this period, the sludge is thickened by the action of gravity, further digested by the microorganisms present in the sludge, and dewatered through drainage of the free water and evaporation. The process is only recommended for dewatering previously digested sludge by aerobic or anaerobic methods, and not for use in the dewatering of primary or mixed sludge. *Among the natural dewatering processes, the sludge lagoons are much less used than drying beds.*

The main difference between this process and the drying beds resides in the fact that evaporation is the principal mechanism of influence in the dewatering process. Percolation has a lesser effect than in the drying beds. The dewatering in the lagoon can be accelerated with the use of devices for the removal of the supernatant water at various levels after the loading of the sludge (Figure 5.6). The use of drains at the bottom is not common practice in drying lagoons, because the sewage sludge has reduced drainability and the risk of pipe blocking is very likely to occur.

When the lagoon is full, it can be put out of operation without the removal of the sludge, thus serving as a solution for final disposal. Another possibility is the removal of the sludge from the full lagoon, allowing its reuse and utilisation as a continuous dewatering unit.

Table 5.11. Advantages and disadvantages of sludge lagoons

| Advantages | Disadvantages |
|---|---|
| • Very small energy consumption<br>• Absence of chemical products<br>• Little sensitivity to the variation of the sludge characteristics<br>• Complementary stabilisation of the organic matter in the sludge<br>• Low requirement of skilled labour<br>• Spare unit in STWs with operational problems in sludge dewatering<br>• Low implementation costs in cases where land is cheap | • Large land requirements<br>• Possible generation of odours of difficult control<br>• Possible pollution of ground and surface waters<br>• Attraction of vectors, mainly mosquitoes and flies<br>• Visual impact |

The main advantages and disadvantages of sludge drying lagoons are listed in Table 5.11.

## c) Centrifuge

Centrifugation is a solids/liquid separation operation forced by the action of a centrifugal force. In a first stage, the sludge particles settle at a velocity much higher than would occur under the action of gravity. In a second stage, compaction occurs when the sludge loses part of the capillary water under the prolonged action of centrifugation. The cake is removed from the process after this last dewatering stage.

Centrifuges are equipment that may be used indistinctly for sludge thickening and dewatering. The operating principle is the same, and it is possible to install the centrifuges in series, the first for the thickening of the sludge and the second for the dewatering. The main types of centrifuges used for sludge dewatering are vertical and horizontal-shaft centrifuges. The main differences are in the type of feeding of the sludge, the intensity of the centrifugal force and the manner in which the cake and the liquid are unloaded from the equipment. Currently, the majority of treatment plants that dewater sludge by centrifugation use horizontal-shaft centrifuges. The semi-continuous feeding of the sludge and relatively lower solids levels in the cake produced by the vertical-shaft centrifuges are some of the reasons for this preference.

Horizontal centrifuges can be classified according to the direction of the sludge feeding and the removal of the cake as *co-current* and *counter-current*. Their main differences reside in the sludge feeding points, in the removal of the centrate (liquid phase removal) and in the direction of the flow of the solid and liquid phases in its interior. In the *co-current* centrifuges, the solid and the liquid phases cross all the extension of the longitudinal shaft of the equipment, until they are unloaded. In the *counter-current models*, the feeding is done in the junction of the cylindrical section with the conical section of the equipment. The solid phase is transported

Table 5.12. Advantages and disadvantages of horizontal shaft centrifuges

| Advantages | Disadvantages |
| --- | --- |
| • Can be used for sludge thickening and dewatering | • Noise and vibration |
| • Low land requirements | • Wearing of some components |
| • Ease of installation | • High energy consumption at the engine start |
| • Operation under high loading rates | |
| • Requirement of small quantities of polymers for conditioning | • Complex and slow adjustments during the start-up |
| • Requirement of low attention from operators | • Requirement of careful maintenance |
| | • High costs in many places (especially when imported) |

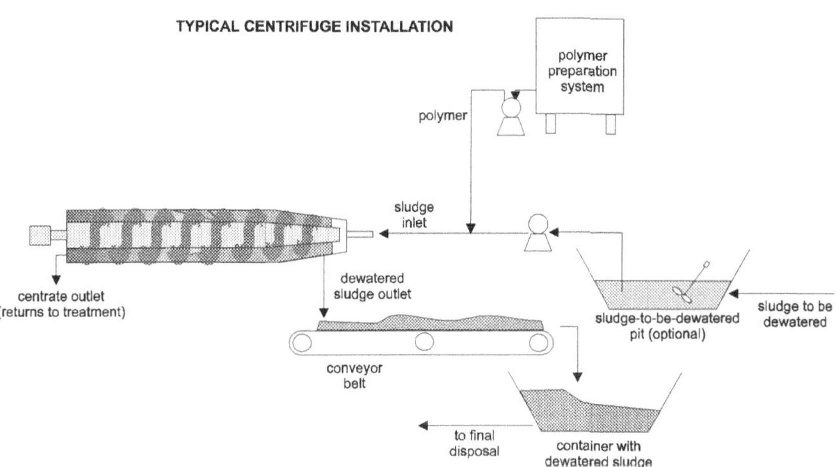


Figure 5.7.   Typical set up of a decanter-type centrifuge

by an screw conveyor to the end of the conical section, while the liquid moves to the opposite direction.

The main advantages and disadvantages of the horizontal centrifuges are summarised in Table 5.12. Figure 5.7 illustrates a typical set up of a centrifuge.

## d) Vacuum filter

Vacuum filters were highly used in industrialised countries for sludge dewatering up until the 1970s. When compared with more modern sludge dewatering processes, *their use entered into decline* due to the high-energy consumption and lower efficiency.

A vacuum filter consists of a rotating cylindrical drum installed with partial submergence in a tank with conditioned sludge. Around 10 to 40% of the drum

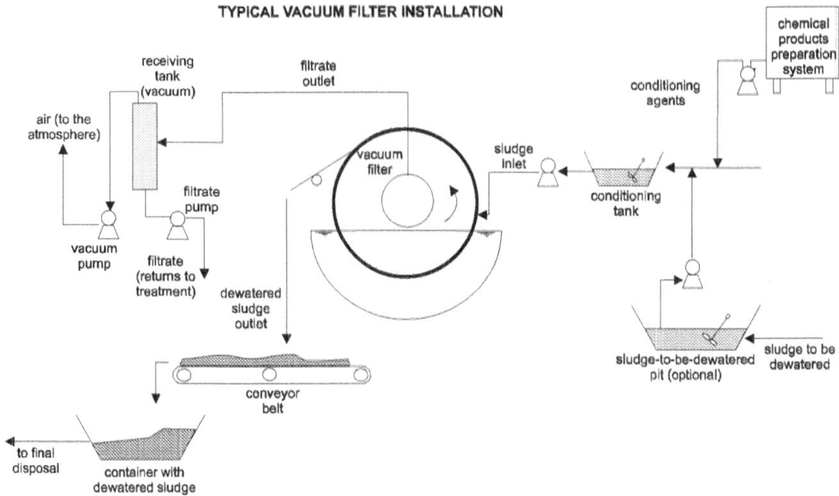



Figure 5.8.   Typical set up of a vacuum filter system for sludge dewatering

surface is submerged in the tank; this part forms the filtration or the cake formation zone. The cake is formed in the outer part of the cylinder, while the filtered liquid migrates to the interior, where there is a vacuum. Next, in the direction of the rotation, there is a dewatering region that occupies between 40 to 60% of the cylinder surface. In the final region of the cylinder that is almost completing the rotation cycle there is the unloading zone. A valve brings the surface of the cylinder to the atmospheric pressure in this region, and the sludge cake is separated from the filtering medium. Figure 5.8 shows a typical set up of a vacuum filter system.

## e)  Filter press

A filter press operates in an intermittent mode, with cycles consisting of sludge loading, filtration, and cake unloading stages. The liquid sludge is pumped into plates surrounded by filter cloths. The pumping of the sludge increases the pressure in the space between the plates and forces the sludge to pass through the filter cloth. The solids are then retained on the filtering medium, forming the cake. Next, a hydraulic piston pushes a steel plate against the other polyethylene plates, making up the pressing. The filtrate (liquid) goes through the filter cloths and is collected by the plate outlet ports. The cake is easily removed from the filter when the pneumatic piston is retreated and the plates are separated. At this moment, the dry cake falls from the plate and can be taken to storage or final destination. Figure 5.9 presents a typical set up of a filter press.

Filter presses were developed for industrial uses and then underwent subsequent adaptations to be used for dewatering sludge. The equipment operates in

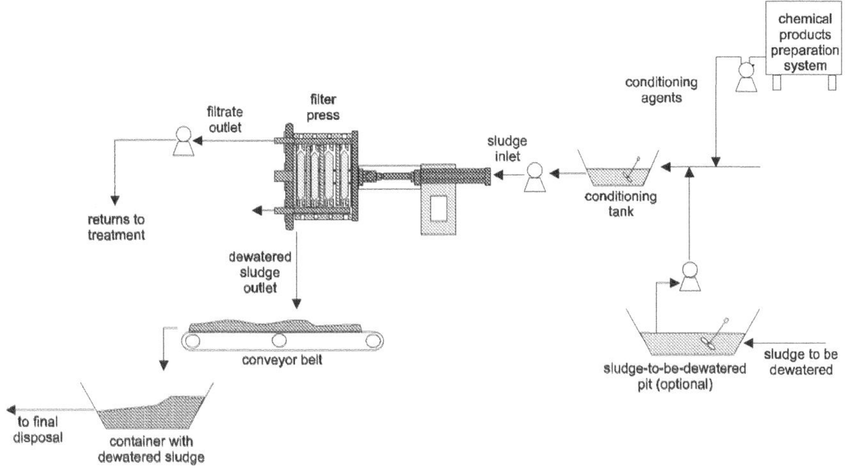



Figure 5.9.   Typical set up of a filter press for sludge dewatering

batch mode, what requires the intervention of trained operators, and has as a main characteristic a high reliability. The principal advantages of filter presses are:

- high concentration of solids in the cake (higher than from other mechanised dewatering processes);
- high solids capture;
- clarified liquid effluent;
- low consumption of chemical products for sludge conditioning;
- reliability.

Currently filter presses are automated, which reduces labour needs. The weight of the equipment, its purchasing costs, and the need for the regular substitution of the filter cloths usually make the utilisation of filter presses limited to medium and large size WWTPs.

## f)  Belt press

Belt presses operate on a continuous mode. The operation process of a belt press can be divided into three distinct stages and zones: (a) zone of gravity drainage, (b) low-pressure zone and (c) high-pressure zone.

The zone of gravity drainage is located at the entrance of the press, where the sludge is applied onto an upper screen and the free water percolates under the action of gravity through the opening pores in the screen. After this, the sludge is directed to the low-pressure zone, where the rest of the free water is removed and the sludge is gently compressed between the upper and lower screens. In the high-rate pressure zone, which is formed by various rollers with different diameters

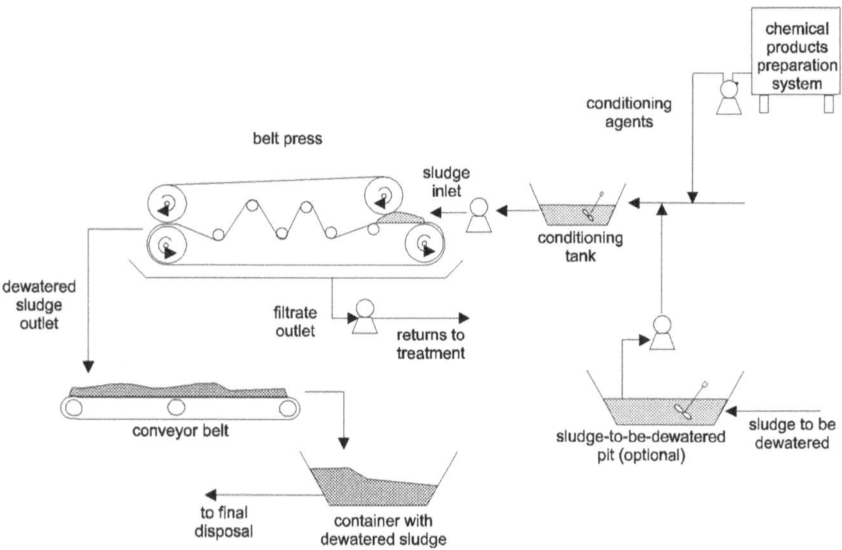



Figure 5.10. Typical set up of a belt press system for sludge dewatering

in series, the sludge is progressively compressed between two screens with the objective of releasing the interstitial water. Finally, scrapers remove the sludge and high-pressure water jets wash the screens. Figure 5.10 presents a typical set up of a belt press system.

Low acquisition costs and reduced energy consumption are the main advantages. However, since the equipment is open, the belt press may have the following disadvantages: aerosol emissions, high level of noise and eventual unpleasant odours (depending on the sludge type). Another large disadvantage of the belt press is the high number of rollers (40–50), which require operational attention and regular substitution.

## 5.8 SLUDGE DISINFECTION

The objective of introducing a sludge disinfection stage in the sewage treatment works is to guarantee a low level of pathogens in the sludge, such that, when it is disposed of on land, will not cause health risks to the population and to the workers that will handle it and also negative impacts to the environment. However, the need to include a complementary sludge disinfection system will depend on the final disposal alternative to be adopted.

The application of sludge in public parks and gardens or its recycling in agriculture implies a higher sanitary level than other disposal alternatives, such as landfills. These requirements can be met by a sludge disinfection process or by temporary restrictions to public use and access.

Some stabilisation processes of the organic matter in the sludge also lead to a reduction of pathogenic microorganisms, producing a sanitarily safe sludge. Others reduce the pathogenic microorganisms to levels lower than the detection limits, after the stabilisation of the organic matter, in a complementary sludge treatment stage. The most important processes are described below.

## a)  Composting

Composting is an aerobic organic matter decomposition process that is achieved through controlled conditions of temperature, water content, oxygen and nutrients. The resultant product (compost) has a high agricultural value as a soil conditioner. The inactivation of the pathogenic microorganisms occurs mainly by the increase of temperature during the highest activity phase of the process. The temperature reaches values between 55 and 65 °C by means of biochemical reactions.

Raw sludge and digested sludge can be composted. Materials such as wood chips, leaves, green waste, rice straw, sawdust, or other structuring agents need to be added to the sludge to improve the water retention, increase the porosity, and balance the ratio between carbon and nitrogen.

Figure 5.11 shows a typical flowsheet of the composting process.

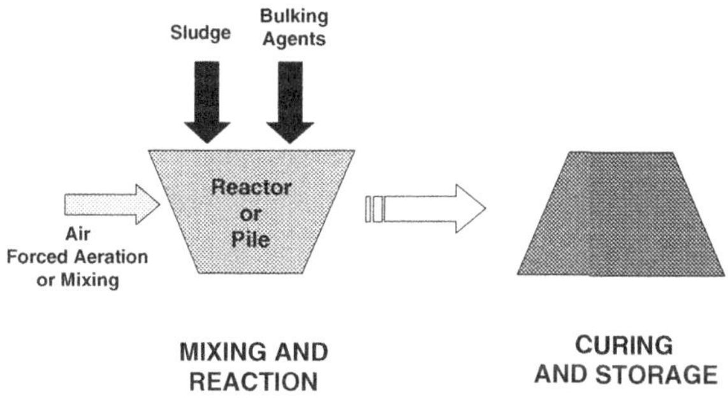


Figure 5.11.    Flowsheet of the composting process

The composting process can be carried out according to the three following main methods (see Figure 5.12 and Table 5.13):

- **Windrows**. Periodical turning, in order to allow aeration and mixture. Detention time between 50 and 90 days.
- **Aerated static pile**. Aeration by perforated pipes from air blowers or exhausting systems. Detention time between 30 and 60 days.
- **In-vessel biological reactors**. Closed systems, with a greater control and lower detention time. Detention time of at least 14 days in the reactor and 14 to 21 days for cooling.

Table 5.13. Comparison between the composting methods

| Composting method | Advantages | Disadvantages |
|---|---|---|
| Windrows | • Low investment cost<br>• Low operation and maintenance costs | • Requirement of large areas<br>• Possible odour problem<br>• Difficulty in reaching the required temperature<br>• Possible mixing problem<br>• High composting period |
| Aerated static piles | • Better odour control<br>• Better conditions for temperature maintenance<br>• Lower reaction time | • Investment required for the aeration system<br>• Moderate operation and maintenance costs |
| Biological reactors | • Small land requirement<br>• High degree of process control<br>• Easiness in controlling temperature and odours | • Higher investment and operation and maintenance costs<br>• Economically feasible only for large scales |

*Source:* Teixeira Pinto (2001)

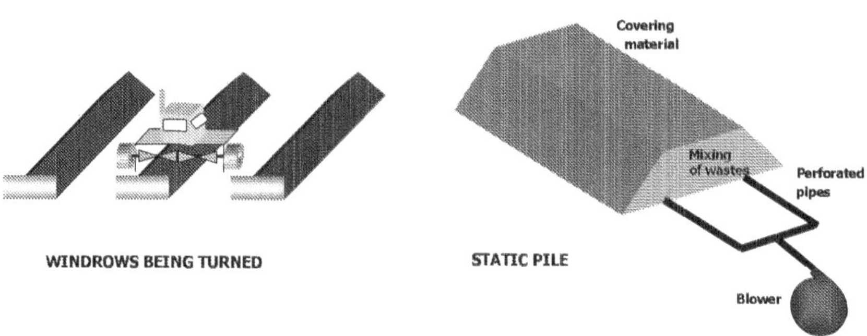


Figure 5.12.　Windrows and aerated static piles (source: Teixeira Pinto, 2001)

## b) Thermophilic aerobic digestion

The process of thermophilic aerobic digestion (also called autothermal digestion) follows the same principles as the conventional aerobic digestion system. The difference is that it operates in a thermophilic phase due to some alterations in the concept and operation of the system.

In this process, the sludge is generally pre-thickened and operates with two aerobic stages, without the need of introducing energy to raise the temperature. Since the reaction volume is small, the system is closed and the concentration of solids in the sludge is higher, the heat released from the aerobic reactions heats the sludge to temperatures higher than 50 °C in the first stage and 60 °C in the second. Due to the temperature increase the pathogenic microorganisms are reduced to

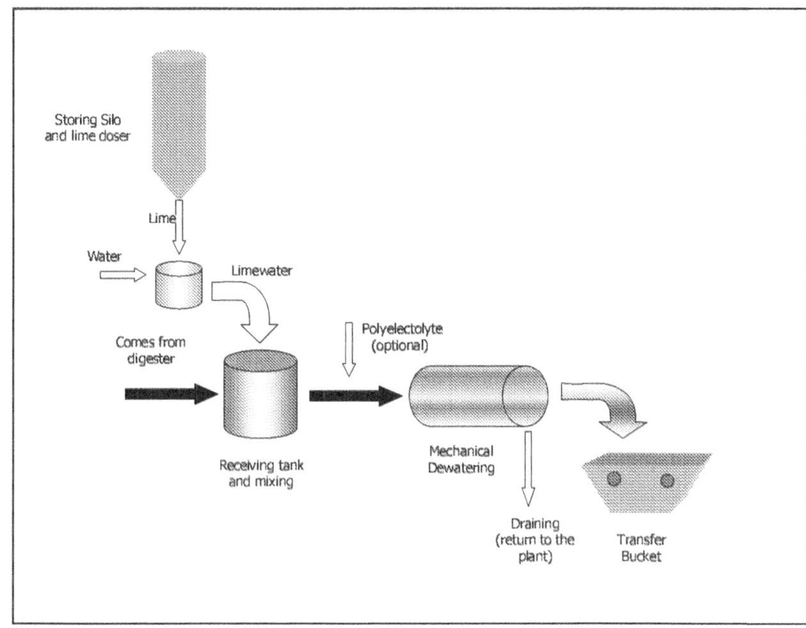


Figure 5.13.   Addition of lime to liquid sludge (lime pre-treatment)

values lower than the detection limits if the sludge is maintained at a temperature between 55–60 °C for 10 days (or 5 to 6 days for reactors in series). The mixing and aeration efficiency are the two most important factors for the operational success of the system.

## c)  Lime stabilisation

Lime stabilisation is used to treat primary, secondary, or digested sludge. When a sufficient quantity of lime is added to the sludge to increase the pH to 12, a reduction of the population of microorganisms (including pathogens) and the potential occurrence of odours takes place. Lime can be added to liquid or dewatered sludges (see Figures 5.13 and 5.14). Owing to the addition of lime, the quantity of sludge to be disposed of increases.

## d)  Pasteurisation

Pasteurisation involves the heating of the sludge to 70 °C for 30 minutes, followed by a rapid cooling to 4 °C. The sludge can be heated by heat exchangers or by hot steam injection. The steam injection process is more commonly used and the sludge is pasteurised in batch to decrease the recontamination risks.

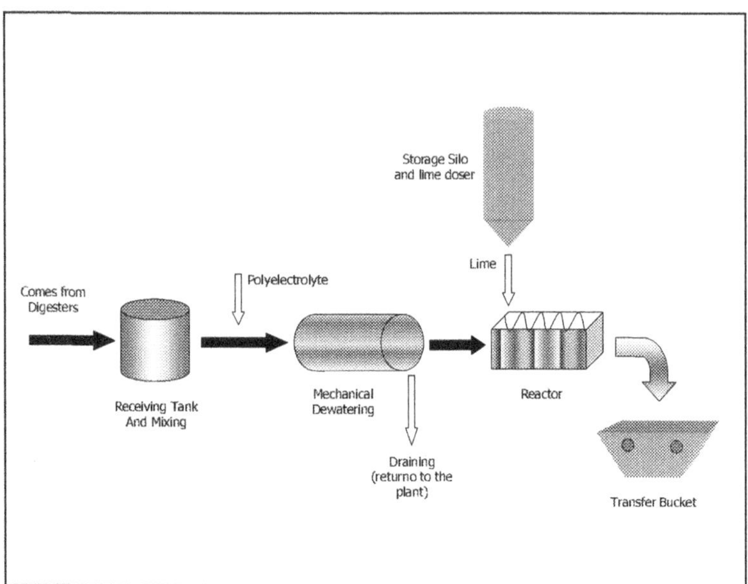


Figure 5.14.　Addition of lime to dewatered sludge (lime post-treatment)

## e) Thermal treatment

Thermal treatment consists of passing the sludge through a heat source that causes the evaporation of the existing moisture in the sludge and consequently the thermal inactivation of the microorganisms. To be economically feasible, the sludge needs to be previously digested and dewatered to a solids concentration in the order of 20–35%, before being thermally treated. The dried sludge has a granular aspect and presents a very high level of solids, in the region of 90–95%.

## f) Comparison between the disinfection processes

Tables 5.14 and 5.15 present a comparison of the characteristics of the various sludge disinfection processes.

## 5.9 FINAL DISPOSAL OF THE SLUDGE

The most commonly applied alternatives for the final disposal of the sludge are summarised in Table 5.16. Table 5.17 lists the main positive and negative aspects associated with each disposal route. Whenever possible, alternatives that bring beneficial uses must be associated with the final disposal.

Table 5.14. Comparison between sludge disinfection technologies – Implementation

| Process | Area | Skilled personnel | External energy | Chemical products | External biomass | Construction cost | O&M cost |
|---|---|---|---|---|---|---|---|
| Composting (windrows/ piles) | +++ | + | +/++ | + | +++ | + | + |
| Composting (reactors) | ++ | ++ | ++ | + | +++ | ++ | ++ |
| Thermophilic aerobic digestion | ++ | ++ | ++ | + | + | ++ | ++ |
| Pasteurisation | ++ | ++ | +++ | + | + | ++ | ++ |
| Lime stabilisation | ++ | +/++ | + | +++ | + | + | ++ |
| Thermal drying | + | +++ | +++ | + | + | +++ | +++ |

*Source:* Teixeira Pinto (2001)
+++: Significant importance   ++: Moderate importance   +: Little or no importance

Table 5.15. Comparison between sludge disinfection technologies – Operation

| Process | Effect against pathogens | | | Product stability | Volume reduction | Odour potential | Observations |
|---|---|---|---|---|---|---|---|
| | Bacteria | Viruses | Eggs | | | | |
| Composting (windrows/ piles) | +++/++ | ++/+ | +++/++ | +++ | ↑ | +++ | Effect depends on mixing |
| Composting (reactors) | +++ | +++/++ | +++ | +++ | ↑ | ++ | Effect depends on mixing |
| Thermophilic aerobic digestion | +++/++ | +++/++ | +++ | ++ | ++ | ++ | Effect depends on the operational regime |
| Pasteurisation | +++ | +++ | +++ | ++ | + | ++ | Must be previously stabilised |
| Lime stabilisation | +++/++ | +++ | +++/++ | ++/+ | ↑ | +++/++ | Effect depends on pH maintenance |
| Thermal drying | +++ | +++ | +++ | +++ | +++ | + | Total stabilisation and inactivation |

*Source:* Teixeira Pinto (2001)
+ + + : Significant importance   ++ : Moderate importance   + : Little or no importance
↑ : Increase in volume

Table 5.16. Main final disposal alternatives for the sludge

| Alternative | Comment |
|---|---|
| Ocean disposal | After pre-conditioning, the sewage is disposed in the sea, through ocean outfalls or barges. Disposal without beneficial uses. |
| Incineration | Thermal decomposition process by oxidation, in which the volatile solids of the sludge are burnt in the presence of oxygen and are converted into carbon dioxide and water. The fixed solids are transformed into ashes. Disposal without beneficial uses. |
| Sanitary landfill | Disposal of the sludges in ditches or trenches, with compaction and covering with soil, until they are totally filled, after which they are sealed. The sewage sludge can be disposed of in dedicated landfills or co-disposed with urban solid wastes. Disposal without beneficial uses. |
| Landfarming | Land disposal process, in which the organic substrate is biologically degraded in the upper layer of the soil and the inorganic fraction is transformed or fixed into this layer. Disposal without beneficial uses. |
| Land reclamation | Disposal of sludge in areas that have been drastically altered, such as mining areas, where the soil does not offer conditions for development and fixation of vegetation, as a result of the lack of organic matter and nutrients. |
| Agricultural reuse | Disposal of the sludge in agricultural soils, in association with the development of crops. Beneficial use of the sludge (which, in this case, is named as a biosolid). |

*Source:* adapted from de Lara et al (2001)

The potential environmental risks or impacts that are related to the sludge disposal alternatives are presented in Table 5.18. The environmental impacts can be more or less complex depending, amongst others, on: (a) quantity of the sludge disposed; (b) physical, chemical and biological characteristics of the sludge and (c) frequency, duration and extent of disposal.

Table 5.17. Advantages and disadvantages of the main sludge disposal alternatives

| Disposal alternative | Advantages | Disadvantages |
|---|---|---|
| Ocean disposal | • Low cost | • Ocean water, flora and fauna pollution |
| Incineration | • Drastic volume reduction<br>• Sterilisation | • High costs<br>• Ash disposal<br>• Atmospheric pollution |
| Sanitary landfill | • Low cost | • Requirement of large areas<br>• Problems with locations near urban centres<br>• Requirement of special soil characteristics<br>• Gas and leachate production<br>• Difficulty in reintegrating the area after decommissioning |
| Landfarming | • Low cost<br>• Disposal of large volumes per unit area | • Accumulation of metals and hardly decaying constituents in the soil<br>• Possible groundwater contamination<br>• Odour release and vector attraction<br>• Difficulty in reintegrating the area after decommissioning |
| Land reclamation | • High application rates<br>• Positive results for the recovery of the soil and flora | • Odours<br>• Composition and use limitations<br>• Contamination of the groundwater, fauna and flora |
| Agricultural reuse | • Large area availability<br>• Positive effects for the soil<br>• Long term solution<br>• Potential as a fertiliser<br>• Positive outcome for the crops | • Limitations regarding composition and application rates<br>• Contamination of the soil by metals<br>• Food contamination with toxic elements and pathogenic organisms<br>• Odours |

*Source:* Lara et al (2001)

Table 5.18. Environmental impacts related to the different sludge disposal alternatives

| Sludge disposal alternative | Potential negative environmental impacts |
| --- | --- |
| Ocean disposal | • Water and sediment pollution<br>• Alteration of the marine fauna communities<br>• Disease transmission<br>• Contamination of elements of the food web |
| Incineration | • Air pollution<br>• Impacts associated with the ash disposal locations |
| Sanitary landfill<br>• Dedicated<br>• Co-disposal with urban wastes | • Surface and groundwater pollution<br>• Air pollution<br>• Soil pollution<br>• Disease transmission<br>• Aesthetic and social impacts |
| Landfarming | • Surface and groundwater pollution<br>• Soil pollution<br>• Air pollution<br>• Disease transmission |
| Land reclamation | • Surface and groundwater pollution<br>• Soil pollution<br>• Odour<br>• Contamination of elements of the food web<br>• Disease transmission |
| Agricultural reuse | • Surface and groundwater pollution<br>• Soil pollution<br>• Contamination of elements of the food web<br>• Disease transmission<br>• Aesthetic and social impacts |

*Source:* Lara et al (2001)

# 6

# Complementary items in planning studies

## 6.1 PRELIMINARY STUDIES

The initial phase of a planning or a design corresponds to the *preliminary studies*. These comprise the overall characterisation of the system to be designed, including a quantitative and qualitative evaluation of the wastewater to be treated, the definition of the treatment objectives and a simple technical–economical screening of the various wastewater treatment processes potentially applicable. This stage is highly important, since the selected alternative will be a result of all the considerations and studies completed in this phase. Consequently, efforts should be directed to obtaining data and subsequently drawing the conclusions, always aiming at the highest possible accuracy and reliability, since the technical success and the economic feasibility of the chosen alternative depend largely on this initial analysis.

Preliminary studies are an integral part of the *planning stage* of the design, which comprise the following fundamental elements:

- *Quantitative characterisation of the influent wastewater* (domestic flow, infiltration flow, industrial flow)
- *Qualitative characterisation of the influent wastewater* (domestic sewage, industrial wastewater)
- *Population forecast studies*

- *Determination of the design horizon and implementation stages*
- *Wastewater treatment objectives and effluent quality requirements*
- *Site selection for the wastewater treatment plant*
- *Initial screening of the treatment alternatives potentially applicable in the situation under analysis*
- *Preliminary design of the most technically promising alternatives* (dimensions of tanks; requirements of power, equipment, resources; flowsheet and plant layout; sludge management)
- *Economical evaluation of the alternatives preliminarily designed*
- *Environmental impact assessment*
- *Selection of the alternative to be adopted and to be subjected to a detailed design, based on the technical and economical analysis*

The population studies and the quantitative and qualitative characterisation of the influent are covered in Chapter 2, while the effluent requirements are discussed in Chapter 3. The selection criteria for the initial screening of alternatives are described in Chapters 4 and 5. Some of the other topics are commented individually in the present chapter. An in depth analysis of these items is not the objective of the chapter, but only to emphasise their importance within the conception and design of the sewage treatment works.

Presented below are short comments on the integration of the points listed above within the preliminary studies phase.

- *Quantification of the polluting loads.* Initially a quantification of the polluting loads needs to be made, based on the quantity and quality of the wastewater. The design population, flows and polluting loads need to be estimated on a yearly (or almost) basis until the end of the design horizon, in order to allow the definition of the staging periods (design stages). (See Chapter 2 for polluting loads and forecasts and current chapter for staging periods studies.)
- *Treatment objectives.* The treatment level and the required removal efficiencies and effluent quality need to be well defined, based on the interaction between the predicted impacts on the water body from the discharge of the effluent and the intended uses for this water body. (See Chapter 3.)
- *Site selection.* Possible sites for the implementation of the treatment plant need to be selected, based on a number of considerations, such as size, geology, topography, ground-water level, flooding level, distance of the intercepting lines, accessibility, neighbouring houses, environmental impact, economics etc. Different sites may be selected, depending on the treatment system to be employed (processes with small or large land requirements). (See relevant sections in this book, related to each treatment process.)
- *Treatment alternatives.* An initial screening of the potentially applicable treatment processes must be undertaken. Based on a global technical analysis, intimately linked to the specificities of the system under analysis, the most promising alternatives are selected for further studies.

- *Process flowsheet*. The flowsheet of each screened alternative to be further analysed should be structured in such a way as to orient the preliminary design stage. The flowsheets should present the main units and flow lines (liquid, sludge, supernatants and recirculations). (See relevant sections in this book, related to each treatment process.)
- *Preliminary design*. The preliminary engineering design is undertaken for the selected alternatives in such a way as to produce data and information to support the subsequent economical analysis. Sub-alternatives may be analysed, such as tank formats, aeration system, sludge treatment options, etc., which may be defined by separate in-parallel comparative technical-economic studies. (See relevant sections in this book, related to each treatment process.)
- *Layout and design of the main units*. Plant layouts, showing in scale the physical arrangement of the units on site should be drawn for the selected alternatives. In order to support the subsequent costs estimates, drawings of the main units should be made, showing the main details that could influence the costs. The preliminary design and the corresponding layouts should be made based on site-specific data, such as topography and geology. (See relevant sections in this book, related to each treatment process.)
- *Economical and financial study*. Based on the characteristics of the main alternatives, a cost estimate for each alternative is undertaken. In many cases it is sufficient to compare only those items which are not common to the options. The economic analysis should take into consideration the construction as well as the operating costs. All of the costs should be brought to the present value, allowing a comparison according to a common basis. (See current chapter.)
- *Environmental impact assessment*. The impacts of the implementation and running of the plant must be taken into account, including the positive impacts resulting from the improvement in the quality of the receiving water body, but also occasional negative impacts associated with the construction, day-to-day operation and occasionally occurring operational problems. Environmental impact assessment (EIA) techniques, such as evaluation matrices, may be employed, and the pertinent legislation must be followed.
- *Selection of the proposed alternative*. The proposed alternative should be the one that offers the greatest advantages from a technical and economical point of view.

## 6.2  DESIGN HORIZON AND STAGING PERIODS

The selection of an appropriate design horizon and its subdivision into staging periods is an item that affects, not only the economy of the plant in terms of construction and operation, but also its performance. These two concepts may be understood as (Qasim, 1985):

- *Design horizon* or *planning period*: period between the initial year and the final year of the plant operation
- S*taging periods:* the time intervals when plant expansions are made

The design horizon of a sewage treatment works should be relatively short, preferably 20 years or less. The design horizon should still be divided into staging periods, in the order of 7 to 10 years. The larger the population growth rate, the more important is the subdivision into stages, the greater should be the number of staging periods and the lower the duration of each stage. High population growth rates are observed in many urban areas of developing countries. On the other hand, very short stages should be avoided in view of the disturbances associated with the almost continuous coexistence with construction works in the plant.

The preliminary studies should be done considering the whole design horizon, in order to allow the estimation of the full land requirements for the plant. However, the detailed design and the construction of the units should be confined to each implementation stage. Some reasons for this are:

- The division in stages is an economically positive factor, which postpones a considerable part of the investments to the future, thus reducing the present value of the construction costs. The higher the interest rates, the greater are the savings.
- For each new stage the design parameters can be reviewed, especially the flows and the incoming loads, as well as the data obtained with the operational experience of the plant itself.
- Over dimensioned units can generate problems, such as septicity in settling tanks (higher than desired detention time), excessive aeration, etc.
- The staging allows the continuous follow-up of the development of the wastewater treatment technology, allowing more modern solutions to be adopted, which may be, in many cases, the most efficient and economical ones.

Therefore, the design of the plant should foresee flexibility for the integration of the existing or first-stage units with future units.

---

### Example 6.1

Carry out a simplified staging study, based on the forecasts of population, influent flow and influent BOD load from the example in Section 2.2.7, which are presented below:

| Year | Served population (inhab) | Average flow ($m^3/d$) | Average BOD load (kg/d) |
|------|---------------------------|------------------------|--------------------------|
| 0    | 24,000                    | 3,888                  | 1,325                    |
| 5    | 47,000                    | 7,477                  | 2,475                    |
| 10   | 53,000                    | 8,409                  | 2,900                    |
| 15   | 58,000                    | 9,179                  | 3,150                    |
| 20   | 62,000                    | 9,820                  | 3,350                    |

---

**Example 6.1 (*Continued*)**

**Solution:**

The following table of percentage population, flow, and load values can be composed, having as a basis the end of the planning period (considered as 100%).

| Year | % of final population | % of final flow | % of final load |
|------|------|------|------|
| 0 | 39 | 40 | 40 |
| 5 | 76 | 76 | 74 |
| 10 | 85 | 86 | 87 |
| 15 | 94 | 93 | 94 |
| 20 | 100 | 100 | 100 |

The staging of this plant does not allow good combinations, because the population growth is not so significant during the design years of the second stage. A possible alternative could be:

| Item | 1st stage | 2nd stage |
|------|------|------|
| Year of stage implementation | *Year 0* | *Year 5* |
| Years covered | *Years 0 to 5* | *Years 6 to20* |
| Duration | *5 years* | *15 years* |
| % of the implementation | *75%* | *25%* |
| Number of modules implemented in the stage | 3 | 1 |
| Total number of modules in the plant | 3 | 4 |

Because the population, flow and load reach in year 5 around 75% of the total final value, the alternative is for the implementation of the first stage comprising 75% of the plant (3 modules, of a total of 4 modules). However, the reach of this first stage is very small, up to year 5 only, when the second stage should enter into operation, until the end of the project (year 20), completing the remaining module (fourth module, in parallel with the others). Although staging is generally advantageous, the benefits in this case are relatively small, because of the need of implementing 75% of the works in the first stage, for only a 5-year period. It is probable that already in year 4 of operation the plant would be in construction works again, associated with the implementation of the units of Stage 2. Therefore, it is unlikely that staging in this plant will be advantageous.

---

## 6.3 PRELIMINARY DESIGN OF THE ALTERNATIVES

There is no need for the elaboration of a detailed design for the technical and economical study of alternatives.

The main objective of the preliminary design is to obtain information to support the technical and economical comparison of the alternatives. The drawing of the plant layout, including the main units, is necessary. In addition, the main dimensions of the units, the area occupied, the earth cut and fill volumes, the concrete volumes, the energy to be consumed, the required equipment and other items judged of relevance in the works in question should be known. Such knowledge serves as the basis for the preliminary cost estimates, which can support the economical study.

For the preliminary design, focus is given to *process* calculations, without the need to deepen in detailing the units and in the hydraulic calculations of pipes and interconnections. *The various design examples presented in this book are at the preliminary-design level.*

## 6.4 ECONOMICAL STUDY OF ALTERNATIVES

The costs for wastewater treatment vary widely with the characteristics of the wastewater, treatment process, climate, design criteria, local conditions and unit local costs for labour, materials, land, and energy.

The cost estimate should comprise the implementation costs (concentrated in time) and the annual operation costs (distributed in time). These costs include (Arceivala, 1981):

- *Implementation costs*
    - construction costs (including equipment and installation)
    - buying or expropriation of the land
    - project costs and supervision, legal taxes
    - interests on the loans during the construction period

- *Annual costs*
    - interests on the loans
    - annual payment for capital recovery
    - depreciation of the works
    - insurance of the works
    - operation and maintenance (O&M) costs of the works

Of the annual costs, the first four items can be considered as fixed since they have to be included if the plant is working or not. In the preliminary economic studies, the costs for construction and land acquisition (implementation) and operation and maintenance (annual) are usually considered.

The present section does not intend to cover the criteria for the making of cost surveys and economical studies. Some simple Economical Engineering methods are presented, which allow the conversion of implementation and annual costs to a common basis that can be used for the comparison between alternatives.

- <u>Present value of a future investment</u>

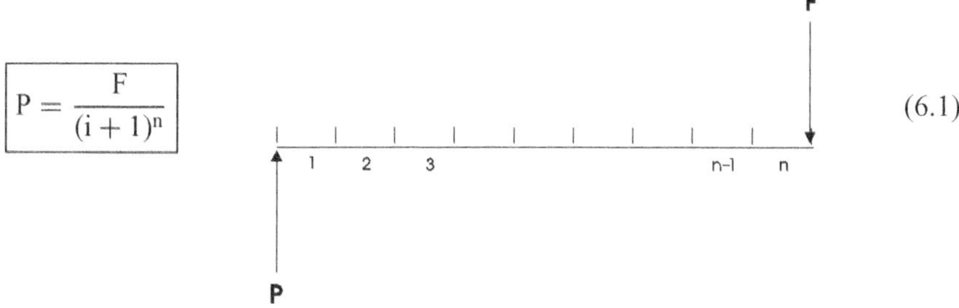

$$P = \frac{F}{(i+1)^n}$$  (6.1)

where:
   P = present value (\$)
   F = future value (\$)
   i = annual interest rate
   n = number of years

- <u>Present value of constant annual expenditures</u>

$$P = A\frac{(1+i)^n - 1}{i.(i+1)^n}$$  (6.2)

where:
   A = annual expenditure (\$/year)

---

### Example 6.2

There are two alternatives for the wastewater treatment in a community, each one with different implementation and operation/maintenance costs. The basic characteristics are:

- Planning period: 20 years
- Interest rate: 11% per year

- <u>Alternative A</u>
   - implementation cost (year 0): US\$ $3 \times 10^6$
   - operation/maintenance cost: US\$ $0.5 \times 10^6$/year

- <u>Alternative B</u>
   - implementation cost (first stage, year 0): US\$ $5 \times 10^6$
   - operation/maintenance cost (first stage): US\$ $0.2 \times 10^6$

---

### *Example 6.2 (Continued)*

- implementation cost (second stage, year 10): US\$ $3 \times 10^6$
- operation/maintenance cost (second stage): US\$ $0.3 \times 10^6$

Based on the lowest present value, indicate the best alternative in economical terms.

**Solution:**

**a) Present value of alternative A**

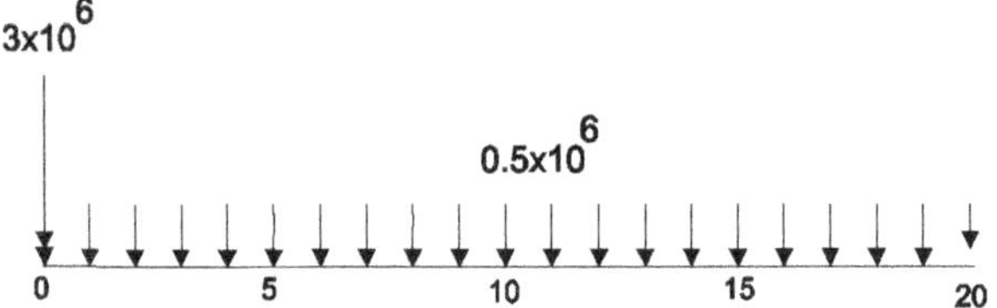


- Present value of the annual expenses (Equation 6.2)

$$P = A\frac{(1+i)^n - 1}{i.(i+1)^n} = 0.5 \times 10^6 \cdot \frac{(1+0.11)^{20} - 1}{0.11 \times (1+0.11)^{20}} = 4.0 \times 10^6$$

- Total present value

Total present value = Implementation cost + Present value of the annual expenses

Total present value = $3.0 \times 10^6 + 4.0 \times 10^6$

**Total present value = US\$ $7.0 \times 10^6$**

**b) Present value of alternative B**

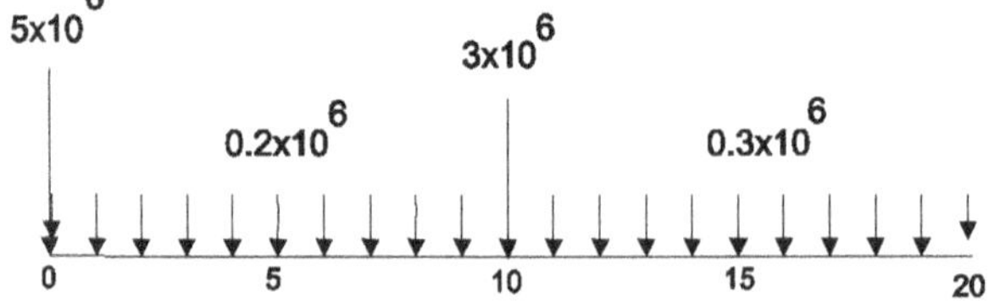


- Present value of the implementation costs of the second stage (Equation 6.1)

$$P = \frac{F}{(i+1)^n} = \frac{3 \times 10^6}{(1+0.11)^{10}} = 1.2 \times 10^6$$

***Example 6.2 (Continued)***

- Present value of the annual expenses of the second stage

Transport the values for year 10 (first year of the second stage) (Equation 6.2):

$$P = A\frac{(1+i)^n - 1}{i.(i+1)^n} = 0.3 \times 10^6 . \frac{(1+0.11)^{10} - 1}{0.11 \times (1+0.11)^{10}} = 1.8 \times 10^6$$

Transport the concentrated value of year 10 to the present value (Equation 6.1):

$$P = \frac{F}{(i+1)^n} = \frac{1.8 \times 10^6}{(1+0.11)^{10}} = 0.6 \times 10^6$$

- Present value of the annual expenses of the first stage (Equation 6.2)

$$P = A\frac{(1+i)^n - 1}{i.(i+1)^n} = 0.2 \times 10^6 . \frac{(1+0.11)^{10} - 1}{0.11 \times (1+0.11)^{10}} = 1.2 \times 10^6$$

- Total present value

Total present value = Implementation cost of the 1st stage + Present value of the implementation of the 2nd stage + Present value of the annual expenses of the 1st stage + Present value of the annual expenses of the 2nd stage

Total present value = $5 \times 10^6 + 1.2 \times 10^6 + 1.2 \times 10^6 + 0.6 \times 10^6$
**Total present value = US\$ 8.0 $\times$ 10$^6$**

**c) Summary of the results**

| Alternative | Present value |
|---|---|
| A | US\$ 7.0 $\times$ 10$^6$ |
| B | US\$ 8.0 $\times$ 10$^6$ |

Based on the lowest present value, alternative A is recommended. In this case, it is more advantageous to have a plant with a smaller implementation cost, even though presenting higher operation and maintenance costs.

The present analysis was made in a simplified manner. Other economical and financial considerations can be included, in order to give support to the study of alternatives.

# References


AISSE, M.M., VAN HAANDEL, A.C., VON SPERLING, M., CAMPOS, J.R., CORAUCCI FILHO, B., ALEM SOBRINHO, P. (1999). Capítulo 11: Tratamento e destino final do lodo gerado em reatores anaeróbios. In: CAMPOS, J.R. (coord). *Tratamento de esgotos sanitários por processo anaeróbio e disposição controlada no solo*. PROSAB/ABES, Rio de Janeiro. (in Portuguese).

ALEM SOBRINHO, P., TSUTIYA, M.T. (1999). *Coleta e transporte de esgoto sanitário*. Escola Politécnica, USP, São Paulo. 547 p. (in Portuguese).

ARCEIVALA, S.J. (1981). *Wastewater treatment and disposal. Engineering and ecology in pollution control*. New York, Marcel Dekker. 892 p.

BARNES, D.; BLISS, P.J.; GOULD, B.W.; VALLENTINE, H.R. (1981). *Water and wastewater engineering systems*. Pitman Publishing Inc, Massachussets, 513 p.

BENENSON, A.B., ed (1985). *Control of communicable diseases in man*. American Public Health Association. 14 ed.

BRAILE, P.M. & CAVALCANTI, J.E.W.A. (1979). *Manual de tratamento de águas residuárias industriais*. São Paulo, CETESB. 764 p. (in Portuguese).

BRANCO, S.M. & ROCHA, A.A. (1979). *Poluição, proteção e usos múltiplos de represas*. São Paulo, Ed. Edgard Blücher. 185 p. (in Portuguese).

BRANCO, S.M. (1976). *Hidrobiologia aplicada à engenharia sanitária*. São Paulo, CETESB. 620 p. (in Portuguese).

CAIRNCROSS, S., FEACHEM, R.G. (1990). *Environmental health engineering in the tropics: an introductory text*. Chichester. John Wiley & Sons. 283 p.

CAMP, T.R (1954). Analysis of a stream's capacity for assimilating pollution – a discussion. *Sewage and Industrial Wastes*, **26** (11), Nov. 1954. p. 1397–1398.

CAMPOS, H.M., VON SPERLING, M. (1996). Estimation of domestic wastewater characteristics in a developing country based on socio-economical variables. *Water Science and Technology*, **34** (3–4). pp. 71–77.

CASTAGNINO, W.A. (1977). *Polucion de agua. Modelos y control*. Lima, CEPIS. 234 p. (in Spanish).

CAVALCANTI, P.F.F., VAN HAANDEL, A.C, KATO, M.T., VON SPERLING, M., LUDUVICE, M.L., MONTEGGIA, L.O. (2001). Capítulo 3: Pós-tratamento de efluentes de reatores anaeróbios por lagoas de polimento. In: CHERNICHARO, C.A.L.

(coord). *Pós-tratamento de efluentes de reatores anaeróbios*. PROSAB/ABES, Rio de Janeiro. p. 105–170. (in Portuguese).

CETESB, Companhia de Tecnlogia de Saneamento Ambiental (1977). *Sistemas de esgotos sanitários*. 2 ed. São Paulo. 467 p. (in Portuguese).

CETESB, Companhia de Tecnlogia de Saneamento Ambiental (1978). *Técnica de abastecimento e tratamento de água*. Vol. 1, 2 ed. São Paulo. (in Portuguese).

CETESB, Companhia de Tecnologia de Saneamento Ambiental (1976). *Ecologia aplicada e proteção do meio ambiente*. São Paulo. (in Portuguese).

CHERNICHARO, C.A.L. (1997). *Princípios do tratamento biológico de águas residuárias. Vol. 5. Reatores anaeróbios*. Departamento de Engenharia Sanitária e Ambiental – UFMG. 245 p.

CHERNICHARO, C.A.L. (coord) (2001). *Pós-tratamento de efluentes de reatores anaeróbios*. ISBN 85-901640-2-0. PROSAB/ABES, Rio de Janeiro. p. 279–331. (in Portuguese).

CHOW, V.T. (1959). *Open channel hydraulics*. Tokyo, Mc Graw-Hill. 680 p.

CHURCHILL, M.A., ELMORE, H.L., BUCKINGHAM, R.A. (1962). The prediction of stream reaeration rates. *Journal Sanitary Engineering Division, ASCE*, **88** (4). July 1962. p. 1–46.

CORAUCCI FILHO, B., CHERNICHARO, C.A.L., ANDRADE NETO, C.O., NOUR, E.A., ANDREOLI, F.N., SOUZA, H.N., MONTEGGIA, L.O., VON SPERLING, M., LUCAS FILHO, M., AISSE, M.M., FIGUEIREDO, R.F., STEFANUTTI, R. (1999). Capítulo 13: Bases conceituais da disposição controlada de águas residuárias no solo. In: CAMPOS, J.R. (coord). *Tratamento de esgotos sanitários por processo anaeróbio e disposição controlada no solo*. PROSAB/ABES, Rio de Janeiro. (in Portuguese).

COUNCIL OF THE EUROPEAN COMMUNITIES (1991). Council Directive of 21 May 1991 concerning urban waste water treatment (91/271/EEC). *Official Journal of the European Communities*, No. L 135/40.

CRESPO, P.G. (1997). *Sistema de esgotos*. Editora UFMG. 1997. 131 p. (in Portuguese).

DAHLHAUS, C., DAMRATH, H. (1982). *Wasserversorgung*. B.G. Teubner Stuttgart. 254 p.

DANTAS, A.P. (2000). *Avaliação limnológica de parâmetros morfométricos em lagos e represas*. Dissertação de Mestrado, Universidade Federal de Minas Gerais, Brazil (in Portuguese).

DERÍSIO, J.C. (1992). *Introdução ao controle de poluição ambiental*. CETESB, São Paulo. (in Portuguese).

DOWNING, A.L. (1978). *Selected subjects in waste treatment*. 3 ed. Delft, IHE.

ECKENFELDER Jr, W.W. (1980). *Principles of water quality management*. Boston, CBI. 717 p.

EMERSON, K. et al. (1975). Aqueous ammonia equilibrium calculations: effect of pH and temperature. *Journal Fish. Res. Board Can.*, **32** (12), p. 2379–2383.

EPA, Environmental Protection Agency (1977). *Wastewater treatment facilities for sewered small communities*. Cincinatti.

EPA, Environmental Protection Agency (1979). *Environmental pollution control alternatives: municipal wastewater*. Technology Transfer, 79 pp.

EPA, Environmental Protection Agency (1979a). *Process design manual. Sludge treatment and disposal*.

EPA, Environmental Protection Agency (1981). *Process design manual for land treatment of municipal wastewater*. Technology Transfer.

EPA, Environmental Protection Agency (1985). *Rates, constants, and kinetics formulations in surface water quality modeling*. 2. ed, 455 p.

EPA, Environmental Protection Agency (1987). *Design manual. Dewatering municipal wastewater suldges*. 193 p.

EPA, Environmental Protection Agency (1992). *Manual. Wastewater treatment/disposal for small communities.* Technology Transfer, 110 pp.

EPA, Environmental Protection Agency (2000). *Manual. Constructed wetlands treatment of municipal wastewaters.* 154 pp.

FAIR, G.M., GEYER, J.C., OKUN, D.A. (1973). *Purificación de aguas y tratamiento y remoción de aguas residuales* [Water and wastewater engineering]. Trad. Salvador Ayanegui j. México, Editorial Limusa, v.2. 764 p.

FERNANDES, F., LOPES, D.D., ANDREOLI, C.V., SILVA, S.M.C.P. (2001). Avaliação de alternativas e gerenciamento do lodo na ETE. In: ANDREOLI, C.V., VON SPERLING, M., FERNANDES, F. (2001). *Princípios do tratamento biológico de águas residuárias. Vol. 6. Lodo de esgotos. Tratamento e disposição final.* Departamento de Engenharia Sanitária e Ambiental – UFMG. Companhia de Saneamento do Paraná – SANEPAR. 484 p. (in Portuguese).

GONÇALVES, R.F. (1996). Aspectos teóricos e práticos do tratamento de esgoto sanitário em biofiltros aerados com leito granular submerso. In: CHERNICHARO, C.A.L., VON SPERLING, M. (eds). *Seminário Internacional. Tendências no tratamento de águas residuárias domésticas e industriais.* Belo Horizonte, 6–8 março 1996. DESA-UFMG, p. 128–143. (in Portuguese).

GONÇALVES, R.F. (coord) (1999). *Gerenciamento do lodo de lagoas de estabilização não mecanizadas.* PROSAB – Programa de Pesquisa em Saneamento Básico. ABES, Rio de Janeiro. 80 p. (in Portuguese).

GRABOW, W. (2002). Water and public health. Theme level paper. In: *UNESCO Encyclopedia of life support systems.*

HELLER, L. (1997). *Saneamento e saúde.* Organização Panamericana de Saúde/Organização Mundial de Saúde. Brasília. 97 p. (in Portuguese).

HELLER. L., MÖLLER, L.M. (1995). Capítulo 3. Saneamento e saúde pública. In: BARROS, R.T.V., CHERNICHARO, C.A.L., HELLER, L., VON SPERLING. M. (eds). Manual de saneamento e proteção ambiental para os municípios. DESA-UFMG & FEAM, Belo Horizonte, Brazil. pp. 51–61. (in Portuguese).

HOSANG/BISCHOF, W. (1984). *Abwassertechnik.* B.G.Teubner Stuttgart. 448 p.

IAWQ (1995). *Activated sludge model No. 2.* IAWQ Scientific and Technical Reports.

IMHOFF, K., IMHOFF, K.R. (1985). *Taschenbuch der Stadtentwässerung.* Oldenbourg, Munique. 408 p.

JOHNSTONE, D.W.M., HORAN, N.J. (1994). Standards, costs and benefits: an international perspective. *J. IWEM* **8**, 450–458.

JOHNSTONE, D.W.M., HORAN, N.J. (1996). Institutional developments, standards and river quality: A UK history and some lessons for industrialising countries. *Water Science and Technology.* **33** (3), 211–222.

JORDÃO, E.P., PESSOA, C.A. (1995). *Tratamento de esgotos domésticos.* 3. ed. Rio de Janeiro, ABES. (in Portuguese).

KLEIN, L. (1962). *River pollution. II. Causes and effects.* London, Butterworths Scientific Publications. v.2.

LA RIVIÉRE, J.W.M. (1980). *Microbiology.* Delft, IHE (lecture notes).

LARA, A.I., ANDREOLI, C.V., PEGORINI, E.S. (2001). Avaliação dos impactos ambientais e monitoramento da disposição final do lodo. In: ANDREOLI, C.V., VON SPERLING, M., FERNANDES, F. (2001). *Princípios do tratamento biológico de águas residuárias. Vol. 6. Lodo de esgotos. Tratamento e disposição final.* Departamento de Engenharia Sanitária e Ambiental – UFMG. Companhia de Saneamento do Paraná – SANEPAR. 484 p. (in Portuguese).

MARAIS, G.v.R. & EKAMA, G.A. (1976). The activated sludge process. Part I – Steady state behaviour. *Water S.A.*, **2** (4), Oct. 1976. p. 164–200.

MARQUES, D.M. (1999). Capítulo 15. Terras úmidas construídas de fuxo subsuperficial. In: CAMPOS, J.R. (coord). *Tratamento de esgotos sanitários por processo anaeróbio e disposição controlada no solo*. PROSAB/ABES, Rio de Janeiro. (in Portuguese).

MATTOS, A. (1998). *Characteristics of effluents from agroindustries*. Personal communication.

MENDONÇA, S.R. (2000). *Sistemas de lagunas de estabilización*. McGraw-Hill. Colombia. 370 p. (in Spanish).

METCALF & EDDY (1981). *Wastewater engineering: treatment, disposal, reuse*. 2. ed. New Delhi, Tata Mc Graw-Hill. 920 p.

METCALF & EDDY (1991). *Wastewater engineering: treatment, disposal and reuse*. Metcalf & Eddy, Inc. 3. ed, 1334 p.

NBR-7229/93 (1993). *Projeto, construção e operação de sistemas de tanques sépticos*. ABNT. (in Portuguese).

O'CONNOR, D.J., DOBBINS, W.E. (1958). Mechanism of reaeration in natural streams. *Journal Sanitary Engineering Division, ASCE*, **123**. p. 641–666.

OPS (Organización Panamericana de la Salud), OMS (Organización Mundial de la Salud) (1999). Sistemas de tratamiento de aguas servidas por medio de humedales artificiales. Colombia, 217 p. (in Spanish).

PEAVY, H.S., ROWE, D.R., TCHOBANOGLOUS, G. (1986). *Environmental engineering*. McGraw Hill. 699 p.

PESSOA, C.A., JORDÃO, E.P. (1982). *Tratamento de esgotos domésticos*. 2. ed. Rio de Janeiro, ABES. (in Portuguese).

PÖPEL, H.J. (1979). *Aeration and gas transfer*. 2. ed. Delft, Delft University of Technology. 169 p.

QASIM, S.R. (1985). *Wastewater treatment plants: planning, design and operation*. Holt, Rinehart and Winston, New York.

RAUCH, W. et al. (1998). River water quality modelling: I. State of the art. *Water Science and Technology*, **38** (11), pp. 237–244.

SALAS, H.J., MARTINO, P. (1991). A simplified phosphorus trophic state model for warm-water tropical lakes. *Water Research*, **25** (3). p. 341–350.

SALVADOR, N.N.B. (1991). Listagem de fatores de emissão para avaliação expedita de cargas poluidoras das águas. In: *Congresso Brasileiro de Engenharia Sanitária e Ambiental*, **16**, Goiânia, Tomo VI-2, p. 3–21. (in Portuguese).

SHANAHAN, P. (1998). River water quality modelling: II. Problems of the art. *Water Science and Technology*, **38** (11), pp. 245–252.

SILVA, S.A., MARA, D.D. (1979). *Tratamentos biológicos de águas residuárias: lagoas de estabilização*. Rio de Janeiro, ABES. 140 p. (in Portuguese).

SILVA, S.M.C.P., FERNANDES, F., SOCCOL, V.T., MORITA, D.M. (2001). Principais contaminantes do lodo. In: ANDREOLI, C.V., VON SPERLING, M., FERNANDES, F. (2001). *Princípios do tratamento biológico de águas residuárias. Vol. 6. Lodo de esgotos. Tratamento e disposição final*. Departamento de Engenharia Sanitária e Ambiental – UFMG. Companhia de Saneamento do Paraná – SANEPAR. 484 p. (in Portuguese).

SOMLYÓDY, L. (1998). River water quality modelling: III. Future of the art. *Water Science and Technology*, **38** (11), pp. 253–260.

STREETER, H.W., PHELPS, E.B. (1925). A study of the pollution and natural purification of the Ohio River. *Public Health Bulletin*, **146**, Washington.

TCHOBANOGLOUS, G., SCHROEDER, E.D. (1985). *Water quality: characteristics, modeling, modification*. Addison-Wesley, Reading, MA.

TEIXEIRA PINTO, M. (2001). Higienização de lodos. In: ANDREOLI, C.V., VON SPERLING, M., FERNANDES, F. (2001). *Princípios do tratamento biológico de águas residuárias. Vol. 6. Lodo de esgotos. Tratamento e disposição final*. Departamento de Engenharia Sanitária e Ambiental – UFMG. Companhia de Saneamento do Paraná – SANEPAR. 484 p. (in Portuguese).

THOMANN, R.V., MUELLER, J.A. (1987). *Principles of surface water quality modeling and control*. Harper International Edition. 644 p.

THORNTON, J.A., RAST, W. (1994). The management of international river basins. In: *Qualidade de águas continentais no Mercosul*. ABRH, publicação no. 2, 1994. 420 p.

VAN BUUREN, J.C.L, FRIJNS, J.A.G., LETTINGA, G. (1995). *Wastewater treatment and reuse in developing countries*. Wageningen Agricultural University, 131 p.

VAN HAANDEL, A.C., LETTINGA, G. (1994). *Tratamento anaeróbio de esgotos. Um manual para regiões de clima quente*. (in Portuguese).

VOLLENWEIDER, R.A. (1976). *Advances in defining critical loading levels for phosphorus in lake eutrophication*. OECD Cooperative Programme in Eutrophication. p. 55–83.

VON SPERLING, E. (1994). Avaliação do estado trófico de lagos e reservatórios tropicais. *Bio Engenharia Sanitária e Ambiental. Encarte Técnico*, Ano III, p. 68–76. (in Portuguese).

VON SPERLING, E. (1995a). Pollution of urban lakes: causes, consequences and restoration techniques. In: *ECOURBS'95*, Rio de Janeiro. p. 18.

VON SPERLING, E. (1999). *Morfologia de lagos e represas*. Belo Horizonte, DESA/UFMG, 137p. (in Portuguese).

VON SPERLING, E., VON SPERLING, M, DANTAS, A.P.(2002). Relationship between trophic status and morphological variables of lakes and reservoirs. *3rd IWA World Water Congress*. International Water Association, Melbourne, Australia, 7–12 Apr 2002.

VON SPERLING, M. (1983). *Autodepuração dos cursos d'água*. Dissertação de mestrado, DES- EEUFMG. 366 p. (in Portuguese).

VON SPERLING, M. (1985). Utilização de gráfico para a avaliação preliminar do aporte de fósforo a uma represa. In: *Congresso Brasileiro de Engenharia Sanitária e Ambiental*, **13**, Maceió. SEEBLA, 1985. (in Portuguese).

VON SPERLING, M. (1987). A study on reaeration in river cascades. *Water Science and Technology*, **19**. p. 757–767.

VON SPERLING, M. (1994). Critérios e dados para uma seleção preliminar de sistemas de tratamento de esgotos. *Bio Engenharia Sanitária e Ambiental. Encarte Técnico*, Ano III, No. 1, Jan/Apr 1994. pp. 7–21. (in Portuguese).

VON SPERLING, M. (1996). Comparison among the most frequently used systems for wastewater treatment in developing countries. *Water Science and Technology*, **33** (3). pp. 59–72.

VON SPERLING, M., CHERNICHARO, C.A.L. (2002). Urban wastewater treatment technologies and the implementation of discharge standards in developing countries. *Urban Water*, **4** (2002). pp. 105–114.

VON SPERLING, M., FATTAL, B. (2001). Implementation of guidelines: some practical aspects. In: FEWTRELL, L., BARTRAM, J. (ed). *Water quality: guidelines, standards and health. Assessment of risk and risk management for water-related infectious disease*. IWA Publishing, World Health Organisation Water Series, ISBN 1-900222028-0. pp. 361–376.

VON SPERLING, M., GONÇALVES, R.F. (2001). Lodo de esgotos: características e produção. In: ANDREOLI, C.V., VON SPERLING, M., FERNANDES, F. (2001). *Princípios do tratamento biológico de águas residuárias. Vol. 6. Lodo de esgotos. Tratamento e disposição final*. Departamento de Engenharia Sanitária e Ambiental – UFMG. Companhia de Saneamento do Paraná – SANEPAR. 484 p. (in Portuguese).

VON SPERLING, M., GONÇALVES, R.F. (2002). Sludge production in wastewater treatment systems in warm-climate countries. *3rd IWA World Water Congress*. International Water Association, Melbourne, Australia, 7–12 April 2002.

VON SPERLING, M., SANTOS, A.S.P., MELO, M.C., LIBÂNIO, M. (2002). Investigação de fatores de influência no consumo per capita de água em estados brasileiros e em cidades de Minas Gerais. In: *Anais, Simpósio Ítalo Brasileiro de Engenharia Sanitária e Ambiental (SIBESA)*, VI, ABES, Vitória-ES, 01–05 setembro 2002. (In Portuguese).

WEF (Water Environment Federation) (1996). *Operation of municipal wastewater treatment plants.* Manual of Practice – MOP 11, 5 ed.

WHO (World Health Organisation) (1989). *Health guidelines for the use of wastewater in agriculture and aquaculture.* Report of a WHO Scientific Group. Geneva, World Health Organisation, 1989 (WHO Technical Report Series, No. 778).

WHO (World Health Organisation) (1993). *Guidelines for drinking-water quality. Volume 1. Recommendations.* WHO, Geneva. 2 ed. 188 p.

WPCF (1990). *Natural systems for wastewater treatment.* Manual of practice FD-16. Water Pollution Control Federation. Alexandria. 270 p.

IWA Publishing's authorised EU representative for General Product
Safety Regulations is Diane D'Arras, 16 rue Duret, 75116 Paris,
France, e-mail: safety@iwap.co.uk.

Printed and bound by CPI Group (UK) Ltd, Croydon, CR0 4YY

12/05/2026

02108880-0003